THE
INVERTED
BOWL

Introductory Accounts of the Universe and Its Life

THE
INVERTED
BOWL

Introductory Accounts of the Universe and Its Life

GEORGE H. A. COLE
University of Hull, UK

Imperial College Press

Published by

Imperial College Press
57 Shelton Street
Covent Garden
London WC2H 9HE

Distributed by

World Scientific Publishing Co. Pte. Ltd.
5 Toh Tuck Link, Singapore 596224
USA office: 27 Warren Street, Suite 401-402, Hackensack, NJ 07601
UK office: 57 Shelton Street, Covent Garden, London WC2H 9HE

British Library Cataloguing-in-Publication Data
A catalogue record for this book is available from the British Library.

THE INVERTED BOWL
Introductory Accounts of the Universe and Its Life

ISBN-13 978-1-84816-503-8
ISBN-13 978-1-84816-505-2 (pbk)

Printed in Singapore by Mainland Press Pte Ltd.

To Tina: In Memorium

And the inverted bowl they call the sky
Whereunder crawling coop'd we live and
die,
Lift not your hands to *It* for help — for *It*
As impotently moves as you or I.

Omar Khayyám (1048–1131)
Persian mathematician, astronomer and poet

[quatrain 16 of the *Rabaiyat* in Edward Fitzgerald's 1889 translation]

Preface

The 1990s was a very exciting period for the planetary sciences because it was then that the first planetary bodies were found orbiting stars other than our Sun. The first discoveries in 1992 involved a pulsar but the orbiting planets were in the range of Earth masses. Unfortunately, these must be dead systems and especially as far as exo-life is concerned. In 1995, however, a body of rather less than Jupiter mass was found to be orbiting the star *51 Pegasus* which is not dissimilar to the Sun. This was exciting because it promised the presence of planetary systems elsewhere in the Universe that were expected to prove similar to our own Solar System. This promise was due for some shocks but more and more systems were discovered and catalogued until today in excess of 300 are known and more are being discovered. These planetary systems might well be expected to support life, perhaps with technological capability.

Bodies of Earth mass cannot yet be detected but technology is being developed that will be able to find them if they are there — and everyone believes they are. If planets like the Earth exist elsewhere around other stars will these bodies also carry living organisms? If so will any be technologically capable? There have been questions in the past whether this might be so and questions why none seem to communicate with the Universe at large. In spite of this negative response it is still interesting to find where such life might be. As a natural part of the Universe such extraterrestrials will be subject to its laws just as we are on Earth. The living Earth then will be an example of what might be found elsewhere so a study of the Earth in its cosmic context will be expected to throw some light on developments elsewhere. This exciting study is what we investigate in this book. We explore the many associated questions and, while we may not offer definitive answers, we will certainly set the questions in a clear universal context. This must involve the cosmos as a whole to gain proper perspective. This all adds up to a most remarkable account of a grandeur beyond our powers of invention.

The ancient Greeks seem to have been the first thinkers who realised the truism that if you want to find out about something then the thing to do is look at it critically and in detail. This is obvious now although it has been forgotten from time to time in the past. This prescription is as true for the Universe as for a garden flower. Unfortunately the Universe is not as easily studied as is a garden flower. A garden flower can be taken inside the garden shed and dissected: obviously we cannot do this to the Universe! The study of the Universe requires the careful interpretation of the readings from observations using the most sophisticated equipment. Generally speaking there is no experimentation where the object can be manipulated in various ways but only the passive observation of events of the greatest grandeur. Equipment and techniques of sufficient power to begin this endeavour have only begun to be available over the last century and new developments are becoming available all the time. Nevertheless our understanding of the Universe is now reaching quite an advanced stage where many important questions can be given answers that are at least safe for the future. Where has this process got to now? This little book is concerned with one aspect of this wide study — the rôle of evolution in the Universe and the exploration to find if extraterrestrial life might be expected to exist elsewhere. Are we likely to be able to make contact with them if they are there?

Observing the Universe has taught us two important things. First, the present phase of the Universe that we now enjoy had a clear origin some 13.7 billion years ago (Bya). We have not got any knowledge of what existed before that nor why the Universe came into existence nor how it happened. The laws of physics that we understand began then. We can presume that something was "there" to start it all off but what we cannot say from our instrumentation. Secondly, the Universe was constructed not to be a simple static thing with its form given once for all at its birth but rather an entity to change and to evolve. There is clear evidence for this in both animate and inanimate forms of matter. The behaviour of inanimate matter is constrained by what we have discovered as physical laws applied without favour but involving a certain degree of chance. The animate matter also evolves in such a way as to be sympathetic to the inanimate environment that it finds at the time. This process has been described by Charles Darwin in his Principle of Natural Selection by Survival of the Fittest. The evolution of the Universe over time is revealed by observing the incoming radiation of the electromagnet spectrum. The visible Universe has been investigated by using ever more powerful optical telescopes but more recently this has been augmented by infrared and ultraviolet light and by radio waves. Earth based telescopes are now augmented by automatic space probes in orbit about the Earth. When we look at the cosmos we are viewing it in the past and the evolutionary pattern becomes clear as we penetrate further and further into space.

The animate past can also be inferred through the study of fossils and more recently using DNA sequencing. The relevance of fossils has sometimes been questioned in the past but they have now become established as an important and reliable tool in the study of the deep and recent past. One surprising result has been that mankind, Homo sapiens, has also arisen as part of the evolutionary process which is still going on. Such a complex evolution in an evolving Universe must surely imply that extraterrestrials, should they exist, will also have arisen in the same evolutionary way as an integral part of an evolving Universe. We explore these aspects of the Universe in the following six chapters and five appendices. Each is designed to be broadly self-contained and, even though this must lead to some repetition of material, it is hoped the reader will tolerate this to avoid the nuisance of having to go searching in other chapters for things that are needed to allow the argument to flow in the chapter being read. Chapter 1 sets down the observations of the planets of the Solar System and the exo-planets that have been found orbiting other stars like the Sun. The changing physical characteristics of the Earth are set down in Chapter 2. The beginnings of like on Earth and the evolution of microscopic creatures is the subject of Chapter 3 while the wider evolution of macroscopic creatures is he subject of Chapter 4. It has now been established that human mankind has evolved according to the principles of evolution and the evidence for this found so far is treated in Chapter 5. This account does not claim to be comprehensive but is meant to offer a very broad indication of the way evolution has acted on Earth over the 4.65 billion years (By) of its history. This is our only model of the working of evolution with animate matter and is meant to suggest how life may evolve on Earth-like planets elsewhere, should they exist. Finally, in Chapter 6 these arguments are drawn together to attempt an assessment of the possibilities for the existence of extraterrestrial life.

The appendices offer ancillary material designed to give perspective to the arguments in the Chapters. Appendix 1 views the broad Universe while Appendix 2 considers some individual members. The ultimate building blocks of everything in the Universe, animate and inanimate alike, are atoms and molecules. Appendix 3 is concerned with the discovery of these entities which we cannot see with our naked eyes. The production of energy is a vital aspect of a living Universe and will be central to the development of technologically capable extraterrestrial life, as is true for us, and the various aspects of energy production are considered in Appendix 4. Finally, in Appendix 5, we consider very briefly the development of mathematics as the language of science and technology. This controls our advances in an understanding of the Universe and the same must also be true of extrater-restrials, should they exist.

This is an elementary introduction to astrobiology and is of the nature of a workbook. The aim has been to include all the many aspects of life in the Universe. It is expected that the reader will wish to extend the arguments to form a personal account of these matters and to this end a Compendium has been added which includes a range of explanations of the various topics involved in the Chapters and Appendices.

The illustrations throughout the book have been assembled from a collection gathered over the years some of whose origins I do not now know. I hope those sources not mentioned will be understanding and not interpret any lapse on my part as discourtesy. It is a pleasure to thank Dr. N. B. Brindle, Reader in Cell Biology at the University of Leicester, for his very helpful comments on certain of the biological aspects of the book. I must also thank Mr D. G. J. Cole, M. Sc., for enlightening comments on various matters but especially concerning Gödel's Incompleteness Theorem. Any errors remaining are, alas, my own. I hope the reader will find the book pleasing to read apart from being interesting and useful as a basis for further studies. Its success will certainly owe much to the professionalism of the members of the Imperial College Press and especially to Kim Tan and her colleagues. Apart from affording me their much appreciated support they have made a most excellent transformation from a raw manuscript to a handsomely finished book.

G. H. A. Cole

g.h.cole@hull.ac.uk

March 2009

Contents

The first chapter is devoted to setting out details of planets. This includes first an account of the Solar System as the one set of star and planets that we continue to explore in growing detail. Then we consider the structure of planets in general making a distinction between a normal planet and a heavier super-planet. This is followed by an account of the exo-planets discovered so far orbiting solar type stars other than the Sun. The orbital and physical details of the orbiting bodies are considered in detail. Four, or perhaps 5, exo-systems are highlighted which appear to have structures not unlike the Solar System and could have Earth type planets. There naturally arises the question of whether there might have developed technologically capable creatures there. This sets the scene for the rest of the book.

Chapter 2.   The Dynamic Earth   39

The Earth is where we have been constrained to develop and live so it is
important to understand its evolution and history. This is the subject of
the present Chapter. The development of continents and oceans, the atmosphere and the general environment are considered. This sets the scene for
understanding the way the environment has affected life.

Chapter 3. Life in Water: The Precambrian 84

The first elementary life appears to have developed in water. It began in very simple ways and remained microscopic for some 4,000 million years. This isn't to be interpreted as total inactivity for the period ends with the appearance of fully viable macroscopic life ready for evolutionary development in the future.

Chapter 4. Life Develops in the Phanerozoic 111

This Chapter is concerned with the remainder of the history of the Earth where evolution gave rise to an enormous range of fantastic creatures but each moving clearly to the next phase. This Chapter considers the evolution between about 534 million years ago up to about 55 million years ago when the very first creatures developed which led directly to hominids and ultimately to Homo sapiens sapiens.

Chapter 5. Hominids — Homo Sapiens 147

Now we come to the evolutionary development of mankind. The different
species and variants within them are set down to establish quite clearly that
we are part of the evolutionary structure. Apparently the structure of the
Universe requires everything to be evolved including to mankind itself.

Chapter 6. A Universe of Exo-Life? 169

The arguments which have been developed in the previous Chapters are
put together in this Chapter to try to assess the likelihood of there being
technologically capable life elsewhere in the Universe. More than this, the
probability is also considered of ways of making some form of contact with it,
should it exist, either indirectly or directly. Finally arguments are developed
to set the existence of living materials in the Universe within the overall
structure.

Appendices

The main Chapters are based on a wide range of detailed information which it would be inappropriate to include as part of the Chapters. This information is collected in five Appendices which support the main Chapters.

Appendix 1.   Discovering the Cosmos   201

This appendix offers a very brief account of the steps that have led to our present view of the Universe as a whole. It is meant to give a broad perspective of the background to the exploration for exo-planets planets and the life that might be associated with them.

Appendix 2.   Further Comments   234

This appendix provides some background to the general properties of planets and the materials of which they are composed. This provides a general background to the observations of exo-planets.

Appendix 3.   The Strange World of the Atom   247

Atoms are too small to be observed and yet they are the building blocks of the Universe. Showing how they were discovered over many centuries is an example of the methods that casm yield information about things unseen.

Appendix 4.   Sources of Energy 264

Advanced technological civilisations must rely heavily on the availability of energy both to have achieved that status and also to maintain and develop it. This would be true of extraterrestrials just as for us. In this Appendix we review the sources of energy available to the Universe as a whole and to technological creatures. Natural sources require a planet to have supported earlier life to provide high density sources so the energy background is also part of the animate evolutionary process.

Appendix 5.   The Language of Science: Developing
Mathematics 296

Mathematics is the language of science and so the language of any quantitative account of the Universe. There have been many misunderstandings in the past about mathematics and its rôle in logic and understanding and this brief Appendix aims to clarify one or two of these issues.

Compendium 315

This compendium contains a wide range of entries the greater majority of which are referred to in the main text. Others are also included as a help to the reader who might wish to pursue matters further. It is hoped that this compendium will be extended by the reader and so form the basis of a comprehensive personal study.

Chapter 1

Planets Orbiting the Sun and Other Stars

The Earth is two things at once. First, it is one of eight planets orbiting the Sun which forms the core of the Solar System. Second, it is the only body on which a vast range of living things is known to exist and through this it is our home. In our quest for other intelligent life elsewhere in the Universe it is of central interest to know whether the Earth is repeated physically in other planets orbiting other stars elsewhere. This chapter is devoted to a review of the observations that are now available of the planets orbiting stars other than our Sun and called here exo-planets.

Some 95% of the stars in the galaxies are very similar to our Sun in composition, mass and radius. They all lie on the same general line of the plot of surface temperature against luminosity which forms the so-called Hertzsprung–Russell diagram (see Appendix 2). This line in the diagram is called the main sequence and involves stars which derive their light and heat from the thermonuclear process of the conversion of hydrogen nuclei into helium nuclei. Most stars do this. They all have broadly the same life span of about 10^{10} y: our Sun is about half way through its life with an age a little below 5×10^9 y. There is firm evidence that the ages of the various members of the Solar System are all close to 4.55×10^9 y and that the Solar System of Sun and planets formed as a single unit. It is generally held that our System probably formed by the collapse of a giant gas cloud. There is nothing to suggest that this System is intrinsically unique so we would expect comparable planetary systems to form about other stars by the collapse of dust clouds. We might expect to find comparable time scales with them as well. While stars like our Sun (solar-type stars) have been observed in the midst of orbiting dust, and so presumed to be in the process of forming planetary systems like our own, no star accompanied by a planetary system had been found up 20 ya.

The situation changed completely in 1992. The first planetary companions of mass a littler greater than Earth were discovered orbiting a pulsar PSR

$1257 + 12.$[1] This was very interesting and even encouraging but pulsars are not the stars likely to support life. Everything was changed, however, three years later by the report of the discovery of a planet of rather less than Jupiter mass orbiting about the star 51 Pegasus. More and more planetary systems have been found orbiting solar-type stars since those very early days. As an example of the rate of discovery of exo-planets, the number known on 10 December 2008 was 333 while on 27 February 2009 this had risen to 342, which is nine discovered in just two months — about one per week. This rate is set to continue. Companion planetary bodies have also been detected about brown dwarfs but these have little interest to us for reasons that will become clear later. Before reviewing these results it is useful first to remind ourselves of the main features of the Solar System because this actually supports life. It is also useful to consider briefly the main features of planets and planetary bodies in general in order to be able to be sure of our interpretation of any bodies that might be found. After that we will explore the exo-planetary systems that have actually been discovered so far orbiting around other solar-type stars.

1.1.   General Features of the Solar System

The Solar System is certainly the one system we know in detail and it was expected, before other systems had been found, that it would ultimately be able to act as a natural standard of comparison for exo-systems orbiting other stars.

1.1.1.   *The planets*

The Solar System is composed of eight planets (four terrestrial planets and four major planets — an interesting divide?) orbiting the Sun together with a range of smaller bodies and dust. The orbits are quite closely in one plane (called the ecliptic) which is essentially the path marked out by the Sun as it moves across the sky during a year. The radii of the orbits of the planets lie between 0.38 AU for the innermost planet Mercury out to a little over 30 AU for the outmost planet Neptune. Orbital data for the planets are collected in Table 1.2. The masses range from 3×10^{23} kg for Mercury to

---

[1]A pulsar is the result of a cataclysmic supernova explosion in which vast quantities of the outer region of the star has been ejected into space. Presumably the observed orbiting planets are the core regions of what were originally Jupiter type gas bodies, the major materials of them having been blown away by the explosion. They are not likely to have any special interest from the point of view of life.

very nearly 2×10^{27} kg for Jupiter, the largest planet. The physical details of the planets are collected in Table 1.1. There is a clear distinction between the masses, radii and densities of the terrestrial planets and those of the major planets. It is seen that the masses of the terrestrial planets are up to a thousand times smaller than those of the major planets but the mean densities are some five times larger. On the other hand, the densities of the major planets are only a half or less than those of the terrestrial planets. These data show that the mean material compositions of the two groups are radically different. While the terrestrial planets are composed of silicate and ferrous compounds, the major planets are composed mainly of hydrogen and helium. Earth, Uranus and Neptune also have hydrogen and oxygen combined as water as not insignificant components. It will be realised from the last but one column in Table 1.1 that Venus is rotating "upside down" while Uranus is rotating on its side. These are understandable arrangements in appropriate circumstances of a rotating planet.

The planets rotate in the same sense in a thin plane about the Sun as a centre with orbital characteristics listed in Table 1.2. The orbital eccentricity indicates how closely the figure of the orbit is to a circular form. It is seen that, with the exception of Mercury, all are below 0.10. Mars has the largest

Table 1.1. Showing the physical data for the planets of the Solar System. The acceleration due to gravity listed in the last column refers to the equatorial value. Data for the Sun are included for completeness.

Name	Mass (kg)	Mean Eq. radius (km)	Mean density (kg/m^3)	Inclination of equator to orbit	Equa. accel. due to gravity (m/s^2)
Terrestrial planets					
Mercury	3.303×10^{23}	2,439	5,340	2°	2.78
Venus	4.870×10^{24}	6,051	5,250	177.3°	8.60
Earth	5.976×10^{24}	6,378	5,520	23.45°	9.78
Mars	6.421×10^{23}	3,393	3,950	25.19°	3.72
Major planets					
Jupiter	1.900×10^{27}	71,492	1,333	3.12°	22.88
Saturn	5.688×10^{26}	60,268	690	26.73°	9.05
Uranus	8.684×10^{25}	25,559	1,290	97.86°	7.77
Neptune	1.024×10^{26}	24,764	1,640	29.6°	11.00
Sun	1.989×10^{30}	696,000	1,408	—	274.0

Table 1.2. Orbital data for the planets of the Solar System (1 AU = 1.4960×10^8 km).

Planet	Orbital eccentricity	Orbital inclination	Mean solar distance (AU)	Local solar const. (kW/m^2)	Obliquity angle
Mercury	0.2056	7.004°	0.3871	9.116	∼0.01°
Venus	0.0068	3.394°	0.7233	2.611	177.36°
Earth	0.0167	0.000°	1.000	1.366	23.439°
Mars	0.0934	1.850°	1.5237	0.588	25.19°
Jupiter	0.0483	1.308°	5.2028	0.051	3.13°
Saturn	0.0560	2.488°	9.5388	0.015	26.73°
Uranus	0.0461	0.774°	19.1914	0.004	97.77°
Neptune	0.0097	1.774°	30.0611	0.001	28.32°

eccentricity after Mercury — primarily due to the gravitational interference of Jupiter. The planets lie within a distance of about 30 AU of the Sun which is about 3.5×10^9 km. The orbits of the terrestrial planets lie within 0.3–3.5 AU while the orbits of the four major planets lie in the range 5–31 AU. These are the general data that we will be searching for in any other planetary systems if we are to reproduce the Solar System there.

1.1.2. *The satellites*

All the planets, with the exceptions of Mercury and Venus, are accompanied by small satellite bodies orbiting around them. These vary in size and mass from bodies fully comparable to our Moon, or a little larger, down to quite small pieces of matter. There are, in fact, 60 known satellites at the present time (very small ones are still being found around the major planets) although seven are the largest. These principal satellites are listed in Table 1.3. It is seen that the ratio of the masses of the satellite to the planet are all rather insignificant numbers except for the case of Earth/Moon where the ratio is still a small number ($\approx 10^{-2}$) but not an insignificant one. The Moon alone will induce significant tides twice daily in the parent planet and this has been suggested as having a special importance for the development of life. We can also notice that the density of the Moon is the largest except for Io although its radius is only fifth down from the largest.

1.1.3. *The asteroids*

The two groups of planets are separated by a region between about 1.8 and 4.8 AU containing small silicate/ferrous bodies in orbit about the Sun. These

Table 1.3. Data for the seven principal satellites of the planets of the Solar System.

Planet	Satellite	Mass (kg)	Mean radius (km)	Mean density (kg/m^3)	Ratio M_S/M_P
Earth	Moon	7.349×10^{22}	1,738	3,340	0.0123
Jupiter	Io	8.94×10^{22}	1,815	3,570	4.71×10^{-5}
	Europa	4.80×10^{22}	1,569	2,970	2.53×10^{-5}
	Ganymede	1.48×10^{23}	2,631	1,940	7.79×10^{-5}
	Callisto	1.08×10^{23}	2,400	1,860	5.68×10^{-5}
Saturn	Titan	1.35×10^{23}	2,575	1,880	2.37×10^{-4}
Neptune	Triton	2.14×10^{22}	1,350	2,070	2.09×10^{-4}

are the asteroids. The largest one, Ceres, was the first to be discovered by G. Piazzi (1746–1826) in 1801 at the Palermo Astronomical Observatory. The physical and orbital data for the largest seven asteroids, each with a radius greater than 150 km, are listed in Table 1.4. These larger bodies are each distinguished by having an essentially spherical surface figure. There are thousands of smaller irregularly shaped bodies within the asteroid belt although the collective mass is not high, perhaps only a fraction of the mass of the Earth. Some are pairs orbiting each other as well as the Sun. They lie in a region where a planet might have been expected. Presumably a planet could not form because the collective mass was too small to come together under the disruptive influence of the gravitational field surrounding Jupiter.[2]

The asteroids are not, in fact, all confined to this gap between the planet groups. Although the overwhelming majority of bodies are in the asteroid belt a large number lie outside with orbits that penetrate the region of the terrestrial planets and the region of the major planets. There are three main groupings, the Atens, the Amors and the Apollos. They are distinguished by their orbits. The Atens have a semi-major axis $a < 1\,\text{AU}$ and an aphelion greater than 0.988. This means they cross the orbit of Venus but lie within the Earth's orbit. The Amors have orbits outside that of the Earth while

---

[2]The distribution of the planets about the Sun is well represented by the curious law ascribed now to the 18th century astronomers Johann Titius and Johann Bode. It was originally intended to extend out to Saturn but also works well for Uranus (unknown at that time). It predicts the value of the semi-major axis a of planetary orbits in astronomical units. The law can be written $a = 0.4 + 0.3 \times 2^m$ with successively $m = -\infty, 0, 1, 2, \ldots$. Although it predicts the planet positions quite accurately out to Uranus it fails completely beyond that. A mean location of the asteroid belt follows for $m = 3$ and this prediction led to the discovery of Ceres (alas, not a planet) by Piazzi. There is still no theoretical explanation for the "law" which remains an empirical curiosity from the past.

Table 1.4.   The data for the seven asteroid bodies with radii in excess of 150 km. This limit is set by the requirement that they should have spherical figures.

Number; Name	Radius (km)	Mean distance from Sun (AU)	Orbital eccen.	Orbital period (years)	Orbital inclination
1 Ceres	457	2.767	0.097	4.61	10.61
2 Pallas	261	2.771	0.180	4.61	34.81
4 Vesta	250	2.362	0.097	3.63	7.14
10 Hygiea	215	3.144	0.136	5.59	3.84
511 Davida	168	3.178	0.171	5.67	15.94
704 Interamnia	167	3.062	0.081	5.36	17.30
52 Europa	156	3.097	0.119	5.46	7.44

the Apollos have orbits that cross that of Earth. The numbers are not large. Some 30 Atens are known but some 220 Apollos. From this it is seen that the space occupied by the planets is an active playground for small objects. Collisions with the Earth are not unknown throughout the history of the Solar System.

1.1.4.   *The Kuiper belt and Oort cloud*

The region occupied by the planets is only a small part of the total System. Outside the orbit of Neptune is another belt of particles in some ways similar to the asteroid belt but much more extensive and of rather different composition. The existence of such an outer belt was first hypothesised by Kenneth Edgeworth (1880–1972) in 1943 and then independently by Gerald Kuiper (1905–1973) in 1951. This Edgeworth–Kuiper (E–K) belt is now known to exist between the general distances 30 and 55 AU from the Sun. It is therefore about 20 times wider than the asteroid belt and probably between 20 and 200 times more massive containing in total some 10 Earth masses. Its composition is quite distinctive. Whereas the asteroid belt is composed of silicate rock and ferrous materials, this E–K belt is composed of frozen volatiles such as water, ammonia and methane in the form of ices with a mean temperature of perhaps 55 K. The central regions are in a state of equilibrium away from the gravitational influence of Neptune. It had been thought for some time that the perturbation of particles within the E–K belt could be the source of comets and especially the short period comets (such as Comet Halley) which have orbits near the plane of the Solar System. Computer simulations have shown that the belt is strongly influenced by the presence of Jupiter and Neptune which impose a series of resonance orbits within the belt.

Interest in the E–K belt has grown since the later 1980s when substantially sized bodies, rather larger than Pluto, were discovered orbiting within the belt. Subsequent discoveries have led to the designation of bodies with orbits outside that of Neptune as trans-Neptune Objects (TNOs). This group includes everything orbiting the Sun in the outer reaches of the Solar System. Very recently the International Astronomical Union has created a new class of object, the Dwarf Planets, to include the smaller spherical bodies of the Solar System. Pluto with its satellite Charon (until then regarded as being a planet with a moon), is now called a dwarf planet together with Eris and its satellite Dysnomia and other discovered bodies such as Haumea and Makemake (and, perhaps rather surprisingly, Ceres of the asteroids so different in composition and so far away). Eris is the largest body found so far in the belt with a mass some 27% greater than Pluto. Other possible dwarf planets are Sedna, Orcus, Quaoar and Varuna but more is needed to be known about them before a decision can be made. These bodies all have masses in the range 10^{20}–10^{22} kg which is similar to the masses of the satellites of the planets themselves. Their densities, however, are less than $2,000 \, \mathrm{kg \, m^{-3}}$. There is an interesting difference of appearance: some bodies are essentially white while others have got a reddish colour presumably due to some organic pollution. The reason for this difference is not known.

Comets are heated when they are near the Sun and eject jets of volatile gases forming a characteristic material tail and a separate ion tail. Each passage near the Sun will heat the body of the comet and release volatile material restricted the time during which the volatiles are there. The end of their life is for them to move as a dead object with all their volatiles used up unless the gravitational attraction causes them to impact a planet or other large object. Since there seems no shortage of comets it follows that there must be a constant source but the Kuiper belt could not reasonably be expected to provide all the variety that are observed. In 1932 Ernst Öpik (1893–1985) made the suggestion that the long period comets originate in an orbiting cloud of matter forming the outermost edge of the Solar System. The idea was revived in 1950 by Jan Oort (1900–1992) when it was realised that some long period comets, incidentally with orbits which can be far away from the ecliptic, had aphelion distances of about 20,000 AU from the Sun. This obviously suggests that there is a source of comets at that distance from the Sun. The hypothesis is that the E–K belt hugging the ecliptic plane is encased in the extensive spherical ball of small ice particles forming a region called the Oort cloud. This spherical cloud is believed to lie between 20,000 and 50,000 AU (or some say as great as 200,000 AU) from the Sun and there is indirect evidence for a central doughnut shaped cloud between 2,000 and 20,000 AU called the Hill's cloud. It must be admitted that there is yet no direct evidence for either the Oort cloud or the Hill's cloud although indirect evidence for them is very strong. It is believed that the cloud contains some 10^{12} bodies of some 1.5 km radius with a total mass of about 3×10^{25} kg. This is some five times the mass of the Earth. The composition is largely frozen methane, ethane, carbon monoxide and hydrogen cyanide. The gravitational effect of the Sun has an effective range of about 1.8 ly for orbiting bodies suggesting the extent of the cloud will be less than this, but it could perhaps be as much as 1 ly. The nearest star, Proxima Centauri, is about 4.2 ly from the Sun.

It is clear that the Oort cloud extends the Solar System far out into what has traditionally been regarded as interstellar space. The result is a gravitational influence on the Oort cloud by the Galactic components and, because the cloud is the centre for the formation of comets with orbits that come near the Sun, the broad influence of the Galaxy extends right down to the inner Solar System. The cloud is only weakly influenced by the solar gravity so the movement of the nearby stars and any distortion of the Galaxy due to rotation will lead to a periodic tidal effect on the cloud. The Solar System is, in this way, not to be regarded as a tight set of bodies isolated in space as was once thought. Stars with planets will generally have a range of dust and gas surrounding them which could well be as extensive. Interstellar space may well, in this way, be a region where stellar systems influence each other and perhaps may even pass material from one system to another. This could have important implications for panspermia.

1.2. The Terrestrial Planetary Bodies

The terrestrial planets form a group with similar compositions and include the Earth where life began. They are characterised by having a definite solid surface. Venus, Earth and Mars each have an atmosphere although Mercury does not now. The terrestrial planets and the Moon show every indication of being composed of the same general spread of chemical elements. Just as the most common chemical element, hydrogen, provides stars so the second most abundant chemically active element, oxygen, also provides characteristic bodies. In fact oxygen combines with a range of chemical elements to form hard mineral oxides which are the basic constituents of terrestrial planetary bodies. The only members known so far lie within the Solar System and these show masses generally with a maximum a little less than 10^{25} kg.

Table 1.5. The principal chemical structures for the crust of the Earth.

Mineral	Composition
Kamacite	(Fe, Ni) (4–7% Ni)
Taemite	(Fe, Ni) (30–60% Ni)
Troilite	FeS
Olivine	$(Mg, Fe)_2SiO_4$
Orthopyroxine	$(Mg, Fe)SiO_3$
Pigeonite	$(Ca, Mg, Fe)SiO_3$
Diopside	$Ca(Mg,Fe)Si_2O_6$
Plagioclase	$(Na, Ca)(Al, Si)_4O_8$
Quartz	SiO_2
Alumina	Al_2O_3
Sodium Oxide	Na_2O
Potassium Oxide	K_2O

There is reason to believe that comparable bodies could exist elsewhere composed of the rarer chemical elements. The principal minerals are oxides of silicon, aluminium, iron and sulphates of iron. These substances take on various forms according to the local temperature and pressure. The principal forms are collected in Table 1.5. A comparison between the compositions of the Earth's crust and the lunar surface are collected in Table 1.6. There is a general similarity between them although a major discrepancy arises in the case of iron. This is actually a very gross comparison. The lunar composition is particularly variable from site to site.

1.2.1. *Different molecular groups*

We might notice finally some affiliations of the chemical elements which have relevance to silicate/ferrous bodies. The various molecules do not combine in even a quasi-random way but according to a pattern set by the nature through the bonds between the atoms. There are three broad categories determined by the electronic configurations and their affinities for different types of crystalline bonds. The groups are:

(1) *Lithophilic elements*. The name comes from the Greek for stone and the elements of this group tend to be associated with oxygen in oxides and silicates. Included here are the more common elements shown below

Table 1.6. A comparison between the surface compositions of Earth and Moon in terms of mineral chemical compositions. It is seen that there is a similarity but the iron content shows a marked discrepancy.

Element	Earth (%)	Moon (%)
Oxygen	47	42
Silicon	28	21
Aluminium	8	7
Iron	5	13
Calcium	4	8
Sodium	3	—
Potassium	2	—
Magnesium	2	—
others	1	9

Table 1.7. Some lithophilic elements.

K	Rb	Mn	Mg
Na	Sr	Be	Li
Ca	Ba	Al	Cr
Mg	Li	Cr	Rare earths

Table 1.8. Some chalcophilic elements.

Cu	Sn	Mo	Cl	Co
Zn	Pb	Hg	Bi	Au

in Table 1.7. The association with stones is because oxides and silicates containing these elements generally form hard rocks.

(2) *Chalcophilic elements.* These elements (named after the Greek for copper) tend to appear as sulphides. A partial list is shown in Table 1.8.

(3) *Siderophilic elements.* These are generally metallic elements (the name of group comes from the Greek word for iron). Examples are listed in Table 1.9.

Table 1.9. Some siderophilic elements.

Ni	Ir	Os
Pt	Au	Pd
	As	

These groupings generally define the molecular composition of a silicate planet.

1.2.2. *Some Venusian data*

The Russian Space Programme was able to perform the very remarkable fete of landing three automatic probes on the surface of Venus at different times. Two, Venera 13 and Venera 14 landed on the surface in March 1981 while the third, Vega 1, landed in 1985. These probes were able to remain active on the surface long enough to send back important information and a photograph (see Fig. 1.1(a)). One probe was able to confirm a very high atmospheric pressure and a very high surface temperature. The other, the most remarkable, was to obtain information about the relative abundances of silicates at two separate sites and to receive a picture from the surface. One site was at a low elevation while the other was on higher ground. These data are collected in Table 1.10. If there were an ocean on Venus the lower location in the valley would be in the oceanic region while that higher up would be on the land clear of water. On Earth this would lead to very different compositions but it is seen from the table that there is very little difference between the compositions of the two sites on Venus with the possible exception of K_2O but this could be instrumental — the measurements were made under extreme conditions and at an extreme distance from the laboratory. The similarity of compositions is important because it is a first indication that the geological processes taking place on Earth and Venus are very different even though the basic material at the formation of the two planets was very similar. Venus had been regarded as a twin of the Earth up until then.

An extended mission to map the Venus surface using synthetic aperture radar (SAR) was made by NASA with the launch of the Magellan space craft 1989 May 4. It was placed in a near-polar orbit on 1990 August 10 and operated until 1994 October 12. During this time, 98% of the surface was recorded with a resolution no worse than 100 m. It was found that rather in excess of 85% of the surface is covered by volcanic flows with the remainder given over to deformed mountain belts. The surface is remarkably flat

(a)

(b)

Fig. 1.1. The surfaces compared. (a) The typically volcanic surface of Venus photographed by the Venera Space Craft (Soviet News). The horizon is visible at the left- and right-hand top corners showing a very flat landscape. The light level at the surface is surprisingly high remembering the very thick cloud layer than surrounds it. (b) The surface of Mars showing rocks and a "soil" surface and a reddish sky (photographed by the NASA Rover). The difference between the conditions of the two surfaces is very clear. The feet of the lander are visible in each photograph.

with over 80% lying within 1 km of the mean radius. This is very clear in Fig. 1.1(a) where the horizon is clearly visible to both the right and left of the picture. The difference in the dynamical behaviour between Venus and Earth appears, in fact, to centre on the differing mechanisms for losing internal heat for the two planets: Venus has a purely volcanic totally dry surface with a temperature of 748 K and an atmospheric pressure of 92 bar while the presence of water on Earth gives rise to a plate tectonic system not yet detected on Venus. The strongly CO_2 dominated atmosphere of Venus is to be expected from a largely volcanic planet and the strong greenhouse properties that follow will account for the high temperature that is found there.

Table 1.10. The chemical compositions at two surface sites on Venus and the mean Earth crust for comparison. There are strong similarities.

Oxide	Relative abundance (% by mass)		
	Venera 14 site	veg 1 site	Mean earth crust
SiO_2	45	49	61
Al_2O_3	16	18	16
MgO	10	8	5
FeO	9	9	7
CaO	7	10	6
K_2O	4	0.2	2
TiO_2	1.5	1.2	0.7
MnO	0.2	0.16	—

1.2.3. *Comments on Mars*

The planet Mars has been the centre of considerable attention over recent years with automatic orbiters photographing the surface in considerable detail. This has culminated now with two surface automatic Rover landers (named Spirit and Opportunity) placed on the surface separately in 2004 January having been launched in mid-2003. These report a planet which is essentially a frozen Earth with only a short period in its early history where it was warm enough to support free surface water. It now has a very thin CO_2 atmosphere and is to all intents and purposes dead. The atmosphere is very thin and a sparsity of greenhouse gases is compatible with the observed low surface temperature.

There is considerable interest in knowing whether microscopic life was ever able to start on the planet and whether fossils exist now. The task of answering this question may be beyond the powers of automatic probes so that answers may have to wait until the proposed human exploration is achieved at some time in the next few decades. In the context of extraterrestrial space travel it is interesting that we are still not able to support an extended mission well beyond the Earth even for a near planet.

1.3. The Characteristics of a Planetary Body

Having looked at the Solar System, the only family of star, planets, satellites and other materials that we yet know in detail, it is interesting to consider the planetary structure more carefully. Our central concern is the occurrence of life on planets and so it is important to be clear about the features that actually make a mass of material of a planet and give it its particular properties.

Like all bodies in the Galaxy the existence of a planet is ultimately the result of the action of gravity. This force collects matter together and compresses it to form a compact body. There is, however, a limit to the degree of compression possible in each case. The body achieves a finite equilibrium radius through the material of which it is made resisting the further compression by gravity. The resistance is offered by the constituent atoms. Each atom consists of an oppositely charged nucleus encased in a cloud of negatively charged electrons.[3] Electrons are Fermi particles[4] with the characteristic of each electron requiring its own "space". Put more technically, each electron occupies one state alone and will resist the constriction of "its" volume by pressure from the outside. This is a quantum property of the electron with no classical counterpart. The whole effect is usually referred to as electron degeneracy. It is possible to make a rough picture by realising that the electron resists the potential energy of gravitation by the kinetic energy of movement — it moves constantly within its cell to apply sufficient pressure to every point of the boundary. It cannot be given a precise location which is compatible with its wave nature. This movement to resist compression turns out to be entirely independent of the temperature so the equilibrium is as strong at the lowest temperatures as at the highest. In this sense the system with the electron is "cold" and a body controlled by atoms can be regarded as a cold body.

Of course the electrons cannot provide an indefinite resistance to gravity and at a sufficient pressure the boundaries of the cell "break" to allow the electrons to move freely outside the atom through the volume of the material, leaving the positive nuclei of the atoms behind. The atom has then become ionised or it is said that pressure ionisation has occurred. One consequence is a lower resistance to compression, that is its compressibility has increased. In simple terms the motion of the electron can increase with increasing external pressure but a limit is reached when the speed becomes comparable to the speed of light which is a natural limit to speed imposed by the special theory of relativity. Atoms have a finite strength and can be broken up if the external

---

[3]In classical theory the electron is taken to be a particle with location and speed but it can also behave like a wave and so fill a space without location, having a frequency of oscillation instead.
[4]The statistics followed by such a particle were developed separately by E. Fermi and P. A. M. Dirac.

pressure is great enough. Applying these ideas to a planet, for the smaller masses the atoms can be regarded as behaving like hard spheres and the equilibrium radius is achieved by the simple packing of atoms under the action of their self-gravity. There is a limiting mass where the gravity is sufficiently strong to cause pressure ionisation and then the packing becomes rather more complicated.

Applying this analysis to the behaviour of a planetary mass of material requires us to identify three separate forces. First, there is the force of self-gravity of the material acting to compress the material and so reduce its volume. Second, there are the electric forces between the protons and electrons in the atom which provide the stability for the atom and again act to pull the material together. Third, there is the electron degeneracy which acts to resist compression. These three forces form a mechanical equilibrium for bodies within a range of masses depending on the precise composition. In taking the arguments further it is convenient to deal in terms of energy rather than directly as forces.

Remember that there are two general types of energy, the kinetic energy of motion and the potential energy of relative position. Forces that depend on a separation distance which varies as the inverse square are special because they are associated with so-called conservative forces whose difference from place to place depends only on the separation distance but not on the actual path taken. Both the electric forces and the gravitational force are of this type. The associated energies in such systems obey a special theorem called the virial theorem. According to this theorem the potential energy in the system has a magnitude of twice the kinetic energy: $2K = -P$ where K is the total kinetic energy and P the total potential energy. The negative sign here refers to attractive energies, such as gravitation. We will now look at the three energies one by one though we will not go through the lengthy task of deriving task of deriving them — they are derived elsewhere.[5]

First, consider the energy working to stop further compression. The body is presumed spherical with radius R_P. Each atom is presumed to have a total mass A (including both protons and neutrons) with Z protons each of mass m_p and so also Z electrons each of mass m_e. The electron degeneracy energy E_K, then, is found to be expressed by the formulae:

$$E_K = \gamma_K \frac{M^{5/3} Z^{5/3}}{A^{5/3} R_P^2},$$

---

[5]See Cole, G. H. A., 2005, *Wandering Stars*, Imperial College Press.

where

$$\gamma_K = a \frac{h^2}{8\pi^2 m_e m_p^{5/3}}$$

and

$$a = 9.8 \times 10^6 \, \text{kg}^{-2/3} \, \text{m}^4 \, \text{s}^{-2}.$$

It is seen that this energy increases both with the total mass of the volume and also with the number of protons present. It also decreases as the inverse square of the radius.

The electrostatic energy, E_e, is given by

$$E_e = -\gamma_e \frac{M_P^{4/3} Z^{7/3}}{A^{4/3} R_P},$$

with

$$\gamma_e = 0.95 \times 10^8 \, \text{kg}^{-1/3} \, \text{m}^3 \, \text{s}^{-2}.$$

Finally, the gravitational energy, E_g, is described by the formula

$$E_g = -\gamma_g \frac{M_P^2}{R_P},$$

with

$$\gamma_g = 6.0 \times 10^{-11} \, \text{m}^3 \, \text{kg}^{-1} \, \text{s}^{-2}.$$

Putting these components together according to $E_K = -2(E_e + E_g)$ gives the final formula for the radius of the body in terms of M_P, A and Z which, of course, specify the mass and chemical composition:

$$\frac{1}{R_P} = \frac{\gamma_e}{2\gamma_K} \frac{A^{1/3} Z^{2/3}}{M_P^{1/3}} + \frac{\gamma_g}{2\gamma_K} \frac{M_P^{1/3} A^{5/3}}{Z^{5/3}}, \qquad (1.1)$$

where

$$\frac{\gamma_e}{2\gamma_K} = 5.9 \, \text{kg}^{1/3} \, \text{m}^{-1} \quad \text{and} \quad \frac{\gamma_g}{2\gamma_K} = 3.0 \times 10^{-18} \, \text{kg}^{-1/3} \, \text{m}^{-1} s^2.$$

This formula is a mass–radius relationship and may look complicated at first sight but remember it does cover the full range of masses from the smallest compatible with an essentially spherical shape up to the largest where the atoms have ionised. It will be realised that there are no thermonuclear processes in what we are describing. We now explore this range of masses.

To begin with, we see that Eq. (1.1) consists of two terms. The first term has got the mass in the denominator and so will dominate the second term when the mass is small. The second term has got the mass in the numerator and so will dominate the first term at high masses. For sufficiently low mass the radius–mass relationship is then

$$R_P \approx \frac{2\gamma_K}{\gamma_e} \frac{M_P^{1/3}}{A^{1/3} Z^{2/3}}.$$

Alternatively this can be written $\frac{M_P}{R_P^3} = (\frac{\gamma_e}{2\gamma_K})^3 A Z^2 =$ constant, where the constant is essentially the mean density of the body. This is the condition met with normal planets such as those of the Solar System. The relation implies that the radius increases as the mass increases — the body becomes bigger as it becomes more massive. This is a natural consequence of "hard" unionised atoms resisting gravity, the constituent material being essentially incompressible.

At the other extreme, for sufficiently large mass, the second term in Eq. (1.1) becomes dominant and then

$$R_P \approx \frac{2\gamma_K}{\gamma_g} \frac{Z^{5/3}}{M_P^{1/3} A^{5/3}}.$$

Now the mass and radius are related by $M_P R_P^3 =$ constant. The size of the body now *decreases* with increasing mass, showing the constituent material to be quite compressible. This is the result of ionisation of the atoms. This extreme dependence of size on mass is not found in planets but is, in fact, found to apply to brown dwarf stars.

The full behaviour of Eq. (1.1) showing the effect of mass on the radius can now be made clear. For small masses the curve of radius against mass is an upward increasing curve: for larger masses it is a downward decreasing curve. There is, therefore, a maximum radius at a particular mass which depends on the composition. This is clear from Fig. 1.2 which shows the plot of planet radius against planet mass: curves for both hydrogen and for iron are shown separately. Also included are the points representing the various members of the Solar System. The maximum radius which applies for a particular mass is very clear. That for an iron body occurs at a higher mass than for a hydrogen body but the maximum radius is smaller.

It is seen that all the planets fit within the maximum and minimum curves with the major planets lying closer to the hydrogen line while the terrestrial planets lie closer to the iron line. This reflects the different compositions of the two groups. There are no members over the maximum radius — but Jupiter is not only the largest planet in the Solar System but it is also very close to the largest planet that is possible without any internal pressure ionisation.

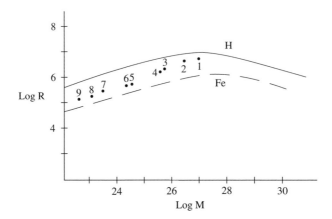

Fig. 1.2. Showing the theoretical relation between mass and radius of a body. Curves for hydrogen and iron are included. Also included are data for the members of the Solar System: 1 — Jupiter; 2 — Saturn; 3 — Neptune; 4 — Uranus; 5 — Earth; 6 — Venus; 7. — Mars; 8 — Mercury; and 9 — Moon.

1.4. Maximum and Minimum Conditions

The conditions of maximum and minimum radius are of importance. In the region of the maximum radius a given radius can refer to one of two masses each on different sides of the maximum. Conditions in this region must be treated very carefully to avoid this ambiguity. It provides a natural upper mass boundary for a planet. Below the maximum lie the true planets with no pressure ionisation. Above is something rather different which we will call super-planets. Although super-planets have internal ionisation conditions are not sufficiently extreme for thermonuclear processes to begin. At about 13 Jupiter masses (about 2.5×10^{27} kg) one enters the region of brown dwarfs with deuterium burning and this marks the upper mass limit for a super-planet.

It follows from Eq. (1.1) that the expression for the maximum mass is;

$$M_{\text{max}} \approx \left(\frac{\gamma_e}{\gamma_g} \right)^{3/2} \frac{Z^{7/2}}{A^2},$$

while that for the associated maximum radius is

$$R_{\text{max}} \approx \frac{\gamma_K}{\sqrt{\gamma_e \gamma_g}} \frac{Z^{1/2}}{A}.$$

Since $A \geq Z$ it follows that R_{max} decreases with increasing A. This means, for example, that the maximum radius is smaller for a silicate body than for a hydrogen one. A comparison with the planets Jupiter and Saturn is:

$$R_{\max}(H) \approx 1.19 \times 10^8 \text{ m}, \qquad M_{R\max}(H) \approx 2.69 \times 10^{27} \text{ kg}$$
$$R(\text{Jupiter}) = 7.15 \times 10^7 \text{ m}, \quad M(\text{Jupiter}) = 1.90 \times 10^{27} \text{ kg}$$
$$R_{\max}(\text{He}) \approx 4.21 \times 10^7 \text{ m}, \quad M_{R\max}(\text{He}) \approx 1.90 \times 10^{27} \text{ kg}$$
$$R(\text{Saturn}) = 6.03 \times 10^7 \text{ m}, \quad M(\text{Saturn}) = 5.69 \times 10^{26} \text{ kg}$$

It seems that Jupiter, taken as a hydrogen body, is very close to the maximum mass. It is clear that Saturn is not a helium body.

The energy associated with the maximum conditions gives insight into the physical circumstances leading to the maxima. Consider the gravitational energy per atom for the maximum conditions. The number of atoms in the body at maximum mass is given by the expression:

$$N_{R\max} \approx \frac{M_{R\max}}{Am_p} = 1.61 \times 10^{54} \frac{Z^{7/2}}{Am_p}.$$

The corresponding gravitational energy is given by

$$E_{R\max} \approx -\frac{GM_{R\max}^2}{R_{\max}} = -4.1 \times 10^{36} \frac{Z^{13/2}}{A^3} m,$$

so the gravitational energy per particle is

$$\mathrm{E}_{R\max} = \frac{E_{R\max}}{N_{R\max}} \approx -2.5 \times 10^{-18} Z^3.$$

For hydrogen this has the value:

$$\mathrm{E}_{R\max} \approx -2.5 \times 10^{-18} \text{ J}.$$

This is extremely close to the lowest ionisation energy for hydrogen, viz. 13.6 e.v. $\approx 2.2 \times 10^{-18}$ J, and can be identified with it within the approximations made in our deductions. Any greater mass will cause ionisation towards the centre. The situation is now clear. There is no ionisation due to pressure within the planets (to the left of the maximum) but there is ionisation within the super-planets (to the right). Of course, thermonuclear processes do not occur in either case. Any planetary object with a mass greater than two or three Jupiter masses must be considered to be a super-planet.

Although we have considered a hydrogen composition specifically the same arguments apply for silicate/ferrous bodies. The mass dividing the planet from the super-planet state is greater than for the hydrogen case as is clear from the formulae above. At the same time the maximum radius is smaller making the mean density high. No bodies of this composition at the maximum radius have yet been found but this doesn't mean they do not exist. The associated acceleration of gravity would be very great making them rather pathological objects from the point of view of living creatures.

We have considered the maximum mass but what can we say about the minimum mass? The minimum mass is taken as that mass for which gravity is just strong enough to cause the mass to assume an essentially spherical shape. For this to be possible the self-gravitational energy per atom must exceed the atomic interaction forces. These can be supposed to have the general magnitude 10^{-19} J. For a mass M of radius R and of composition Am_p the energy per constituent atom is given by the expression

$$\varepsilon \approx \frac{GM^2}{R}\frac{Am_p}{M} = \frac{GMm_pA}{R}.$$

We now require that $\varepsilon \geq\approx 10^{-19}$ that is

$$\frac{M}{R} \geq \frac{5.98 \times 10^7}{GA}.$$

This can be written in terms of the mass alone if the density ρ is specified. Then

$$M \approx 4.62 \times 10^{11} \left(\frac{3}{4\pi\rho G^3 A^3}\right)^{1/2}.$$

The minimum mass depends on the composition. For a silicate body it is of the order 10^{15} kg while for a largely ice body it is of the order 10^{19} kg. Bodies with masses below these values will be irregular and the internal structure will not be controlled by gravity. This is the case for the smaller asteroids in the Solar System and for the smaller satellites. The larger Kuiper belt objects (such as Pluto) will just fit into the lowest mass category and be spherical bodies.

1.5.   Planetary Bodies: Cold Bodies

We have seen that it is the mechanical equilibrium between two forces which control a planetary mass. These are the self-gravitation of the total mass on the one side and its incompressibility on the other side. It is really a question of atomic packing. There is no requirement that the body orbit the Sun. It could, indeed, orbit a planet (when it becomes a natural satellite) but it could actually be entirely free and still satisfy our criteria. Such a wide range of masses with a common basis, encompassing planets and satellites, can usefully be given a single title. We will call them planetary bodies. In particular this will cover collectively all bodies which are intrinsically spherical and whose constituent atoms are not ionised. There is no intrinsic heat component in the equilibrium because the degenerate energy is independent of temperature so it is natural to regard a planetary body as a cold body.

The result of packing under the control of gravity has a very important consequence. Because, by definition, the gravitational force is stronger than the forces between the atoms within the interior it will determine the relative arrangements of the atoms and molecules. The heavier ones will sink towards the centre forcing the lighter ones towards the top. The result in practice will be an interior differentiated by density. With the silicate/ferrous composition, the ferrous components (iron, nickel, cobalt and so on) will concentrate at the centre to form a core encased in a silicate mantle. Outside everything will be a crust of silicate materials. Whilst the interior will be closely spherical the surface can be somewhat irregular because the self-gravity there is weak. Indeed, there will be a boundary where the gravitational energy is comparable to the gravitational energy. This theoretical division has a real example in the interior of the Earth shown in Fig. 2.1. The body may be encased in an atmosphere and this can be very extensive. While the atmospheres of Venus, Earth and Mars are relatively small components of each planet those for the major planets are very substantial. While the central cores in each case are most likely composed of a few Earth masses of ferrous/silicate materials, the atmospheres of Jupiter and Saturn will be composed of hydrogen with some helium while those of Uranus and Neptune will contain some water as well. This makes a Jupiter mass body a small version of the Sun but with a distorted cosmic abundance of the chemical elements. The pressures in the deep atmospheres will ensure liquid-like densities for the mantles but with a relatively thin top layer of more usual gas densities. The silicate/ferrous cores will be liquid with a broadly thin boundary with the mantle but it probably academic whether or not we regard the nonferrous/silicate components as part of a mantle or just part of a very complex atmosphere.

1.6. Methods Used to Detect Bodies of Planetary Mass

We now pass to the more general issue of the search for planetary bodies orbiting stars other than our Sun. We are used to looking at the members of the Solar System through telescopes and having direct contact with them by means of automatic space vehicles but new techniques are necessary when we wish to explore the environs of other stars that are so far away from us. The difficulty is partly because the planetary bodies appear to be so small at their large distance from us but also because, when viewed from the Earth, the light from the central star will totally obscure the light reflected from any planets that may be there. In general terms the light from the central star will typically be a million times brighter than that of the planet. There is the possibility in some cases of shielding off the light from the central star (see Fig. 1.7) but this is still rare and in general the exploration of possible planets around other stars, the exo-planets, has involved indirect methods of observation. Fortunately there are a number of alternative ways to do this. Although it will not be relevant to us now we might remember that the very first planets to be discovered were not around solar-type stars but around a pulsar and were discovered through the radio wavelength emissions.

1.6.1. *Motion of the central star*

Johannes Kepler gave the analysis for the motion of a single planet about the Sun in his famous three laws of planetary motion which can be extended immediately to apply to the motion of an exo-planet about a star. These laws are expressions of the conservation of energy and angular momentum of the pair of bodies in their motions. They describe the limiting case of a very small body orbiting a very massive one where the massive one does not move. This may be a very good approximation for a planet such as the Earth about the Sun but the central Sun (star) in fact responds to the motion of the planet even if only in a small way. Actually when one body orbits another under the action of their mutual gravity the combined motion is a synchronised movement of both bodies in elliptic orbits about the centre of mass of the pair. The orbit of the lighter planet will be the larger but that of the heavier star need not be insignificant and can be open to detection. The motions are shown in Fig. 1.3. The central feature is the movement of each with a common time period. The rate of the movement of the planet and the star is such as to make the (hypothetical) line joining the body to the star sweep out equal areas of the orbit in equal times. These laws in fact apply only approximately if there is more than one planet due to interactions between the planets. The contributions are separate so the effects of more than one planet can be represented as a super-position of the individual effects.

   To gain an idea of the relative scale of the orbits, for the case of Earth/Sun the major axis of the dotted orbit is 1 AU long whereas that of the Sun (solid orbit) is about 500 km long and well within the body of the Sun. The major gravitational influence in determining the position of the centre of gravity of the System is, however, Jupiter. The centre is calculated to be at a distance of about 7.41×10^5 km from the solar centre which places it at a distance of about 4.4×10^4 km above the solar surface. The semi-major axis for the solar orbit (solid curve) will therefore be closely to 7.41×10^5 km. This is to

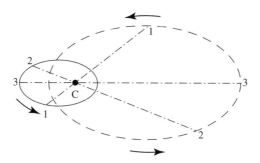

Fig. 1.3. Showing the elliptic orbits of a central star (solid line) and a planet (dotted line). The motions are synchronous with the corresponding points 1, 2 and 3 of the orbits marked. C is the centre of mass of the pair of bodies.

compare with the distance of 7.75×10^8 km as the size of the Jupiter orbit (dotted curve) or a factor of about 1,000 larger.

It would be best to measure the scale of the dotted orbit directly but the planet is too faint to be seen. Instead one must try to measure the size of the solid orbit and then infer the remaining information. The measurement is the observation of the very small wobble of the star. This proved too difficult a task until 1995 when advances in spectroscopic techniques made it possible using the Doppler shift technique. The actual measurements are of the Doppler shifted motion of the star. This is in synchronous motion with the planet so once the stellar motion is measured the planetary path can also be found using Kepler's laws of planetary motion. From the Doppler effect, as the star in its orbit approaches the observer on Earth the light is moved to the blue: as it moves away at the other end of its orbit the light is red shifted. By linking the shifts to speeds a curve of speed against time can be drawn. The form of the curve gives information about the ellipticity of the orbit. An exactly circular orbit gives an exactly sinusoidal speed–time plot because the orbital speed is constant. Such a plot is shown in Fig. 1.4. One time period is shown in the figure.

As an example of the analysis let us consider a single planet orbiting a star in a circular orbit, which is the simplest case. The details of the argument follow from Kepler's laws. The conservation of energy for the planet is expressed by the speed in the orbit v_p in terms of the planet mass M_P, the semi-major axis of the orbit a and the constant of gravitation G. This expression is

$$v_p = \sqrt{\frac{GM_p}{a}}.$$

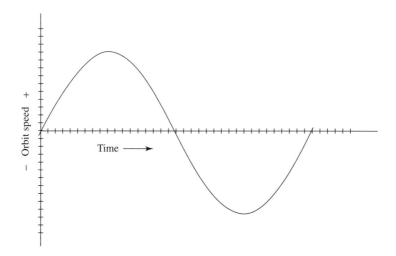

Fig. 1.4. A schematic representation of orbit speed against time for a planet in a circular orbit. The curve is a simple sine curve.

Kepler's third law expressing the conservation of angular momentum of the planet in the orbit is written as

$$a^3 = \frac{GM_o}{4\pi^2}T^2,$$

where M_o is the mass of the star and T is the period of the orbit (which from Fig. 1.3 is obviously the same for the planet and for the star). Now combining this with the statement of the conservation of momentum

$$M_p v_p = M_o V_o,$$

where V_o is the actual speed of the star in its orbit. The Doppler measurements allow the mean speed K of the star to be measured which is different from V_o. Unfortunately the orientation of the axis of the planet is not known so the speed cannot be measured unambiguously. If the angle i denotes the inclination of the orbit to the observer ($i = 90°$ when the orbit is in the same plane as that of the Earth — the ecliptic plane). Explicitly, $K = V_o \sin i$ so that $V_o = \frac{K}{\sin i}$ then the measurement gives $K = V_o \sin i$. The conservation expression is actually

$$M_o v_o = M_o V_o \sin i.$$

These formulae are combined to provide an expression for $M_p \sin i$. Unfortunately this is not M_p itself. Remember, $\sin i \leq 1$ so this method allows only a minimum value of the mass of the planet to be deduced if the mass of the star is known alternatively. The stellar mass follows reliably from aspects of the theory of stellar structure. It might be expected that the inclination angle will be in the neighbourhood of 90° so the calculated mass which would make the deduced planet mass close to the actual mass. The formulae lead to the expression for K

$$K = \left(\frac{2\pi G}{T}\right)^{1/3} \frac{M_p \sin i}{(M_p + M_o)^{2/3}}.$$

The extension to the more general case of an elliptic orbit offers no problems. Then it is found that the measured orbital speed of the star is linked to the other variables according to the expression including the ellipticity of the orbit ε:

$$K = \left(\frac{2\pi G}{T}\right)^{1/3} \frac{M_p \sin i}{(M_p + M_o)^{2/3}} \frac{1}{(1 - \varepsilon^2)^{1/2}}.$$

For a circular orbit $\varepsilon = 0$ while other values are obtained from the speed–time curve.

The speed in an elliptic orbit is not constant. It has, according to Kepler's second law of planetary motion, a region (near perihelion closest to the star) where the speed is greater and the aphelion region (at the other end) where it is less. In the speed — time plot this gives a region where the changes are faster than the other region where they are slower. The effect is a distorted sine curve in which the degree of distortion reflects the degree to which the orbit differs from the circular. This behaviour is shown schematically in Fig. 1.5.

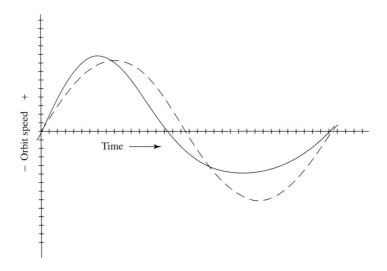

Fig. 1.5. A schematic representation of the speed–time curve for a planet in an elliptic orbit. The dotted curve shows the curve for the corresponding circular orbit.

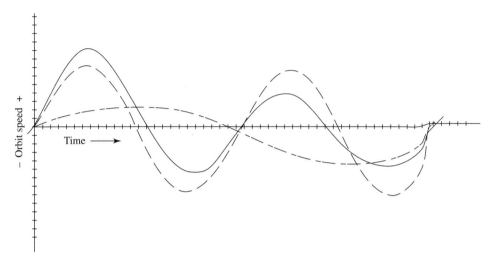

Fig. 1.6. A schematic representation of the orbit curves for a multiple planet system. The two dotted lines represent the circular paths of two orbiting planets while the solid curve is the resultant that the observer measures.

An extension of the analysis allows more than one orbiting body to be recognised in principle. The measured curve can get difficult to interpret and the results a little ambiguous because the orbital characteristics are not known at the beginning. It is realised, of course, that the different planets will influence the central star differently with their masses and this can act as a useful guide. The basis of the approach is shown in Fig. 1.6 where the solid measured line hides the two orbiting planets represented by the broken lines. The dominant effect is the simple dashed line while the curve for the second planet is the dash–dotted line. The two components could be identified by what amounts to a Fourier inversion.

The great majority of the exo-planetary systems found so far have resulted from the use of this Doppler approach. Of the 333 planets found to date (December 2008) 263 planetary systems have been found by the spectral shift involving 307 planets with 31 multiple systems which is about 13% of the total observed so far.

1.6.2. *Direct transit*

Disadvantages of the Doppler method are that it does not provide any information about the orientation of the orbit or about the radius of the planet itself. If the mass and radius of the planet were known the mean density could be calculated and this would give some indication of the composition. It would also provide a useful confirmation of the authenticity of the Doppler method — one can never cross check too much. The radius of the planet can be deduced directly for those star–planet systems where the plane of the planet orbit is parallel to that of the Earth about the Sun. As the much less luminous planet moves across the face of the bright star it will cut off a proportion of the light radiated and its presence can be detected if this so-called transit can be observed from the Earth. Again, the effect is very small (the observed luminosity of the star may be reduced by as little as 1 part in 10,000 or less) but yields full information of the planet. The observation, then, is the light curve for the star with periodic minima as the planet moves in its orbit. The result is direct information of the unseen planet.

The quantity of light lost is a measure of the area of the planet. Presuming it to be essentially a sphere the cross-sectional area, A, allows the radius, R, to be inferred because $A = 4\pi R^2$. The periodicity of the light curve gives the period of the planet in its orbit. Knowing the mass of the star allows the semi-major axis of the planet orbit to be found from which the definition of the centre of mass and the known mass of the star allows the semi-major axis of the of the stellar ellipse to be deduced. This essentially specifies the complete system. Unfortunately the number of systems so aligned to show a transit is very limited but those that are available add confidence to the general approach. As to numbers, 54 planetary systems have been found so far by this method involving 54 planets with no multiple systems. It is, perhaps, surprising that no multiple systems have been found — one might have expected five or six from the experience of the Doppler approach. There are plans to launch automatic probes to observe such transits and more especially NASA launched the probe code named Kepler for this purpose during March 2009. It will take some time before any clear results can be expected.

1.6.3. *Gravitational micro-lensing*

It is a result of general relativity that matter attracts electromagnetic radiation as if it were mass — energy and mass are related. This was first demonstrated by Sir Arthur Eddington (1882–1944) in the solar eclipse of 1919 when it was found that a star actually hidden by the Sun in purely geometrical terms was nevertheless visible during the period of darkening. The Sun had moved the trajectory of the light round to be visible to the observer. The same effect applies in other cases and is available for detecting planets

around stars. It has been used successfully on a number of occasions so far
providing eight planetary systems with one multiple system.

The approach is the analogue of diffraction in the behaviour of light round
an obstacle but with gravitation being the essential component in this gross
astronomical case. The magnification comes from the increased angle of the
incoming light caused by an obstructing mass. Suppose a distant star is
obstructed by a strong gravitational source so that in geometric terms it
cannot be seen from the Earth. The gravitational attraction of the strong
source can bend the light round to become visible from Earth. The view from
Earth will be a characteristic gravitational "diffraction" pattern revealing the
presence of the hidden star. Now suppose the strong gravitational source is
orbited by a planet. The effect is to distort the observed diffraction pattern in a characteristic way that can be interpreted in planetary terms. The
method has the merit that the detection is independent of the mass of the
planet which should make it especially useful in the detection of Earth mass
bodies.

1.6.4.   *Direct imaging*

As Earth based telescope apertures become ever larger and space craft ever
more sophisticated there is the possibility of photographing a planetary
object directly. This hope is being realised to a limited degree. Very recently
the planetary system HR 8799 has been photographed by artificially obscuring the radiation from the central star. This is a system with three planetary
bodies b, c and d, 130 ly from our Sun. The system is shown in Fig. 1.7. The
central pattern is the obscuration of the light from the central star to allow
the planets to be seen. The planets have masses of several Jupiter masses
and so are not directly relevant to our studies but could well be the largest
members of a system which contains at least one body of Earth mass. The
photograph is significant, however, in being among the first to show exo-
planets directly in orbit about a star.

A comparison between the results the various methods have provided
so far is made in Table 1.11. It is seen that the overwhelming number of
discoveries has been made using the Doppler approach but this situation
is likely to change substantially in the future as the other techniques gain
in sophistication. The timing methods are linked to pulsars and have been
included in the table for completeness. The lack of multiple systems in the
case of transit measurements could be the result of different inclinations of
otherwise hidden planets.

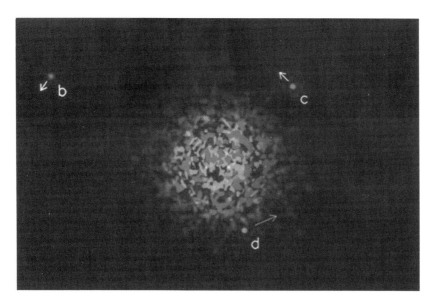

Fig. 1.7. The three planet system HR 8799 photographed in infrared light using the Keck 10 m telescope on Hawaii. The three planets, labelled b, c and d, cover essentially the same spatial range as Sun–Neptune in the Solar System. The central patch is the obscuration of the light of the star (C. Mamis *et al.*, NRC, Canada).

Table 1.11. This gives a list of exo-planets found using the different methods of detection. The pulsar systems at the bottom of the table are added for completeness. The numbers refer to 1 March 2009.

Method	Results
Doppler shift	270 planetary systems 316 planets 33 multiple systems
Transit measurements	58 planetary systems 58 planets No multiple systems
Microlensing	7 planetary systems 8 planets 1 multiple system
Imaging	9 planetary systems 11 planets 1 multiple system
Timing methods (pulsars)	4 planetary systems 7 planets 2 multiple systems

1.7.   Observed Exo-planets

There had been speculation for the last several hundred years that planets could exist around other stars like our Sun. Of course they would be expected to be composed of orbiting Jupiter-type and Earth-type planets with a general distance pattern similar to the Solar System (plane orbits contained within a circle some 40 AU in radius). If other copies of the Solar System exist why, then, should they not be inhabited by thinking creatures like us? The idea had been developed to a high degree by science fiction writers while the (still unexplained) phenomena of flying saucers even suggested to some that this was reality. The mind is the fertile source of much imagination.

This was the situation up until the early 1990s when two extraordinary things happened. In 1992 Aleksander Wolszczan and Dale Frail announced the discovery of three planets around pulsar PSR $1257 + 12$. These planets were some 5 AU from the planet (roughly the distance of Jupiter from our Sun) and had a mass of a few Earth masses. Here were the long sort after planets but sadly they were not orbiting a normal star but a pulsar that had gone through a supernova experience. The conditions on them were unlikely to be anything like on Earth. Then in 1994 the second system was found PSR B 1620-26B with an orbiting planet of mass $2.5\,M_J$ ($M_J = 1.98 \times 10^{27}$ kg). These were discovered through timing in radio emissions.[6] The real breakthrough from our point of view came in 1995 when Michel Mayor and Didier Queloz announced the discovery of a planet in a circular orbit around the star 51 Pegasus, a solar-type star. The single planet is referred to as 51 Peg b. The measured speed was $K = 53$ m/s. The discovery was soon confirmed by Macey and Butler. The corresponding curve is shown in Fig. 1.8. This discovery looked much more interesting although the observation actually gives unexpected results as we will now explain.

The single planetary member orbiting the central star is calculated to have a minimum mass of $0.5\,M_J$, where $M_J = 1.98 \times 10^{27}$ kg is the mass of Jupiter. This is $M_p \geq 7.92 \times 10^{26}$ kg which is a body with of mass midway between the masses of Jupiter and Saturn. The sinusoidally shaped speed–time curve shows the orbit to be almost exactly circular. So far, so good. There could well be accompanying Earth-type orbiting members which are not heavy enough for the observations to reveal — or could there? The big surprise is the value of the semi-major axis, a, of the orbit: the measured

---

[6]Two further systems have been found by timing methods: V391 Peg b with a planet of mass $3.2\,M_J$ discovered in 2007 and in 2008, HW Vir b,c with masses, respectively, of $19\,M_J$ and $8\,M_J$. The stellar surface temperatures are too low to be of interest in our discussion.

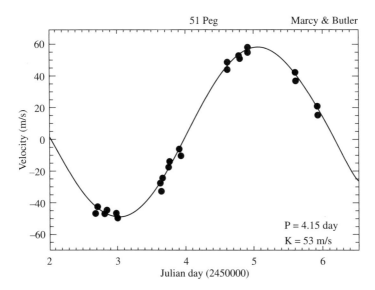

Fig. 1.8. The velocity–time curve for 51 Pegasus discovered by Mayor and Queloz of the Geneva Observatory. The curve shown is due to Marcy and Butler. This curve should be compared with Fig. 1.4.

period of $T = 4.15$ days corresponds to a value $a = 0.05$ AU $= 5.96 \times 10^5$ km. For comparison, the planet Mercury has an orbit of mean radius 0.3 AU $= 4.47 \times 10^7$ km and Jupiter a mean radius close to 5 AU $= 7.5 \times 10^8$ km. Apparently, the planet is uncompromisingly close to the star. Assuming the planet radius is similar to Jupiter/Saturn, calculations suggest the "surface" temperature will be in excess of 1,100°C. It seems that it is, nevertheless, possible for the planet to retain a substantial atmosphere even under these extreme conditions. As a comparison with the measured time period there are indications that the rotation period of the star 51 Peg is of the order 30 days (similar to our Sun). This knowledge confirms that the measured time period is of the orbit and not of the star itself. This is a strange system and these features of a small distance between the star and its planet has proved to be a feature of many of the 300 systems or so found so far (see Fig. 1.9).

Since the initial discovery in 1995 ever more exo-planets have been found and more are being found every month. At the present time (August 2007) 333 planets have been found of which 25 are in multiple systems. Radial velocity techniques have discovered 236 of these planets which include the 25 multiple systems. Transit measurements account for 23 planets, micro-lensing for 4 planets while imaging also accounts for 4. These are the numbers now but it must be realised that the numbers are growing by a few every month. There are probably just enough planets available to provide reliable

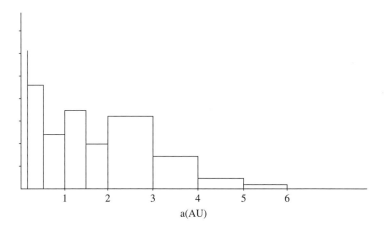

Fig. 1.9.   The observed distribution of semi-major axes for the exo-planets up to 6.0 AU.

generalised information but it is always possible to find new and contradic-
tory data in the future. The data associated with these planets show many
unexpected things and we set out these details now in graphical form.

(a) The first extraordinary thing is the semi-major axis of the planets.
The majority of these are very small while only rarely are larger values
encountered. Of the total number of planets no fewer than 123 have semi-
major axes less than 0.75 AU which is 1.12×10^8 m. A further 51 have got
semi-major axes less than 1.5 AU (2.2×10^8 m) while only 9 are greater than
3.0 AU and 1 greater than 4 AU. For scale comparison the semi-major axis
for Jupiter is 5 AU. The full set of data is shown in Fig. 1.9. It is seen from
the figure that there is a concentration of planetary orbits with very small
major axes in the range smaller than those of Mercury and Venus about the
Sun. The figure does not include data beyond 6 AU because such data are
not numerous at the present time. The indications there suggest that the
number of systems with larger orbital sizes decrease strongly with distance.

(b) The eccentricities of the planetary orbits are also very different from
those that would have been expected. A schematic plot is given in Fig. 1.10.
A circular orbit is associated with a small value of the eccentricity, ε, and
for the Solar System the plot would be confined to very small values. Surprisingly, only 28% of the orbits have eccentricities in the range 0–0.05 while
only a further 15.9% have orbits in the range 0.05–0.1. This means that
56.1% of the orbits are markedly elliptical. Indeed, the figure shows some
orbits have values close to unity which means highly elliptical obits with the
planets moving very close to the star at perihelion. In these circumstances
the planet and star almost touch providing the most extraordinary surface

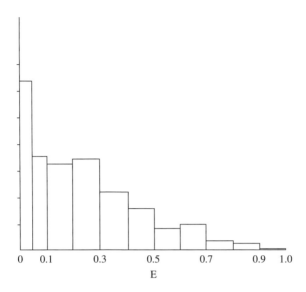

Fig. 1.10. A schematic plot of the eccentricities, ε, of 226 observed exo-planets.

conditions for the orbiting body. This represents a most remarkable series of observations.

(c) The spread of planetary masses is of importance and is shown in Fig. 1.11. The distribution is seen to favour the smaller masses but larger masses are certainly represented. In fact although some 28% of the planetary masses are 1 Jupiter mass or lower (the lowest mass measured so far is $0.016\,M_J = 3 \times 10^{25}$ kg), 17.4% have a mass between 1 and $2\,M_J$ and 9% have a mass between 2 and $3\,M_J$. This still leaves over 40% with masses above $3\,M_J$. It is interesting that there seems to be a gap in masses around 12–$13\,M_J$ which may divide the super-planets from the brown dwarfs. It is generally believed that whereas brown dwarfs form by condensation planets form by accretion. The apparent gap might be a demonstration of the dividing line between the two processes.

(d) Further Comments

There are one or two further comments that can be made about the present state of the observations.

The masses of the central stars show a spread about the solar mass. The smallest mass is $0.3\,M_o \approx 6 \times 10^{29}$ kg which is close to the minimum mass for hydrogen burning. The greatest mass so far is about $2.6\,M_o \approx 5.2 \times 10^{30}$ kg. The central temperature for such a star will allow more than simple hydrogen burning. Of the stars measured to date, about 74% have closely the solar mass while about 26% have a lower mass and about 9%

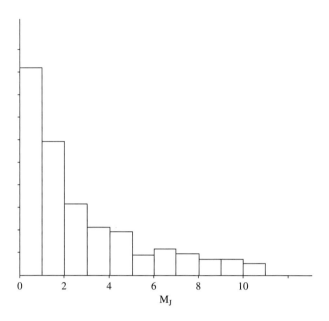

Fig. 1.11.   The distribution of observed exo-planet masses: $1\,M_J = 1.98 \times 10^{27}\,\text{kg}$.

have a higher mass. This could reflect the background distribution of stellar masses for the overwhelmingly most numerous group of stars in the Galaxy.

The composition of a star is described by its metallicity. The composition is divided into hydrogen/helium and the remaining heavier elements. The heavier elements are, for historical reasons, called metals. For the Sun the metals defined this way account for about 2% of the mass. The metallicity of the Sun is taken as unity and that of other stars is compared with it on a logarithmic scale. Of the stars associated with exo-planets examined so far about 66% have a higher (positive) metallicity while the remaining 34% have a lower (negative) metallicity. The proportion of heavier elements in the central star has implications for the planets that might orbit about it.

The distances of the stars are also relevant for our central interest in advanced life and communications. Most lie within 500 pc of the Earth while a few lie further away. These are large distances suggesting the corresponding planetary systems are widely separated.

1.8.   Relevance for the Occurrence of Advanced Life

The data that we have discussed will have relevance to the problem of isolating possible sites for advanced life. The first requirement for a stable evolutionary development is a stable planetary base within the habitable zone (where liquid water can exist freely) of the star in question.[7] In particular the surface temperature must remain sensibly constant and appropriate to allow the necessary chemical processes to proceed at near their maximum efficiency. This implies a closely circular planet orbit. The fact that not many systems meet these requirements was surprising because it reduces the possibility of finding a ready home for technologically capable life. Microscopic life is much less choosy. We must keep these thoughts in mind in the chapters that follow and especially for Chap. 6 where we will attempt to summarise our arguments and reach some form of conclusion about the status of life in the Universe.

1.9.   Summary

1. Several different major components of the Solar System were considered which can be viewed collectively as a presenting concentric regions of different chemical compositions.

    The region 0–5 AU from the Sun contains the terrestrial planets and the asteroids. These are composed primarily of oxygen (the second most abundant chemical element) in combination with the less abundant chemical elements and principally carbon, aluminium, silicon, magnesium, sulphur and iron. These form the basic rocks. They are polluted by other elements such as the rare earths, tin, bismuth, silver and gold among others. Separate from the oxide of iron, pure iron with cobalt and nickel is also present in quite large quantities. There is also a small percentage of the radioactive elements and especially the long half life uranium ^{235}U. There will also have been short period radioactive nuclei early on in the history of the terrestrial planets but these will by now all but decayed. The interiors of the planets, including our Moon, are differentiated according to the density giving the three layers core/inner core, mantle and crust, separated by sharp core/mantle and sharp mantle/crust boundaries. The atmospheres of Venus and Mars have got carbon dioxide as a principle component. The masses are not high — perhaps 10^{16} kg for Mars and 10^{21} kg for Venus. The atmosphere of Earth alone has a high percentage of oxygen (some 21% by volume) and of nitrogen (essentially the remainder) with a total mass of perhaps 10^{19} kg. There are also some very minor components such as argon and neon (see Fig. 2.5).

---

[7]This is sometimes referred to as the Goldilocks zone after the nursery rhyme of the three bears — neither too hot nor too cold but just right.

The Earth is unique in the inner Solar System in having a substantial quantity of water in an ocean covering. It will be remembered that this is a combination of the most abundant and next most abundant active chemical elements. The mass is approximately 10^{21} kg. This is about the same mass as an average comet (presuming there is such a thing) but this is a coincidence. The isotope ratio of oxygen is not the same in the two cases. The Earth is also unique in the Solar System in having an orbit which is in the middle of the range of those which allow liquid water to exist freely on the surface. Because water is essential to life as we understand it, this is the so-called habitable zone around the Sun. Venus is right on the inner edge while Mars is right on the outer one. It is important for life that a planet in this zone has a solid silicate surface and is not gaseous.

The region 5.0–30.0 AU includes the giant (major) gaseous planets. The first two, Jupiter and Saturn, are very large hydrogen bodies with some helium while each of the remaining pair, Uranus and Neptune, has got in addition significant quantities of interior water. Each planet has a heavy core composed of terrestrial rock/iron-type materials of several Earth masses. It might appear that the cores acted as a nucleus for the gravitational accretion of hydrogen and helium during the initial formation processes 4.6 Bya.

The major planets are also accompanied by a very grand range of orbiting satellites. About 60 are known so far and very small ones are still being discovered. The grandest are Io, Europa, Ganemede and Callisto of Jupiter, Titan of Saturn and Triton of Neptune. These are of comparable size and mass to our Moon and these seven bodies constitute the Principal Satellites. Of these Moon and Io are rock/iron bodies but the remaining five, all of the major planets, are different. These are essentially ice (water ice) with an admixture of rock and iron. The proportion of rock varies but is generally of the order of a few percent for most of them. This places water ice in this planetary belt with a likely mass rather less than an Earth mass.

In the region 30–100,000 AU the predominant component is water ice although other ices will be present such as ammonia and methane ices together with some silicate/ferrous dust.

2. Next the general characteristics of a planet were reviewed and it was found most convenient to move to the atomic scale. Two new concepts emerged. The analysis applies to masses quite generally and is not confined to those

orbiting a star but includes also the heavier bodies orbiting a planet rather than the Sun. This suggested calling all the masses planetary bodies.

3. A plot of planetary radius against planetary mass shows a rising curve which reaches a maximum radius at a mass after which the radius decreases again. The rising curve comes from the simple packing of atoms essentially as hard spheres. Beyond the maximum radius the gravitational energy acting on the atoms causes pressure ionisation though there are no thermonuclear processes active. The masses below the maximum, which is about 3 Jupiter masses ($\approx 6 \times 10^{27}$ kg) refer to planetary bodies but those above are super-planets.

4. Next the methods used to detect exo-planets were considered. Most systems detected so far have come from observing the motion of the central star by means of a Doppler shift.

5. The properties of the observed exo-planet systems were reviewed and some comments made about the relevance of these properties to the existence of exo-life. Many observed exo-planets are associated with very small semi-major axes giving a micro-scale to the system. Again, many systems comprise a single large super-planet and not the complex forming the major planets of the Solar System.

Further Reading

1. Cole, G. H. A., 2006, *Wandering Stars*, Imperial College Press, London.
2. The discoveries of exo-planets are listed and kept up to date by Professor Jean Schneider of Observatoire de Paris, 92195 Meudon, France on the website: http:/exoplanet.eu/.

Note added in proof:

In 2009 September the exo-planet system CoRoT–7 was reported, detected by transit measurements. It is at the distance 489 ly (150 pc). There are two orbiting planets of masses a few times that of the Earth. The details are:

Star: CoRoT–7a: mass ≈ 0.93 M$_{Sun}$; radius ≈ 0.87 R$_{Sun}$;
surface temperature ≈ 5275 K; metallicity ≈ 0.03; age $\approx 1.5 \times 10^9$ years

For the planets (inclination of 80 degrees):

CoRoT–7b: mass ≈ 4.8 M$_{Earth}$; semi-maj. axis ≈ 0.0172 AU; eccen. ≈ 0.
CoRoT–7c: mass ≈ 8.4 M$_{Earth}$; semi-maj. axis ≈ 0.046 AU; eccen. ≈ 0

These are the lowest planet masses yet observed orbiting a main sequence star but are below the limits for gaseous compositions. They are almost certainly silicate/ferrous like the Earth though this is not known from the observations. Unfortunately the orbits are anyway too close to the central star to sustain higher life forms.

This system of terrestrial-type planets is interesting because the observed planetary masses are comparable to those of the central regions of the major planets of the Solar System. It is very likely an illustration of a hazard for gaseous planets in very close orbit around a star. The planet can be disrupted and ultimately destroyed by gravitational tidal forces, but at first leaving the core. This is the opposite of pulsar planets where the cores of gaseous planets in larger orbits have been laid bare by a catastrophic stellar explosion. It shows that great care must be taken in assessing the circumstances and age of orbiting bodies with Earth-type masses.

Chapter 2

The Dynamic Earth

The Earth is where we live. It is the platform where life first emerged whatever its origin, then developed and now lives into the future. Our experience during our life time tells of the Earth as an essentially beautiful but unchanging backcloth to our activities and this has been true for essentially all recorded history. There have been some minor upsets but nothing sufficiently serious up until now to stop us regarding the world as a benign and static backcloth for our development without our having to make special reference to it.[1]

Once an accurate figure for the age of the Earth became established[2] it was interesting to ask whether the Earth had always been as it is now and if not what the past conditions had been like. However can this be found? Certainly there is no photographic evidence — or is there, perhaps, something equivalent? The possibility of finding out about something of the past, however meagre, followed the realisation in the 18th century of the significance of rocks and the fossils they contain. If this evidence can be interpreted correctly there is a possibility of piecing together something of the Earth's early history. Although this is a chance study, since finding fossils is not something that can be guaranteed, nevertheless it soon became clear that the Earth environment has changed very drastically several times over the history of the planet. This has had catastrophic effects on the flora and fauna when the changes occurred. We see now that the Earth has never been a constant unchanging place over the longer time span. It so happens that conditions have remained very much the same over the last half a million years but this can be expected to change in the future. Its changes affect the environment as well and, because this provides the back cloth for life, in its turn it also affects life. Consequently, in reviewing the history of life

---

[1]The recent recognition of the action of industrially based global warming came as a shock.
[2]We will find later that the age of the Earth is closely 4.56×10^9 years, written 4.56 By.

on Earth, it is important to trace in parallel the ways the surface of the Earth and its atmosphere have changed over its history. As a matter of fact, this is a good time to approach these matters because palaeo-data are now beginning to have a precision and reliability not known before.

The main problem of these studies in the past has been assigning precise dates to individual rocks. More or less accurate dates were ascribed to individual rock layers at one place but it had proved very difficult to correlate the precise age of one rock structure with another one elsewhere. There has been a rule that rocks that contained the same fossils must be of similar age. This allowed rocks at different places to be put on a *relative* time scale. Unfortunately this identified rocks of essentially the same age but it did not tell precisely what that age actually was. Two techniques have been perfected to overcome this problem. Their application has provided an unbelievable wealth of material.

Almost all rocks contain radioactivity of some sort. Measuring the radioactivity in the material allows us to estimate with some accuracy the age since the rock formed as it is now. The first results began to appear during the early decades of the 20th century. This revolutionised the problem of supplying a date for the geological formations but created some consternation because ages were found that were orders of magnitude greater than those accepted at that time. A more spectacular technique was to come. Virtually all rocks are at least weakly magnetic. Their magnetism was frozen in at the time of their formation and so is evidence of the magnetic conditions locally at the time of their formation deep into the past. This remanent magnetisation of the rocks, as it is called, has led even to an assessment of the latitude and longitude of the rocks at the time of their formation. Applied first very tentatively in the 1950s it revealed an astonishing story of the migration of the continents across the face of the Earth with a time scale provided by the radioactive dating.

The idea that the continents had moved relative to each other was first proposed by Alfred Wegener (1880–1930) in the early years of the 20th century having realised that the contours of the west coast of Africa and the east coast of South America were such a remarkable fit together that, he conjectured, this could not be simple coincidence. The rocks, flora and fauna on each side of the South Atlantic are also remarkably similar. He presented similar evidence to suggest that all the land masses had once at some time in the past formed a single entity which he called Pangea.[3] He lacked firm

---

[3] A more restricted form of the same idea also occurred to Alfred Russel Wallace (1823–1913) rather earlier who defined a line (the Wallace line) in the Malay Archipelago north of which life

evidence to support his theory which was largely ignored at the time but acceptance came much later with the applications of the two techniques, radioactive dating and remanent magnetisation, which confirmed that these movements had indeed happened. Our understanding of the history of the surface of the Earth was totally changed by these measurements and the foundations laid for our present representation of the history. There have, of course, been problems to overcome which will be considered later.

The state of the palaeo-atmosphere and the palaeo-climate at different times in the past is of immense interest and can be found with some reliability in some circumstances although the very earliest times are the most difficult to have any certainty about them. The climate determines what grows and flourishes and here again microscopic fossils of organisms and ancient flora obtained by drilling on land and in the oceans continue to give invaluable information. For instance, the type of flora can allow inferences to be made of the local temperature. It is true that it is necessary to link fossils with modern equivalents to be sure of such inferences but this is most often possible after about 550 Mya. Some information is available before that and it has proved possible to determine the conditions during some periods of the early Earth. The palaeo-atmosphere has also begun to be studied by the relatively new micro-technique of analysing small bubbles of ancient air contained in some rocks and more recently in Antarctic ices. In the present chapter we outline the conclusions drawn from this wide range of observations. More information is accumulating all the time so we will attempt no more than to provide a basis for the reader's further study.

2.1. The Geological Divisions

The geological time scale of the changing surface is most conveniently expressed using divisions based on the changing rock structures that are found now. This geological grid also acts as a very convenient framework for correlating different aspects of the history of life. The geological list is set down in Table 2.1.

The structure of Table 2.1 is the result of a very great deal of study and debate. It was the 17th century Danish doctor Nicholas Steno of Copenhagen (1638–1686) who was the first to realise that rock strata (layers) were laid down in an ordered sequence and reflect the fact that the conditions of the

---

is closely similar to that of Asia but south of which it is much more like that of Australasia. He proposed (correctly) that the two sides of the line had separate histories but he did not know what they were.

Table 2.1. The geological divisions of the history of the Earth as accepted by the International Commission on Stratigraphy (Kya = thousand years ago).

Eon	Era	Epoch		Time scale
Precambrian 4,550–543 Mya	Haden Archaean Proterozoic	Palaeo-proterozoic (2,500–1,600 Mya) Mesoproterozoic (1,600–900 Mya) Neoproerozoic (900–635 Mya) Ediacaran (635–543 Mya)		4,550–3,800 Mya 3,800–2,500 Mya 2,500–543 Mya
Phanerozoic (543 Mya to the present)	Palaeozoic	Cambrian (543–490 Mya) Ordovician (490–443 Mya) Silurian (443–417 Mya) Devonian (417–354 Mya) Carboniferous (354–290 Mya) Permian (290–248 Mya)		543–248 Mya
	Mesozoic	Triassic (248–206 Mya) Jurassic (206–144 Mya) Cretaceous (144–65 Mya)		248–65 Mya
	Cenozoic	Tertiary (65–1.8 Mya)	Paleocene (65–54.8 Mya) Eocene (54.8–33.7 Mya) Oligocene (33.7–23.8 Mya) Miocene (23.8–5.3 Mya) Pliocene (5.3–1.8 Mya)	65 Mya– present
		Quaternary (1.8 Mya– present)	Pleistocene (1.8 Mya–10 Kya) Holocene (10 Kya–present)	

environment remain stable for a period of time after which they change, sometimes dramatically. He also made the implication that the associated animate fossilised matter can also be expected to change with the environment and so with the rock forms. It follows that rocks with the same fossils are of the same age period wherever they may be. Again, one rock layer over-laying another will be younger than the rock below it. All this sounds pretty obvious now but at that time it was very original. Although this comparison provides only relative information and does not lead to an absolute time scale it led, nevertheless, to a classification of the rocks which also provided at least a rough classification of the order of the changes of life over the history of the Earth. Later techniques have fully confirmed Steno's approach and have also allowed an absolute time scale to be identified.

Many, but not all, fossils are unique to particular geological epochs so it is natural to base the animate calendar on the stratigraphic scale. For animate use it is appropriate to divide the times scale into two time periods, called eons, of very unequal length. The first is called the Precambrian and spans the large time interval from 4,550 Bya to 543 Mya.[4] This covers some 88% of the present age of the Earth. The second period (called the Phanerozoic) covers the remaining of 12% of the age being the much shorter period from 543 Mya to the present day. This recent eon has been packed with activity and includes the evolution which has seen the appearance of advanced living forms and especially includes the appearance of Homo sapiens. By contrast, the Precambrian was regarded as devoid of fossils until the last decade or so after which a few microscopic forms have been found in rocks but nothing larger.

2.2.   The Formation and Isothermal Structure of the Earth

None of the rocks of the very earliest Earth have survived to the present day due to the continuously recycled surface, but some knowledge of the conditions about the time of the formation of the Earth can be inferred from a variety of other data. Various pieces of evidence, some direct and some indirect but mainly using radioactivity, place the age of the Earth as closely 4,550 My (4.55×10^9 y or 4.550 By). Evidence from the rocks on Earth, and especially at Barberton in South Africa and the Isua Greenstone belt in Western Greenland, allow us to take the measured age back only to about

---

[4]Note we use the obvious shorthand notation Bya for billion years ago (that is one thousand million years ago) and Mya for million years ago. Similarly, By stands for billion years and My for a million years.

3,800 By. At that time there was certainly an ocean and the rocks are largely volcanic. The Earth clearly existed before that. There is some evidence of elementary fossil bacteria in rocks at Barberton with an age of about 3.2 By. These pieces of information give us a minimum measured age for the Earth but not an actual age. For this we must look elsewhere.

It is firmly accepted now that an early Moon was orbiting the Earth from its formation — indeed, it is most likely that Earth and Moon shared a common origin. The NASA Apollo Moon landings of the 1960s allowed the lunar surface to be studied in situ and also for rocks to be brought back to Earth for laboratory study. It soon became clear that much of the lunar surface now is essentially the original pristine surface and that the oldest radioactive date for the rocks can give the age of the Moon directly and, on current thinking, also that of the Earth. The age of the Moon deduced by this method is closely 4.540 By so this becomes a lower estimate of the age for the Earth itself. There is further support for the same age in the study of meteorites that have fallen to Earth. The results are significant. The meteorite that landed at the Mexican village of Pueblito de Allende on 8th February 1969 (and so referred to as the Allende meteorite) is the oldest found so far with an age of formation of precisely 4.560 ± 2 By. It contains some microscopic diamonds with a strange signature which apparently indicates an extra-solar origin. Because it is of the same age as the other components of the Solar System, the composition of the Allende meteorite is taken as representative of that from which the Solar System was formed. More than this, the age of Moon/Earth is taken now to be the age of the Solar System itself. It is generally accepted that the initial trigger that led to the formation of the Solar System was a supernova explosion of a star which caused a cloud of molecular gas nearby to condense. On this basis the Sun will be only marginally older than the planets, with a theoretical age approaching 5.0 By.

2.2.1.   *The early earth*

There are clear indications that the Earth had a very fiery beginning with a molten surface resulting initially from the accretion of the materials that formed it. The heat energy here comes mainly from the transfer of kinetic energy of the material falling inwards to random energy as the accreted material is brought to rest in the implosion: there is also a contribution from the consequential compression of the material. The near surface layer will have cooled quickly from the molten by radiation into space and calculations suggest that a "solid" surface will have formed after a few hundred million years and indeed that a layer some 200 km thick could have cooled to essentially solid form in about 200 My. This would mean that the Earth had a solid surface of sorts by about 4.350 Bya. This estimate is consistent with the discovery of Zircon crystals from the Jack Hills of the Narreyer Gneiss Terrane, Western Australia, which have been dated with ages between 4.4 and 4.3 By. Obviously, a continental crust region must have formed by that time. The ^{18}O values in the crystals are high and there are micro-inclusions of SiO_2. This is consistent with a granitic volcanic source under water which is another evidence that there will have been some form of liquid water surface covering at that time. With water present it follows that the near surface temperature must have been appreciably below 100 °C so that water could condense and form at least a shallow covering for the planet. These rather high temperature conditions at the surface were supported by high internal temperatures within the Earth and these would give rise to extensive volcanic activity as a mechanism for losing internal heat. The level of radioactivity would have been at least double the present value. This estimate follows from the radioactive decay of the elements known now supplemented by contributions from short lived radioactive elements which by now will have all but decayed away.

It is believed that the atmosphere actually associated with the formation of the Earth was largely composed of hydrogen with helium but, due to the relatively small mass of the Earth and its high surface temperature, these gases will have been lost very quickly into space. Volcanoes release a range of gasses, most of them noxious to us. The main components will have been broadly CO_2 and CO together with sulphur compounds, nitrogen (probably in the form of ammonia) and a great deal of water. The surface waters would have produced a substantial water cloud structure in a basically CO_2 atmosphere. It is these components that would have formed the first stable atmosphere. Continuing volcanism would have provided ever more water giving a water layer on the surface of steadily increasing thickness. Evaporation will have produced rain and the rain would have been contaminated with carbolic acid resulting from the interaction between H_2O and CO_2. There would also have been some mineral input from rain running off what land there was. This process is called weathering and will have been active quite early on. The very earliest picture must be of a wide range of very active modest sized volcanoes protruding from hot lagoons and shallow seas partially obscured by steam. There is no clear indication of what the depths of the seas might have been at this early time. The lava issuing from the many volcanoes spread out on the surface show every sign of having extruded under water (see Fig. 2.1) forming ever more land and the first concentration of

Fig. 2.1. Irregular volcanic rock formed underwater 3.5 Mya in the Barberton, South Africa. The random structure is typical of molten rock cooled by water as it is extruded at very high temperature from a volcanic nozzle. It is clear that this region, though now land, was under a considerable cover of water at this early time in the past.

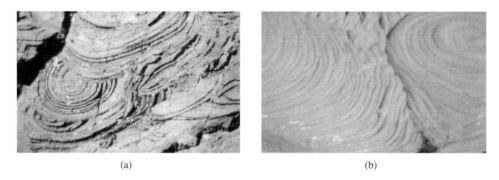

(a)   (b)

Fig. 2.2. (a) An ancient volcanic mud pool fossilised in rock. The rock is in Barberton, South Africa and is 3.5 By old. (b) For comparison a modern active volcanic mud pool from New Zealand. This comparison shows that the same mechanism for losing heat was operative then as now. This does not of itself, however, prove that plate tectonics had already developed at this early time.

hard, dry land formed at this time. It will have contained lava mud pools and an ancient example is shown in Fig. 2.2(a) with a modern one for comparison (Fig. 2.2(b)). If the land stayed a unit as time went by it is called a craton. There could have been more than one such formation. This will have included the Yilgarn craton containing the Zircon crystals that we have mentioned already. The material issuing from the many volcanoes over time will have formed growing quantities of land around the cratons to form the continents — the first continents will have been rather small but will date from about this early time.

2.2.2. *Internal differentiation*

The material of the initial condensation to form the Earth was a random mixture of silicates and ferrous perhaps with some water ice. Chemical processes converted the initially simple material into a more complex molecular form providing a range of materials of different densities mixed together. This was not mechanically stable and separated slowly under gravity into chemically distinct layers. The hot plastic interior allowed the most dense ferrous materials to sink into the interior so forcing the less dense materials upwards towards the surface. Thus began the chemical differentiation of the interior which is the central feature of the Earth today and, incidentally, of the other terrestrial planets and of the Moon. The increasing pressure with depth leads to the appearance of different forms of the basic molecular aggregates providing a rich mineralogy. An equilibrium was eventually established which can be described in terms of an hydrostatic equilibrium. The result is the "three layer" Earth with a heavy (ferrous) core forming about one half of the central region encased in a mantle with a thin relatively low density crust enclosing the whole. The core itself has been found to have a structure with a solid inner component (the inner core) and a liquid outer core. This is shown schematically in Fig. 2.3.

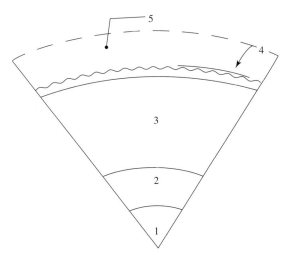

Fig. 2.3. Schematic cross-section of the Earth. Region 1 is the solid ferrous inner core; region 2 is the liquid ferrous outer core; region 3 is the silicate mantle; region 4 is the crust; and region 5 is the atmosphere. The core occupies closely half the radius from the centre. The equatorial radius is 6,378 km and the mass is 5.974×10^{24} kg.

2.3.   Internal Thermal Balance

The dynamical behaviour of the Earth has been driven by the processes leading to a loss of initial internal heat. There is a gradient of temperature through the Earth from the central core region where the temperature is estimated as being about 6,000 K to the surface where it is rather less than 300 K. This gives a very small mean gradient throughout the Earth of a little less than one degree per kilometre. This is about 0.001 K per metre (or 10^{-5} K per cm) which is a very small gradient indeed. In fact the gradient is not uniform and is much greater than this in certain regions inside (for instance where there is a concentration of radioactivity) but this detail can be neglected now. There are two separate contributions to the gradient. First it must be realised that there would be a gradient of temperature anyway (even for body without independent heat sources) due simply to the increasing compression with depth below the surface. The temperature gradient which mirrors the compression is the so-called adiabatic temperature gradient. Being a compression effect it does not of itself give rise to a mean flow of heat. The internal material would be in a stable hydrostatic equilibrium under this gradient alone with no resultant flow of material. There is, however, a separate heat flow throughout the volume due to a source of heat inside partly from the formation of the planet and partly from presence of internal radioactive elements. Heat is then moved to the surface, down the temperature gradient, certainly by thermal conduction but, if this is not sufficient to carry the energy, heat is also carried by the direct convective movement of the material itself. This advection to the surface using material motions results in a set of complex convection currents throughout the volume which breaks the condition of strict mechanical equilibrium. This disequilibrium also affects the surface with dramatic consequences for living materials. In spite of this it may still be possible for the interior to be in a state of quasi-hydrostatic equilibrium.

The precise conditions inside will provide the stability of a self-correcting system through the action of irreversible processes. The convective motion of the fluid will be restricted by the retarding action of viscosity, a friction force. This arises ultimately from the resistance to change of shape of molecular arrangements due to the action of the attractive forces between the molecules. It is an inner friction which tends to oppose change. The effect depends on the temperature which affects the kinetic energy of the molecules. Increasing temperature allows the molecules to move further apart which reduces the attractive force between them. The higher the temperature the greater is the intrinsic molecular movement and so the lower is the force resisting change of the configuration. This means the viscous drag is lower. The effect of temperature is very marked and can be represented by an exponential expression. Here is the basis of a controlling mechanism.

There is a limit to the quantity of heat that can be conducted from the hotter to the colder materials through a given temperature gradient. This again is determined by nature of the interatomic forces. Beyond this limit the static conduction of heat must be supplemented by the physical movement of the fluid carrying heat through the gradient — this free convection of fluid is often called advection. The reason for this is that the density of the hotter material lower down is less than that of the colder material above which is unstable in the presence of the gravitational field. Movement of the fluid will supplement the simple conduction for the transport of energy and act to sustain mechanical equilibrium. The movement of fluid, however, will introduce an opposing viscous force so there is a tendency for the motion to be stopped if it is too weak. This brake prevents small movements of the fluid. It means the convection cannot begin for a very small excess density gradient in the fluid and a critical value must be passed before motion can take place. When motion does occur equilibrium can be established between the force driving the convective flow and the force of viscosity opposing the motion. Motion will occur when the driving force is the larger of the two. The crucial factor is the strength of the viscous force which depends on the local temperature.

The material forming the mantle of the planet is, in fact, highly viscous and will not be like the normal fluid (like water) that we might first think of. Over short time scales it will behave as an elastic solid as is shown by its ability to carry both longitudinal and transverse elastic seismic (earthquake) waves. Over the longer term, however, it will flow slowly as a very viscous fluid. As the temperature builds up across a small volume the viscosity will decrease allowing the fluid to convert heat along the direction of the gravitational field. This release of heat will lower the temperature and so increase the viscosity and so the resistance to the flow. The flow will be reduced and so the temperature will rise. The result is a reduction of the viscosity allowing more movement of the fluid. The action of this control mechanism will take place over a considerable period of time, perhaps many thousands or even millions of years. In this way the interior will be maintained in a state of quasi-equilibrium as the internal temperature store gradually reduces. Even though there is a mean heat gradient throughout the interior the magnitude will be sufficiently small for thermal equilibrium to be effectively achieved everywhere locally to a high degree of approximation. This gives rise to the condition of local thermodynamic equilibrium which allows the arguments to true equilibrium thermodynamics to be applied to high approximation.

2.3.1. *Thermal forces*

The thermal balance described in the last sub-section can be expressed in more general terms. This can usefully involve a dimensional approach which puts the argument on a semi-quantitative basis. To do this we find expressions for the convection and viscous forces and then compare them.

Buoyancy force: The force of buoyancy acts when there is a difference of density between a small volume element of fluid and its neighbourhood, but in the presence of a gravitational field. This is sometimes in engineering called alternatively natural convection or free convection. The density of the hotter fluid will be less than that of the cooler surroundings. Consequently, the volume of the hotter fluid will be greater than that of the cooler fluid and the amount will depend on the volumetric coefficient of expansion, β, of the fluid. Suppose the density of the hot fluid is ρ_f while that of the cooler immediate surroundings is ρ_s. This means that $\rho_s - \rho_f > 0 \equiv \rho$. Suppose the temperature difference ΔT is maintained across the fluid: this gives the density difference $\beta \Delta T$. The buoyancy force, which acts against the gradient of the field, is then written $\rho g \beta \Delta T$ per unit volume of the fluid and where g represents the acceleration of gravity.[5] If the temperature is held across a distance L the temperature gradient is $\frac{\Delta T}{L}$ and the buoyancy force F_B is written then

$$F_B = \rho g \frac{\beta \Delta T}{L} L$$

per unit volume of fluid.

Viscous force: The viscous force acts to stop the motion and so is opposed to the movement due to buoyancy. This force will depend upon the action of shear viscosity which, in the simplest form considered here for a Newtonian fluid, is proportional to shear on the fluid due to the movement. The proportionality is closed by a constant of proportionality called the shear viscosity and denoted by η. For our fluid element the shear stress is denoted by $\frac{v}{L}$ where v is the velocity of the fluid. The viscous force acts across a unit area L^2, say, so the viscous force per unit volume (to match the buoyancy force)

---

[5]This effect will be absent in space where g is closely zero. It means, as one example, that a candle or a match will not burn in a space craft.

is written

$$F_V = \eta \frac{v}{L} L^2 \frac{1}{L^3} = \eta \frac{v}{L^2}.$$

For convection to take place we must have $F_B > F_V$ or

$$\rho g \frac{\beta \Delta T}{L} L > \eta \frac{v}{L^2} \quad \text{that is} \quad \frac{\rho g \beta \Delta T L^2}{\eta v} > 1.$$

This condition involving the various parameters is to be satisfied if convection is to occur. If it is not satisfied there will be no convection.

2.3.2. *Dimensionless numbers*

It is most convenient in general discussions to replace this form of the criterion for convection by another involving dimensionless numbers. These represent the ratio of forces acting on the fluid and so have not got a dimension. There are several numbers that are relevant to us now.

Reynolds number, Re. This represents the ratio of the force of fluid inertia and the force of viscosity. Explicitly

$$Re = \frac{\rho v L}{\eta} = \frac{v L}{\nu},$$

where $\nu = \frac{\eta}{\rho}$ defines the kinematic viscosity.

Prandlt number, Pr. This compares the viscosity to the thermometric heat capacity which measures the strength of the buoyancy motion to keep going. Heat energy can be lost by thermal conductivity specified by the thermal conduction in the fluid with the thermal coefficient κ. The heat reservoir is the heat capacity ρC_P so the ratio of the outflow of heat to the capacity of the reservoir is given by the ratio

$$\alpha = \frac{\kappa}{\rho C_P}$$

called the thermometric conductivity. The ratio of the kinematic viscosity and the thermometric conductivity gives the ratio of the viscous and thermal forces. This defines the Prandtl number

$$Pr = \frac{\eta}{\rho \alpha} = \frac{\nu}{\alpha}.$$

Rayleigh number, Ra. This follows from the equation of the last sub-section

$$\frac{\rho g \beta \Delta T L^2}{\eta v}.$$

This expression is rearranged to introduce explicitly the numbers Re and Pr. The expression is rewritten successively as

$$\frac{\rho g \beta \Delta T L^2}{\eta v} = \frac{g \beta \Delta T L^3}{v^2} \left(\frac{v}{vL}\right) = \frac{g \beta \Delta T}{v^2} \frac{1}{Re}$$

$$= \frac{g \beta \Delta T L^3}{\alpha v} \left(\frac{1}{Re}\right) \frac{\alpha}{v} = \left(\frac{g \beta \Delta L^3}{\alpha v}\right) \left(\frac{1}{Re}\right) \left(\frac{1}{Pr}\right).$$

The total expression has no dimensions and the components Re and Pr are dimensionless. Therefore the terms in the first bracket must also be dimensionless. This defines the Rayleigh number Ra

$$Ra = \frac{g \beta \Delta T L^3}{\alpha v} = \frac{g \beta}{\alpha v} \left(\frac{\Delta T}{L}\right) L^4,$$

where the second form includes explicitly the mean temperature gradient across the region. The numerical value of this number tells whether convection is present or not. In general terms convection can be expected for $Ra > Ra_c$ where the critical value Ra_c is usually of the order 10^3.

2.3.3. *Models*

The dimensionless numbers we have deduced are not just compact ways of describing a physical situation. The arguments leading to the numbers did not specify a particular physical system and they will, in fact, apply to any that show the same physical features irrespective of size or time scale, that is, to dynamically similar systems. They can allow real situations to be realistically simulated by models according to the principle of dynamic similarity. The only restriction is how accurately we can model the real situation on the laboratory scale. This is the basis of modelling which plays so important a role in physics and in engineering. As one example, it is seen in action when a model aeroplane flying in the sky could according to our eyes be a real aeroplane but far away. In engineering the method allows models, such as of aeroplanes, bridges, barriers, towns and ships, to be tested as part of the design process. It turns out in practice that the method has some limitations. For instance, the bolts holding the skin of an aeroplane cannot be modelled precisely. Often more than one number is involved (for instance in a ship model a number for riding waves and a number to represent friction of the water) and it may be impossible to model both numbers simultaneously. Nevertheless, the method does give insights into the behaviour of a particular system. This means in our case that the numbers can apply equally to a body like the Earth or to a physically similar model in the laboratory.

Our interest now is with free convection in the Earth's interior. An important quantity is the Prantl number which compares heat conduction with viscous drag. This estimate for the Earth is of the order $Pr \approx 10^{40}$, a very large number. It shows that viscosity is overwhelmingly the most important quantity in the interior. The effect of temperature, then, will be on the strength of the viscous opposition to motion. Although we do not wish to go into details here we can notice a problem of taking the material for the model. Where can we find a material of such a very high Prandtl number? It turns out that molasses treacle at about room temperature is an acceptable model material on the laboratory scale. It is not appropriate to enter into the details of the experiments here except to say that they all indicate that there is convective motion in the mantle region and in the crustal region. This is quite consistent with what we see at the surface. It is possible that the connective motion could be strongly turbulent over some regions. With internal heat energy on this scale it is clear that the surface will show significant changes over time.

2.4. Geochronology: Measurement of Rock Radioactivity

All rocks, or almost all, contain radioactive atoms which undergo spontaneous radioactive decay. In this an unstable atom decays through the emission of neutrons, or protons or β-particles by the nucleus. The decay in a particular case is characteristic of that particular nucleus with a rate of decay which is entirely independent of the external circumstances but determined by the nuclear structure itself. Once the constant decay rate is known the amount of the decay product (determined by a mass spectrograph) in a particular rock can then be used to determine the time since the rock achieved that particular form. The decay constant, usually denoted by λ, is easily measured for each chemical element. Let N_o, be the number of atoms present in a sample of the rock initially and let N_t denote the number of the same atoms in the sample after time t. The law of radioactive decay is the statement that the number of atoms, δN, which decay during a time element, δt, is proportional to the number, N, there at the beginning of the interval. This is written

$$\delta N = -\lambda N \delta t,$$

where the negative sign shows that the number of atoms that have not decayed is decreasing with time. Over a time interval τ the number that will have decayed is given by

$$N_\tau = N_o \exp\{-\lambda \tau\},$$

where exp denotes the exponential function. It is conventional to indicate the time scale of the decay by specifying the half life, which is the time, $T_{1/2}$, for half the initial atoms to decay. Setting $N_\tau = \frac{1}{2}N_o$ we

$$T_{1/2} = \frac{\ln 2}{\lambda} = \frac{0.693}{\lambda},$$

where ln refers to the natural logarithm. It is seen that the half life is shorter the greater λ which indicates a rapid decay process.[6]

This is the basic analysis but it can become more complicated. For instance, there may be a radioactive chain where the initial atom decays to a daughter atom which itself decays to a further daughter and there may be several such steps. There is no problem of principle but the analysis is repeated and can be a little complicated.

What has been said so far assumes the decay process is unique but there are examples of atoms which decay in more than one mode. One example is potassium $^{40}_{19}K$. There is an 89% probability that this will decay to the calcium isotope $^{40}_{20}Ca$ by the emission of an electron but the remaining 11% probability is for decay to the Argon isotope $^{40}_{18}Ar$ by the capture of an electron. Each component of this bifurcated decay is treated as for the simple case but combining the two adds complication.

The methods rely on being able to measure the quantity of the remaining daughter atoms and this can lead to uncertainties unless very careful safeguards are applied. If the daughter product is a gas there is always the possibility that the rock is porous and some gas escapes. Rocks are, in fact, not able to contain gases above a so-called closure temperature which is characteristic of each rock. Closure temperatures vary widely. For instance, the closure temperature is $540\,°C$ for Hornblende, $750\,°C$ for Zircon but about $350\,°C$ for feldspar. The age determination then is that referring to the time that the rock cooled to the closure temperature. The case of the decay of Uranium and Thorium is special in that the different isotopes decay to different isotopes of lead according to the scheme $^{238}U \rightarrow {}^{206}Pb; {}^{235}U \rightarrow {}^{207}Pb; {}^{232}Th \rightarrow {}^{208}Pb$. There is always the possibility that some lead was present initially. This problem can be alleviated if age determinations of a given rock sample are made using more than one dating method.

Separate dating methods are often possible using different parent/daughter combinations. Some useful for dating purposes, other than Uranium

---

[6]This rule applies more generally. This expression will also tell what rate of interest will cause money to double in a given time (for example 3% interest will lead money to double in about 23 years) as but one example.

and Thorium, are

$$^{87}\text{Rb} \rightarrow {}^{87}\text{Sr} \qquad {}^{187}\text{Sm} \rightarrow {}^{183}\text{Nd} \qquad {}^{40}\text{K} \rightarrow {}^{40}\text{Ca} \qquad {}^{40}\text{K} \rightarrow {}^{40}\text{Ar}$$

$$^{39}\text{Ar} \rightarrow {}^{39}\text{K} \qquad {}^{176}\text{La} \rightarrow {}^{176}\text{Hf} \qquad {}^{187}\text{Re} \rightarrow {}^{187}\text{Os} \qquad {}^{14}\text{C} \rightarrow {}^{14}\text{N}$$

The problem is to select an element with an appropriate half life but we will not delve into that here.

These various methods have allowed the ages of the rocks to be found to great precision and so have enabled the rock sequences to be dated unambiguously since the method was pioneered by Lord Rutherford (1871–1937) and his students during the early years of the 20th century. It now forms a standard method.

2.5. Measurement of Remanent Magnetisation

A surprising development during the 1950s was the use that can be made of the fact that rocks generally have a small magnetism called remanent magnetism. This arises because virtually all rock materials have got some ferrous material within them. These act as very small magnets and will line themselves up in sympathy with the local intrinsic field of the Earth. As the material solidifies, if volcanic, or is compressed, if sedimentary, these elementary magnets remain frozen in the orientation provided by the Earth's field at the time. This means the rock is magnetised. But more than that, it is magnetised in the direction of the field at the time of its deposition and this initial magnetisation is locked into the rock. Any change in the conditions over geological time will be reflected in magnetic measurements so any changes in the local field can be detected.

The major part (over 80%) of the Earth's intrinsic magnetic field has a dipole form with magnetic north and south poles. Now these are close to, though not coincident with, the geographic poles which mark where the hypothetical rotation axis cuts the Earth's surface. The north pole of the magnet points north which makes the magnetic pole there a south pole. The hypothetical line through the Earth joining the measured north and south poles is called the magnetic axis: for the Earth this points in the opposite direction to the rotation axis so the Earth's magnetic field is said to be anti-parallel. The mean orientation of the elementary magnets in the rock was along the lines of force of a dipole at the time the rock formed. By measuring this direction accurately one is able to find both the direction of the poles in the past and the position on the Earth for the specimen — especially the latitude. Several interesting discoveries were made when this general method was applied to the rock magnetism using a sensitive magnetometer.

Perhaps the most surprising discovery was that the magnetism of the Earth is not fixed but that its polarity has "flipped" from time to time throughout geological history. There is not a true period of change but the polarity suddenly changes with roughly a million years or so as the time scale. In some instances it was much more while in others it was less. The time of change is geologically instantaneous — of the order of 10,000 y. There are no intermediate states although there can be cases where the field begins to flip but quickly returns to its initial condition. The characteristic condition is when the dipole field slowly decays and then builds up again but with the opposite polarity. There have been roughly equal times in parallel or anti-parallel alignments over the history of the Earth with a slight preponderance for the parallel alignment. The Earth's dipole field is decaying at the present time. The last reversal to the present polarity occurred about 700,000 ya.

Plotting the magnetic data for a continent showed unequivocally that the north pole wandered in time relative to the geographic pole — and presumable the south pole did the same in unison. The rotation axis which determines the geographical poles will have remained unchanged in space. The same pattern was found for each continent but the different curves did not fit one over another as they might be expected to do. This disconcerting situation was put right and all the data could be reconciled if the continents were moved one with respect to another ultimately allowing for a perfect match (see Fig. 2.4). The remanent magnetism has shown the movement

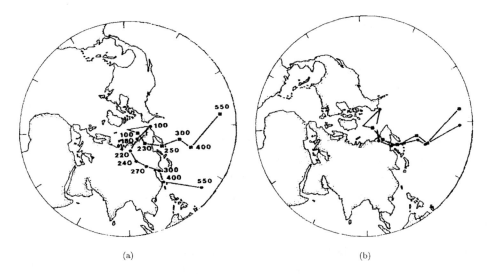

(a)   (b)

Fig. 2.4. An example of magnetic pole wander using data from Europe and North America. (a) The raw data showing a wide disparity of pole positions using the present continental positions. (b) Showing how the data become consistent if the relative positions of North America and Euroasia are adjusted appropriately. (The numbers are in millions of years.)

of the continents over the history of the planet and the results correlate with other information such as that provided by the fossil flora and fauna. Remanent magnetism has proved an indispensable tool in finding the details of the early Earth.

The polarity of the Earth is used today by birds and some insects as a coordinate system for navigation. A reversal of polarity would reverse their directions and no field, even temporarily, would leave them at a loss. Presumably these variations over the last 500 My have had these effects but there is no actual evidence of what difficulties, if any, these effects provided for life.

2.6. The Land Surfaces — the Development of Continents

The convection pattern within the Earth has very important consequences for the surface structure. The surface region itself at the present time has a thickness of about 8 km thickness under the oceans and a mean of about 20 km under the continents.

2.6.1. *Plates and plate tectonics*

This crustal region does not form a giant slab covering the interior but rather a series of a dozen or so smaller independent "plates" that form the surface. The global plate system is represented pictorially in Fig. 2.5. The plates move one with respect to one another with very important consequences. In their motions new plate material is formed in a global mountain region called the mid-ocean ridge system, the material swelling up from the mantle below the crust in the ridges. This 20,000 km range of mountains is almost entirely under the deep oceans. The mountains are high — reaching perhaps 2 km or more at their highest peaks — and the Earth is encircled by them (see Fig. 2.6). The ridge system exuded hot magma which moves away to form new ocean crust. The size of the Earth is not increasing so the increasing ocean floor must be absorbed back into the body of the Earth somewhere, a process called subduction. This occurs at plate boundaries where one plate is pushed under another. The whole process is shown schematically in Fig. 2.7. One expanding plate presses on a neighbour and one of the two plates at the boundary subducts under the other. The subducted material returns to the mantle ultimately to be recycled but only after some refinement. More especially, the subduction takes some ocean water down with the plate and this has an important consequence. The presence of water lowers the melting point of the magma and reduces its density. The result is an up welling of magma above the subduction region which bursts at the land surface in the

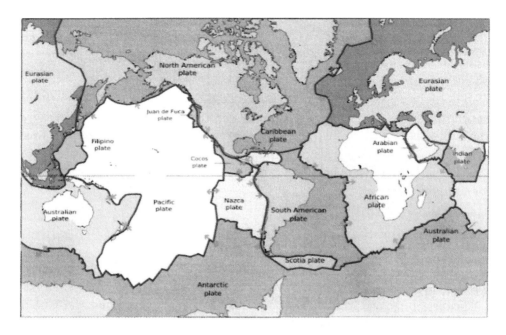

Fig. 2.5.   The global mid-ocean plate system set against the present continents.

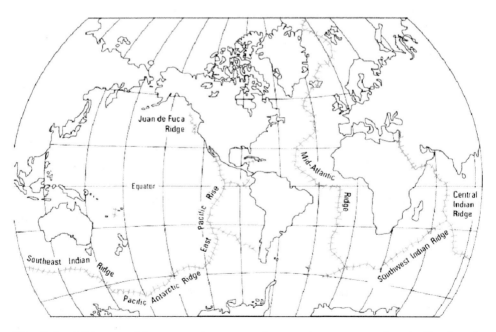

Fig. 2.6.   Mid ocean ridge system is a range of mountains encompassing the Earth with a length of some 50,000 km and a height that can reach 3 km.

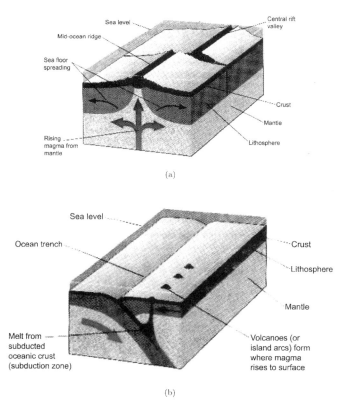

Fig. 2.7. Sketches representing (a) the production of ocean floor at a mid-ocean ridge and (b) the subduction of ocean floor under a continental plate producing an ocean trench.

form of a series of volcanoes. The volcanic magma produced this way accumulates to form new continental land which is not subsumed back into the mantle. The consequence is that land is continually produced and the continental area of the Earth increases steadily. A schematic representation of the ridge system is shown in Fig. 2.7 where material from the mantle is seen forcing itself out to form new sea floor. New material forces that recently released away from the ridge so the ocean floor is continually expanding. The expansion is symmetrical about the ridge. It is this process that drives the plate system. There is no information about the workings of these processes in the early history of the Earth although we have seen there is fossil evidence of black smokers in rock in South Africa that is about 3.5 By old (see Fig. 2.8). This does not confirm that the plate system was operative then but makes it very likely. The oldest evidence for the plates is 2.5 Bya in China. Presumably the initial solidified surface material was broken up on a larger scale by material brought up by convection from below. Certainly collections of basalt and metamorphic rocks appeared (cratons) which moved about the

(a)   (b)

Fig. 2.8. (a) A modern active black smoker from the mid-ocean ridge in the Pacific. (b) A fossil black smoker in rocks in Barberton, South Africa. The opening is about 5 cm across at its widest part. The plug is composed of ancient minerals in the process of being ejected. The rock is 3.5 My old showing an ocean ridge system was under water and operative then as now. Life may have started beside such smokers.

surface under the action of plate movements to coalesce and then to break away. The area of the plates is quite large. The present areas of the seven main plates are listed in Table 2.2.

We can ask whether these same processes have taken place in the past. The evidence there suggests they go back a long way. There is evidence that plate tectonics were operative at least 2.5 Bya and probably even further back. There is another evidence. There are regions on the underwater mountain ridge system ridge where gases and super-heated water are ejected from below into the overlying water in the form of a black plume: these are called black smokers (see Fig. 2.8(a)). They have a very special interest for us because there is a range of characteristic fauna associated with them.

Table 2.2. The present areas of the seven principal surface plates of the Earth. This covers essentially the full surface area.

Plate	Area (million sq. km)	Plate	Area (million sq. km)
Pacific	103.3	Antarctic	60.9
North America	75.9	Australian	47.2
Eurasia	67.8	South America	43.6
Africa	61.3		

There is evidence of ancient smokers in rock 3.5 By old (see Fig. 2.8(b)). It might reasonably be inferred that ridge conditions applied then but this does not prove that this is associated with a plate tectonic system though many workers believe this was so.

2.6.2. *Ancient continents*

Two cratons are significant here: one is the Pilbara craton of Western Australia and the other the Kaapvaal craton of South Africa. These two regions are among the oldest that exist now on Earth. Radiometric data, including palaeo-magnetic and geo-chronological, show that these two cratons came together some 3.6 Bya to form the first super-continental land mass (small as it probably was but it was the biggest at the time) and remained in an essentially contiguous arrangement until about 2.8 Bya when they moved away from each other. This must show that the plate tectonic processes were active very early on. Certainly the geological activity will have been greater then than now — for instance we know that then the surface heat flow was some three times the present value and the viscosity of the material will have been appropriately lower. Presumably the plate structure resulting from the enhanced activity will have been more complex than now and working to a shorter time scale although there is no direct evidence of this. This would suggest that the rate of formation of land was initially quite high but quickly became reduced leaving the land area to increase roughly proportionately to the age of the planet or perhaps more precisely to the level of volcanic activity.

It appears that around 3.475 Bya there was an especially large outpouring of lava called the Komatii formation. This was comprised of a magnesium rich material which enhanced the land area quite significantly. The total land area was still quite small — probably much less than that of Australia today — but it did form a super-continent for its time (about 3.0 Bya in the early Archaean) if by this one means a construct that included essentially all the land area. This is the earliest accredited continent and has been called Ur. A more extensive land mass formed later, at about 2.7 Bya, by the coming together of new land masses with the cratons of this Archaean period. This also followed very extensive volcanic activity. In particular, the cratons comprising: the modern North America and Greenland, and called now Laurentia; the Baltic craton Baltica (which forms modern Scandinavia and the Baltic); Western Australia and Kalahari all came together. This accumulation of material is called Kenorland. Palaeo-magnetic studies suggest

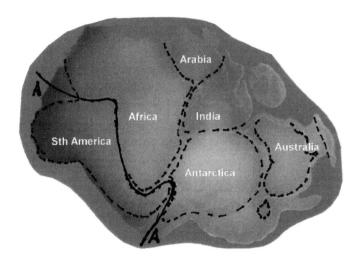

Fig. 2.9. The structure of the island continent Gondwana about 500,000 ya which lay in the southern hemisphere. Modern regions are indicated. The line A–A shows the path of the split that will ultimately form the South Atlantic Ocean. The south polar region is just below the marked Antarctica.

this was in low latitudes. This is included in Table 2.3 together with later formations up to, and including, Pangeaea.

Gondwana began to form during the era 570–510 Mya during the Cambrian. It comprised what are now S. America, Africa, India, Australasia, Antarctica and Arabia. It occupied the southern hemisphere from essentially the pole to the equator and was really quite a stable configuration. It lasted in various guises until about 200 Mya when it began breaking up to form the components of what is now the Southern Hemisphere. Pangea was formed from the joining of Gondwana and the northern combined terrain of Eurasia and Laurentia as is clear from Fig. 2.10. This dates from about 300 Mya and lasted until about 180 Mya when it began to break up. It is still in the final stages of this disintegration. Extrapolation of the modern plate tectonic structure suggests that it might reassemble as a single unit again in about 250 My in the future. The two units which formed Pangea have shown remarkable stability. It is difficult to estimate the rate of production of continental crust over the history of the Earth but it is likely that some 25% of the current crust had formed 2.5 Bya which rose to 80% 1.5 Bya (Fig. 2.11).

The effect of plate tectonics on the land distribution has been decisive in the past and the associated volcanic activity has influenced the development of life in fundamental ways. There is evidence, though some would say not especially compelling, to suggest that plumes of mantle material can rise from as far down as the core/mantle boundary some 3,000 km below the surface

Table 2.3. A partial list for which there is evidence of the most significant super-continents of their day that have existed during the history of the Earth. Pangaea was the greatest and its dissolution is still going on. The land will likely come together again to form a second Pangea in about 250 My time.

Name	Beginning (Bya)	End (Bya)	Comments
Ur	3.0	2.8	First established early Archean super-continent. Probably smaller than present day Australia.
Kenorland	2.7	2.1	The break up led to great rainfall which depleted the CO_2 in the atmosphere. Banded iron structures show increase of oxygen from 0.1% to 1% of the atmosphere. Methane then decreased (decomposes to CO_2 and H_2O). No greenhouses gases and the weak early Sun led to glaciation which lasted some 60 My.
Nena (Hudsonia)	1.8	1.5	Found by palaeo-magnetic evidence.
Rodina	1.1	0.75	Probably lay south of the equator. Composed of eight continents including Laurentia, South America, Australia, Antarctica and Africa craton. Kalahari (S. Africa), Congo (W. central Africa and S.E. South America were separate.
Pannotia	0.6	0.54	Eight continents that formed Rodina reassembled. Baltica and Amazonia drifted off to form Panthalassic Ocean — the Palaeo-Pacific.
Gondwana	0.57	0.18	Stayed very much a unit and spreading to form the Southern Hemisphere today.
Euramerica	0.38	0.3	This was a simple merging of Lauentia and Baltica
Pangea	0.3	0.18	Contained all the land around 250 Mya. It spread essentially from pole to pole. Break up is still proceeding — India hitting Asia, East African Rift and the movement north east of Australia. It is likely to reform in some 0.25 G years time.
Laurasia	0.3	0.60	Joined Pangea and then broke away almost as the same unit of cratons forming the modern Northern Hemisphere.

and slowly rise to eventually break surface in the form of volcanoes. It is generally accepted that some features of the modern surface are the result of massive plumes of lava rising from the core/mantle boundary. One example is the volcanic structure which forms the extended islands of Hawaii. Here the temporal structure of the islands has a linear geometry which implies that the rotation speed of the Earth's core is a little different from that of the main body of the planet. Another quoted example is Iceland, where the Atlantic Ridge breaks surface to be seen on dry land, though the existence of this plume has recently been disputed.

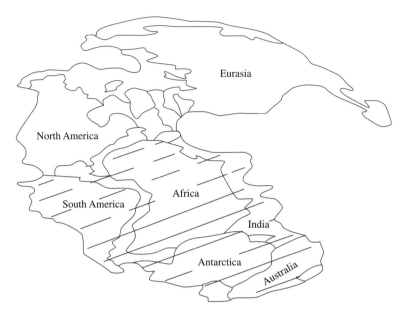

Fig. 2.10. The super-continent of Pangea about 300,000 ya. The various areas are marked which are well known features today. The cross-hatched region is Gondwana which joined with northern Laurasia to form the total continent. The ultimate split was north–south as Americas/Eurasia moved away to give the modern two block continental structure.

2.7.   The Surface Temperature

The principal source of heat for the surface of the Earth is the radiation from the Sun. The radiation from a very confined heat source is spread over the ever expanding surface of a sphere about the source as the radius R of the sphere increases. The area of the surface is $4\pi R^2$ which makes the intensity on the surface decrease as the inverse square of the distance from the source, which is like $1/R^2$. The radiation received by a body will heat it to some temperature T so the heated body itself becomes a source of heat radiation into space. In this way the thermal relation between a planet orbiting a star (for us the Sun) in a stable orbit which determines the surface temperature of the planet becomes an equilibrium between the primary radiation from the Sun and the induced radiation from the planet. The Sun and the Earth are each very close to being black bodies which means they emit and accept all frequencies of the radiation. There is a simple relation between the quantity of heat radiated by a black body and its temperature which was first identified empirically[7] by Jožef Stefan (1835–1893) and later given a theoretical

---

[7]Stefan deduced his relationship from the experimental measurements of the Irish physicist John Tyndall (1820–1893) who was Professor at, and later Superintendent of, the Royal Institution in London.

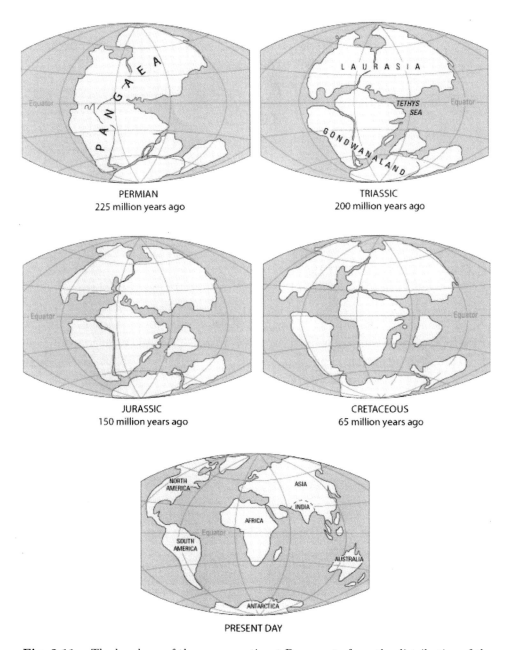

Fig. 2.11. The break up of the super-continent Pangaea to form the distribution of the continents familiar today.

basis by his student Ludwig Boltzmann (1844–1906). This so-called Stefan–Boltzmann radiation law states that the total quantity of heat radiated is proportional to the fourth power of the surface temperature. The constant of proportionality is the Stefan–Boltzmann constant of radiation, traditionally denoted by $\sigma(= 5.67040 \times 10^{-8}\,\mathrm{J\,m^{-2}\,s^{-1}\,K}$ in SI units), and refers to the radiation from unit area of the source of heat in a direction perpendicular to the surface. Any nonparallel radiation can be accounted for by applying a reduction by the appropriate angle.

Let the total radiation emitted by the Sun be denoted by S: this is called the solar luminosity. This is reduced with distance away from the Sun according to the inverse square law and at a distance R equal to the mean distance of the Earth in its orbit is reduced to $S_E = \frac{S}{R^2}$ per unit area perpendicular to the direction of the Sun. This is called the solar constant. It is now known from measurements in space that it is not strictly a constant but the deviations are quite small. For the whole Earth the quantity of received radiation is $\pi R_E^2 \frac{S}{R^2}$, where R_E is the mean radius of the Earth. Not all this radiation is actually absorbed by the Earth. A certain proportion is reflected by clouds and particles in the atmosphere and even by the material of the surface itself. The proportion of radiation reflected away, and which never enters the thermal regime of the Earth, per unit area is called the albedo and is denoted by A. This reflected radiation is of amount $\pi R_E^2 A \frac{S}{R^2}$ for the whole Earth and this is to be subtracted from the total radiation entering the region of the Earth. The radiation H_S actually received by the body of the Earth from the Sun is therefore given by

$$H_S = \pi R_E^2 \frac{S}{R^2} - \pi R_E^2 A \frac{S}{R^2} = \pi R_E^2 (1 - A) \frac{S}{R^2}.$$

This radiation raises the temperature of the surface of the Earth which we denote by T_E. The Earth now radiates as a black body at this temperature according to the same Stefan–Boltzmann law so the quantity of heat H_E radiated away is given by

$$H_E = 4\pi R_E^2 \sigma T_E^4.$$

For a stable orbit there is the heat balance $H_E = H_s$ and this equality leads to the expression for the equilibrium temperature at the Earth's surface

$$T_E = \left[\frac{S}{4\sigma R^2} (1 - A) \right]^{1/4}.$$

As a check it is seen that the temperature is lower the more distant the Sun, as it must be and lower the higher the value of the albedo. The heat received by the surface per unit area per second is called the irradiance and is measured in watts per square metre ($\mathrm{W/m^2}$). The alternative designation is the radiation received per hour and this is expressed in kilowatt hours ($\mathrm{kW\,h}$).

The solar luminosity is $S = 3.827 \times 10^{26}$ W at the present epoch. This has risen from about 1.5×10^{26} W for the early Earth and will become higher as the Sun ages. The surface is destined to get hotter.[8] It has been found that the luminosity changes by very small amounts on a daily/monthly basis which is consistent with the radiation being due to a random emission of photons from the interior. The intensity of radiation received in the vicinity of the Earth is actually not strictly constant because the distance from the Earth to the Sun varies in its slightly elliptic orbit. The Earth is nearest to the Sun in January and farthest in July. The corresponding values are: January $S = 1412$ W/m^2 while in July $S = 1321$ W/m^2. This is a variation of nearly 7%. The mean value of the luminosity is $S = 1367$ W/m^2 with a 3.4% variation between the beginning and the middle of the year. January is the summer period in the Southern Hemisphere and the temperature is marginally higher there than for summer in the Northern Hemisphere. The South Pole is, by contrast, colder than the North Pole. Taking the Stefan–Boltzmann radiation constant as $\sigma = 5.670 \times 10^{-8}$ J m^{-2} s^{-1} K^{-4}, the equilibrium temperature over the Earth's surface is given as $T_E = 256$ K. Surprisingly this is 17 K below the freezing point of water. In this state the Earth would be frozen over and would be incapable of supporting life other than perhaps the most extreme microscopic forms but we know this is not so. The surface is encased in an atmosphere which raises the temperature above the freezing point enabling water to lie on the surface in liquid form. This is the natural greenhouse effect. It seems this has been the case from the very earliest times even though the heat from the Sun was rather lower then than now. Stronger volcanic activity earlier on will have provided more greenhouse gases which would have retained heat at the surface level. The level of warming gases would fall as the volcanic activity decreased.

2.8.   The Climate

It has been possible to gain some knowledge of the past climate by various indirect methods. The distribution of fossil remains show deserts, tropical soils and coal fields show the presence of early forests. Flora and fauna which are sensitive to temperature, such as mangrove swamps, palm trees, alligators, sharks and other fish are an invaluable aid. The result is the overall estimates collected in Table 2.4 up to the time that Pangea began to

---

[8]This is a natural effect quite separate from the so-called global warming. In about 1 By the surface will be sufficiently hot to evaporate away the water so the seas will disappear. Life could not survive.

Table 2.4. Possible climates up to the break up of the super-continent Pangea.

Period	Climate	Continents
Cambrian (540–520 Mya)	Intermediate temperature south pole probably warm and temperate.	Land spread in southern hemisphere. Most of the land arid.
Ordovician (480 Mya)	Probably warm temperate; cool south pole.	Land still spread mainly in southern hemisphere.
Late Odivician (440 Mya)	Climate cools; south pole cold; Other land still arid.	Africa and S. America in south covered by ice.
Silurian (420 Mya)	Glacial conditions near S. pole.	Coral reefs in southern arid belt.
Early Devonian (400 Mya)	Land still largely in southern hemisphere — generally temperate climate. South pole cool temperate.	Arid conditions over much of land.
Mid-Devonian (380 Mya)	Land plants grew profusely in equatorial rainy belt which formed coal. Arctic Canada on the equator.	Australia, Siberia and N. America lay under warm shallow seas. Continents began to move.
Late Devonian (360 Mya)	Cool S.Pole with some glaciation in Amazon basin at S. Pole. Tropical rain forests in Canada and S. China see coal formed.	Land mass concentrates to begin formation of Pangea.
Early Carboniferous (340 Mya)	Tropical rain forests move to Newfoundland and W. Europe. Southern hemisphere begins to cool.	Land continues to concentrate and centre of gravity moves northwards.
Mid-Carboniferous (320 Mya)	Rain forests cover tropical central land: North and south bounded by deserts. Ice cap formed at S. Pole and moves northwards.	Land still moves northwards: Tethys Sea formed just south of equator.
Late Carboniferous (300 Mya)	Temp. range increases — cold at both pole, then cold temperate strip, the arid while centre is tropical.	Ice cap now covers S. Pole.
Early Permian (280 Mya)	Large fraction of southern hemisphere covered by ice; glaciers cover much of southern hemisphere and moving northwards.	Land pushes northwards: Tethys now large enclosed sea; Land extending from south to north poles.
Late Permian (260 Mya)	Southern ice cap disappeared but north pole ice accumulates. Sudden temperature rise at boundary with Triassic: perhaps 25 °C.	Movement of land northwards continues: rain forests replaced by deserts: China crosses equator.

(*Continued*)

Table 2.4. (*Continued*)

Period	Climate	Continents
Triassic	Interior hot and dry. High temperatures — probably highest ever through period.	Pangea forms. Tethys Sea opens up. Land spreads from North to South. Poles warm and temperate.
Early Jurassic	Interior of Pangea very hot and arid. Monsoon wins over some regions — China lush.	Land forms single solid mass.
Late Jurassic	Global climate change — ice in polar regions.	Pangea begins to break up.

Table 2.5. Possible climate and the distribution of continents post-Pangea.

Period	Climate	Continents
Early Cretaceous	Cooler-ice at poles during the winter season. Cool temperate forests cover poles.	Channels separate regions: Africa/S. America together; Eurasia separate; N. America separate; Antarctica/India/ Australia together. Africa and S. America separate. India breaks away with separate Madagascar. Australia moves north.
Late Cretaceous	Rather warmer than today's. No ice at poles. Dinosaurs migrate with seasons.	Africa and S. America in place. Australia essentially in place.
Palaeocene	Climate warm — no polar ice.	N. and S. Atlantic oceans widening. India just south of equator. Madagascar in place as today.
Early Eocene	Very warm — alligators at N. Pole.	India joins with EuroAsia. N. and S. Americas joined.
Late Eocene	Climate still warmer than today but ice now forms at S. Pole.	
Oligocene	Temperature much as today. Ice at S. Pole but not at N. Pole.	Continents much as we know them today
Miocene	Climate similar to now	Central America land bridge now firm.

break up while the post Pangaea period is collected in Table 2.5. It is found that the climate has swung from hot and dry to cold with snow and ice. Of the two, the Earth has more often been hot than cold. The mean surface temperature now is about 19 °C but for much of its history the temperature seems to have been around 22 °C or perhaps rather more. There was a period at the Permian–Triassic boundary when it was near to 25 °C. This was also a time when there was a massive extinction of life on Earth and the two things may have been related. The period when the dinosaurs flourished appears to have been particularly hot. It may well be that the extraordinary rise in temperature (the greatest that is known so far) was due to excess production of methane gases by the very large vegetarian dinosaurs. A fall of temperature will have coincided with their demise as a result of the lowering of the production of the greenhouse gas.

The particularly cold spells seem to have occurred at geological boundaries. The Ordovician–Silurian boundary and the Carboniferous Permian boundary were particular cold times. A cooling between the Jurassic and the Cretaceous produced temperatures only slightly below those we are familiar with today. The temperature appears to have dropped significantly at the end of the Tertiary Period and have been recovering slowly since. We are at the present in a warming period. Left to natural forces the present climate would be expected to last for 15,000 y or so but the effect of the man made enhanced greenhouse effect makes the future more difficult to predict. Today a significant quantity of methane is produced by animals and especially domesticated farm animals. This can be expected to grow as the number of such animals increases as a consequence of the demands of the growing human population. The human population itself also adds to these emissions.[9]

2.9.   The Atmosphere

The atmosphere raises the surface temperature and so allows life to survive. There is the natural greenhouse effect. Although the Earth's atmosphere started with an alien composition, the current composition is in crude terms 78% nitrogen, 20% oxygen and 2% of trace elements. The oxygen has arisen primarily from living creatures as will be seen in Chap. 3. A detailed inventory is given in Table 2.6. The CO_2 component is currently rising due to industrial pollution.

The atmosphere is divided into layers formed under gravity according to local temperature. A little more than 99% of the atmospheric gases lie below 32 km. There are four main layers and an outer cover:

(i) *Troposphere* (between 0 and 12 km). This layer contains 75% of the atmosphere and is characterised by a falling temperature with height

---

[9] Reverend Thomas Malthus (1766–1834) saw food as the essential controlling factor for population growth. Perhaps other things, like temperature rise and pollution, will prove more significant.

Table 2.6. The components of the Earth's atmosphere (US Geological Survey).

Element	Abundance (mass %)
N_2	78.084
O_2	20.947
Ar	0.934
CO_2	0.0314
Ne	0.001818
He	0.000524
CH_4	0.0002
Kr	0.000114
H_2	0.00005
Xe	0.000008

of about 6.5 °C/km. This is often called the lapse rate and results from the upward motion of the air from the warm ground. Here lies the jet stream, a current of fast moving air which blows towards the east.[10] This is the region of high flying commercial flights. The top of this layer is called the tropophase where the temperature is essentially constant for a kilometre or two at about −60 °C.

(ii) *Stratosphere* (between 12 and 50 km). The ozone layer rests here which absorbs the solar ultraviolet (UV) light causing a slight rise of temperature. Ozone is produced and destroyed here by the UV light. The top of this layer is called the stratopause where the temperature has a local maximum of about −40 °C.

(iii) *Mesosphere* (between 50 and 80 km). This is the coldest part of the atmosphere with a temperature of around −100 °C. A region of very low gas pressure, it is in this region that meteorites burn up to be seen as shooting stars. The top of this layer is the mesopause where the temperature has a local minimum of about −180°C. The cooling is due to the interaction between infrared radiation and CO_2 molecules.

(iv) Thermosphere (from 80 km upwards to about 1,000 km). The gas pressure here is virtually zero giving a vacuum better than any generally obtainable in the laboratory. The temperature here can reach up to 2,000 °C and beyond due to the absorption of ultraviolet radiation producing photodissociation and photoionisation of oxygen. There are two distinct divisions.

---

[10]Its actual existence was discovered during the Second World Pacific War by high flying bombers which appeared to be flying backwards with respect to the ground due to the strong wind although a strong jet flow had been observed over Japan in the 1920s by the Japanese meteorologist Wasaburo Ooishi.

(a) *Ionosphere* (between 80 and 550 km). The gas is ionised by the incoming X-rays and ultraviolet radiation to form a plasma. This reflects
radio waves and so allows world wide radio communication. It is a
complicated layer with three sub-layers denoted by D, E and F and
we can do no more here than note that fact. They are controlled by
the magnetic conditions on the solar surface apart from the electromagnetic radiation. The D and E layers disappear on the night side
of the Earth but the F layer is a permanent feature night and day.

(b) *Exosphere* (550–1,000 km) The air here is very thin indeed with an
indefinitely large mean free path between interactions with neighbours. The atoms are, therefore, essentially in free linear trajectories
and the fastest moving (with the greatest kinetic energy) can escape
from the Earth's influence. This is actually a very small fraction of
the whole in any time interval. This is the region of artificial satellites
and of the International Space Station.

(v) *Magnetosphere* (beyond 1,000 km). Encasing the whole atmosphere is
this region controlled by the Earth's magnetic field but influenced by
the solar wind. It is the source of charged particles from the Sun which
collide with atmospheric atoms to produce auroral displays centred on
the north and south magnetic poles. The density of atoms becomes
sufficient to provide awesome displays at a height of about 80 km.

2.10.   Energy Absorbed from Solar Radiation

Not all of the incoming radiation is accepted. On the average about 35% of
it is reflected straight back into space (this is the albedo when expressed as
a fraction of incoming radiation and denoted by A). This figure varies from
place to place: for instance, fresh snow has a high albedo (some 0.85) but for
farmland it could be as low as 0.09. Of the 65% of the incoming radiation
(that is $0.65 \times 1367 = 885 \, \text{W/m}^2$) there is an unequal division between the
surface and the atmosphere. About 80% is absorbed by the surface (which
is approximately $0.8 \times 885 = 708 \, \text{W/m}^2$) leaving the remaining 20% to be
absorbed by the atmosphere (that is $177 \, \text{W/m}^2$). These are average values
and will differ over the surface depending on the composition of the surface
and the latitude (and so the tilt of the surface relative to the direction of
the Sun).

The atmospheric gasses provide a greenhouse effect to a greater or lesser
degree. CO_2 and CH_4 are especially important in this connection. The atmosphere raises the temperature from 256 K quoted above to about 290 K which

is actually found with the atmosphere. This is the natural greenhouse effect so vital for life on Earth. The general effect is for the atmosphere is trap some of the infra-red radiation emitted by the Earth keeping it close to the surface rather than be radiated away into space. This has the same effect as a lowering of the albedo factor A so that the factor $(1 - A)$ is increased. In fact, the effective value of the albedo is a critical factor in the surface temperature. A gas traps radiation to a greater or lesser extent but some gases are more effective than others. The most effective are methane (CH_4), carbon dioxide (CO_2) and water vapour (H_2O). Of these, water vapour in the form of clouds is perhaps the most effective. On the average the cloud cover is some 50%. The Earth's average surface temperature is about $19\,°C$, which is $290\,K$, so the atmospheric effects are vital if any advanced life is to be able to form and develop.

The precise cause of cloud cover was not known until quite recently. It was realised that ice crystals collected as clouds which could precipitate to pass water (rain) to the surface. This will carry atmospheric chemicals to the surface. Data from Earth satellites first became available during 1979 and it soon became apparent that an increasing local surface temperature is not necessarily accompanied by changes in the temperature of the low atmosphere. This is a finding that requires a clear explanation. Various observations have led to the recognition that the precise extent and nature of the cloud cover is the answer.

2.11. Astronomical Factors Affecting the Atmosphere

Although clouds generally form as a collection of water crystals acting as the nuclei for further condensation, it has become clear that ions in the atmosphere can do this as well. (One might think of the Wilson cloud chamber but this is not a good analogue because it relies on the behaviour of supersaturated fluid under compression which is different from the atmospheric problem.) The major source in the upper atmosphere is cosmic ray showers. Cosmic rays are charged particles with essentially the solar abundance of the elements and have origins outside the Solar System, possibly even as far away as in the nuclei of active galaxies. Being charged their trajectories are affected by magnetic fields and consequently are affected by the magnetic properties of the Sun and especially of the solar wind. The solar wind is subject to solar "weather" and so varies in a way difficult to foresee. As the ray particles strike atmospheric atoms they produce the nuclei and this process is dependent on the cosmic ray entry characteristics. These are affected by the magnetic conditions on the Sun. Satellite data show a clear correlation between cosmic ray intensity and the fraction of cover by low clouds. It would seem that there is a link between global temperature and long term solar activity. The mechanism causing this link is still under investigation and it is not possible yet to make a quantitative prediction of these effects.

2.11.1. *Milanković cycles*

Not only are conditions on the Sun variable but the orbital characteristics of the Earth are also subject to change. This is because the Earth behaves as if it were a spinning top under the gravitational influence of the other members of the Solar System, and especially of the pair Jupiter and Saturn. There are several effects. This is rather obvious to us now but it was not in the 19th century when this situation was first pointed out. The first ideas were proposed by the Scottish scientist James Croll (1821–1890) in the 1860s initially basing his calculations on the formulae developed by the French scientist Le Verrier (1811–1877) in his analysis of the planetary orbits that led to the discovery of Neptune. This followed independently the work by the French mathematician Joseph Adhemer (1797–1862) who attempted to link the occurrence of ice ages with details of the Earth's orbit. The modern work was undertaken by the Serbian civil engineer and geophysicist Milutin Milankovic (1879–1958).[11]

The Orbit: According to Kepler's laws of planetary motion the orbit of a small body around a more massive one is a stable plane ellipse. The small body performs the same closed orbit continually in a fixed time period and with a fixed eccentricity. The effect of other bodies also orbiting the central massive body is to perturb the simple Keplerian orbit. While it leaves the magnitude of the semi-major axis unaffected it will open the orbit to a greater or lesser extent and also change the eccentricity to become a periodic property and not a constant. These changes affect the intensity of radiation received by the planetary body and so affect its surface temperature.

At the present time the mean value of the eccentricity of the Earth is $e = 0.028$. This varies due to the gravitational interactions with the other planets from 0.005 (an almost circular orbit) to 0.058. The present difference in the solar distance at perihelion and aphelion is 5.1×10^6 km or some 3.4% and is associated with a 6.8% difference in the incoming solar radiation. The main components of the changes are due to interactions with Jupiter and Saturn and this provides change with a period of 413,000 y.

---

[11]He was a practicing engineer and an expert on concrete before he began to undertake scientific work.

Other components of the interactions give periods of 95,000 and 136,000 y but the effects on the eccentricity are small. The effects of the orbital changes can be very important: when the orbit is at its most elliptical the radiation received at perihelion (which is winter in the northern hemisphere) is 28% greater than at aphelion. Such variations lead to differences in the lengths of summer–winter and autumn–spring.

The plane of the orbit also varies with time. The plane of the Sun and the planets collectively forms an invariable plane against which the plane of the Earth (which determines the ecliptic). The period of oscillation is closely 100,000 y. The invariant plane very probably contains a level of dust and debris.

Obliquity: The axis of rotation of the spinning Earth does not make an invariant angle with the ecliptic. At the present time the angle between the normal to the ecliptic and the rotation axis is $23°26'$ but in varies between $21°6'$ and $24°30'$. The period of the oscillation is closely 41,000 y. The fact that this angle exists is the reason for the seasons and differences between the seasons in the north and south hemispheres. The smaller this angle the less are the differences between the seasons. There are, apparently, differences in the seasons that change periodically and which have consequential effects.

Precession of the axis: The obliquity angle precesses about the direction of the normal to the ecliptic. This is the well-known characteristic of a spinning body under external constraints. The constraints for the Earth are the gravitational interactions between the Earth and the Sun/Moon and separately the planets (especially Jupiter and Saturn). The direction changes periodically with a period of 26,000 y. These changes change the insolation at points on the surface periodically.

Resumé: It has been seen that the physical relation in distance and orientation between the Sun and the Earth changes with time with a range of periods. These will lead to a corresponding range of climate variations that is not easy to predict in detail. In very general terms we can say that a cooler summer with tend to allow the Earth to enter the winter at a lower mean temperature than a warmer summer and so will be conducive to a greater accumulation of ice and snow. If a region of the Earth enters the winter with snow and ice already present that associated with the winter to come will compact this more. This situation could act as a trigger for a move towards a colder mean temperature. There is an urgent need to place these arguments on a more quantitative footing.

2.12. Effects of Volcanoes

There are times when the Earth's internal heat breaks through to the surface with particular consequences. Any break in the surface will allow hot molten magma to exude through together with gases and vapours of various kinds. In this way a magma reservoir below the surface can be transferred to the surface as a mound or as a spread of new land. Continual escape of lava and gases at one region will form a mountain over a period of time and will inject gases continually into the atmosphere. This is the most general description of a volcano and of its general behaviour. Volcanoes can occur on the continental surface or on the sea bed. There is evidence that such activities have taken place on Earth from the earliest times in its history. Indeed, it is likely that volcanic activity was much stronger with a shorter time scale earlier on the history of the Earth than now.

Volcanoes are associated with plate tectonic compressions (where one plate is subducted under another) or expansions (where new plate material forms at a ridge). Some are also the result of a rising lava plume reaching the lithosphere region from the lower mantle. An important component in these processes is the presence of water and especially so in subduction processes. Water acts as a lubricant, lowering the magma melting temperature and so facilitating the flow. There is also the possibility of a volcanic eruption with a substantial reservoir of liquid magma as a source.

The composition of lava differs with different volcanic structures. The central difference is different proportions of silica (silicon dioxide, SiO_2). This appears especially in quartz and opal. There are four divisions:

(1) A felsic lava which contains a high percentage of silica, typically >65%. This shows high viscosity and may contain trapped volatiles as gases. This makes it rather explosive covering a wide area with fine ash which may also penetrate high into the atmosphere.
(2) An intermediate silica composition called andesitic with a silica composition between 52% and 63%. This is very characteristic of subduction zones such as the volcanoes around the Pacific rim.
(3) Mafic lava contains between 45% and 63% silica. It contains a high proportion of magnesium and iron (hence its name from MAgnesium and iron F). This has a low viscosity and this lava can be associated with a higher temperature than the others. It occurs at mid-ocean ridges where two ocean plates are pulled apart. Magma rises in a pillar form to plug the gap. This material is found on the ocean floor. This is also the material forming shield volcanoes such as those forming Hawaii which arises

from a lava plume from deep in the mantle. It is also found in continental flood basalts.

(4) Lavas with less than 45% silica are called ultramafic, or sometimes komatite. They are rare now but are very prominent in rocks before the Protozoic (2500–542 ya). These are hotter than the mafic lavas and were associated with an Earth of much higher heat flow than today, perhaps three times as great. There are several different types of eruption.

(a) One set are steam generated and are called Pheratic.
(b) High silica lava tends to be explosive. An example is rhyolite.
(c) Low silica lava like basalt can be effusive.
(d) Pyroclastic flows are particularly deadly.

The flow is a fluidised conglomerate of hot gas and rock called tephra. The gas is usually at a temperature anywhere in the range 100–800 °C and travels at high speed which may be as high as 700 km/h (about 450 miles/h). The gas stays close to the surface and shows a speed which is dictated by the surface gradient. Larger flows tend also to have the higher speed. The effect of such a flow can be seen in the ancient Roman cities of Heraculanium and Pompei.

Although the spread of lava is certainly a feature of volcanic activity there are also effects on the atmosphere. The gases plume upwards mixing with the atmosphere. The most abundant components are H_2O in the form of vapour, CO_2 and SO_2. Other possible components are HCl, HS and HF. There is often small traces of H_2, CO, halogens and organic compounds and volatile metal chlorides. There is also a large quantity of pulverised rock. The largest eruptions inject these various compounds high into the atmosphere, perhaps as high as 16–32 km (which is 10–20 miles). For perspective, this range is several times the height of Mt Everest.

The presence of these compounds in the atmosphere leads to a series of special consequences. He, HCl and HF are quickly absorbed by water vapour producing an acid mixture — this falls as acid rain. The ash falls to Earth over a period that might be measured in days or weeks depending on the quantity of ash and the sizes of the constituent particles. This can add to the nutrients of the soil and so enhance its agricultural quality. Another immediate consequence of SiO_2 is its conversion to H_2SO_4 which very quickly condenses in the atmosphere to form crystals of sulphate aerosols. These reflect radiation at the top and the bottom — they at one and the same time reflect away incident solar radiation increasing the albedo and also retain radiation emitted by the surface. This means that the effect is to cool the troposphere but heat the stratosphere. It has been recorded in various recent volcanic eruptions that the net effect is to cool the surface.

The CO_2 released provides carbon for various biochemical cycles. Volcanic eruptions release on the average between 1.3×10^9 and 2.3×10^9 kg of carbon per year which is about 1.3×10^{-9} Earth atmospheres. The sulphate aerosols also promote complex reactions on the crystal surfaces. Among others, one is to produce chlorine monoxide (ClO) which attacks ozone, O_3. Ozone is very effective in preventing ultra-violet light from reaching the surface and forms a layer in the upper atmosphere. This is reduced or even destroyed by the aerosols. The aerosols also promote the formation of high cirrus clouds which also increase the albedo and so work to reduce the surface temperature.

There is one aspect of volcanic eruptions that must also be included here. This is the occurrence of underwater earthquakes and those near oceans. These could be associated with tectonic plate subduction such as that which occurs generally in the Pacific Rim. A vertical movement of the ocean floor which lifts water can produce a wave train which travels under surface gravity. The principle wave may be very high and have a large wavelength: the height can be up to 150 m, though 50 m is more usual, and the wavelength can be up to 100 km. These extreme values still leave the wave unrecognised at sea but will cause great devastation when it bursts on an ocean beach. They will spread a flood over substantial distances inland with the associated devastation. These waves are called tsunami, a Japanese word (literally meaning a harbour wave though they are not associated with harbours alone). The name reflects the frequent occurrence of such waves off the coast of Japan. Similar wave patterns will be formed by large landslides into the ocean. These again can result from volcanic mounds which become too large to be stable. There has always been a substantial population living on the low land at the water's edge so tsunami waves have been a constant threat to flora and fauna over the centuries. The effects on modern cities can only be imagined.

In summary, it is clear that volcanic activity can have the most profound effects on the surface and so does more than influence the atmosphere to produce vivid red sunsets.

2.13. Ice Ages

For periods over the history of the Earth substantial regions of the surface have been covered by ice and snow which has spread from the poles towards the equator. The mean temperature of the Earth is then lower than the average and ice sheets with glaciation advance to become more extensive. These periods are generally known as ice ages. There have been several such events during the history of the Earth and each time (at least so far!) each ice age has passed away but this need not always be so. We are used to having ice at the geographical poles, which is a feature of ice ages, but again there have been periods when the poles were completely free of ice. In fact, Antarctica is actually a very large continent at present buried under ice but there were periods when it was a lush wilderness.

Ice ages can be recognised by their characteristic "fossil" remains which define the nature of ice. The most obvious sign is the spread of rock debris carried over distances by ice and dropped where if fell when the ice melted. This can leave some strange, and sometimes large, pieces of rock in geologically strange places. Other features are scoured out valleys and etchings in rock. Typical landscapes are found over Northern Europe and North America. While identifying these landscapes might seem easy in fact erosion and the wear and tear of history make this less so with age. The colder climate favours flora and fauna that likes the cold but not that that does not. If these characteristics can be identified then conditions at a particular time can be found. This approach will work if the fossil record is stable for sufficient time for a range of relevant fossils can be found now which may be for several million years. A third effect is the proportions of particular chemical isotopes present in sedimentary rocks and cores at a particular time. Periods of ice change the flora and fauna and so the isotopic ratios because living materials tend to absorb the lighter isotopes. Any extinction of life will, therefore, leave a higher proportion of lighter elements in the mix of elements that are laid down.

All these indicators are subject to substantial errors. One major difficulty is the dating of different deposits. Comparative studies involving different sites and continents are consequently subject to considerable uncertainties making global analyses difficult to come by. Luck also plays a very major role here, as is other aspects of geo-studies, in accumulating the raw data to be interpreted. The situation is different over the last few million years because data derived from the drilling of continental, oceanic sediments and ice cores are open to a much more stringent analysis. These studies have fully confirmed relationships between ice ages and crustal phenomenon such as glacial morains, glacial erratics and glacial drumlins. Once established for the more recent periods of ice these can be taken as indicators for past ice ages.

Using these various techniques it has been established that there have been at least four very major ice ages and several lesser ones during the geological history of the Earth. We can summarise these as follows:

The earliest is believed to have occurred during the period about 2,700–2,300 Mya. This was during the Proterozoic Eon.

A well established period of ice was from 850 to 630 Mya (during the appropriately called Cryogenian period) when ice covered substantial regions of the Earth. Permanent sea ice extended even as far as the equator at some places. It has been suggested that this produced a snowball Earth where essentially the whole surface was ice bound. This is perhaps the only time discovered so far when the whole surface was possibly covered in this way. The ice was broken eventually by the heat of volcanic activity.

There was a substantial ice age during 450–420 Mya.

The next ice age seems to have been during 360–260 Mya.

The present period is in an ice age which began about 30 Mya and is now slowly coming to an end.

It is seen that these periods of ice cover periods of several hundreds of million years. They were not periods of entirely static conditions. The extent of the ice increases and decreases giving the ice cap a pulsing appearance. The times when the ice has its maximum extent are called a glacial maximum; those when the ice has a minimum extent are called interglacial. The Earth is in an interglacial at the present time.

The cause of ice ages is still not agreed nor the reasons why they have so far dispersed. Two very important influences are ocean currents and atmospheric currents. The former is controlled by the distribution of the continents while the latter is affected by mountain building (orogenic behaviour). The effects of ocean currents can be quite important — one has only to picture the effects on the British Isles if the Gulf Stream current were cut off. In the extreme case, growth of a super-continent will make ocean currents irrelevant but the placing of the land in latitude will be important. The effect of mountains is to force the air currents upwards and so to levels of lower temperature. An altitude is reached where the saturation temperature for water vapour in air occurs with precipitation as the result. The associated clouds alter the albedo locally. In this way the climate is affected by surface effects.

2.14. External Impacts

Interplanetary particles contain a great deal of kinetic energy. A mass M kg travelling at v m/s has kinetic energy $E = 1/2Mv^2$. As an example a mass of 10,000 kg travelling at 10 m/s carries kinetic energy $E = 5 \times 10^5$ J. The binding energy between two atoms is of the order of magnitude 10^{-19} J so there is sufficient energy in the travelling mass to separate 5×10^{24} paired atoms. Were the mass to strike the land surface of the Earth (or any planet for that matter) its kinetic energy would turn to heat which would vaporise the immediate surface throwing it in the air when it would land some distance away. In more familiar terms it would cause a crater, the larger the heavier the mass and the greater its speed of impact. If instead it struck the ocean it would cause a large wave and, if the impact were energetic enough, a substantial tsunami. The associated high temperature would evaporate a substantial quantity of sea water which would first become thick clouds and then very heavy rain laden with dust. The impact on land will cause local havoc. Apart from the crater, material will be sent into the atmosphere in a way similar to a volcanic eruption. It will return to Earth over a period of days or weeks blotting out the Sun for a period depending on the magnitude of the impact. The very smallest particles could remain in the atmosphere for a year or more. This will reduce the intensity of light reaching the surface and so the surface temperature. The presence of substantial quantities of dust in the atmosphere could be catastrophic to life.

If the body is of large mass the impact will lead to a very substantial loss of kinetic energy which will appear as heat. The heat blast may reach very high temperatures, sufficient to burn all vegetation and creatures for a considerable distance from the impact site. This itself will have a very significant effect on life. One example of a world catastrophe along these lines is believed by many paleantologists to have occurred 65 Mya with the demise of the land dinosaurs and other species.[12]

2.15. Summary

This chapter is concerned with the Earth bringing our swift survey of the inanimate world we live into a close. One interest has been the recognition of effects that could affect the stability of populations and especially of very

---

[12]Other paleontologists feel the observed crater at Chicxulub in Mexico is nowhere near sufficiently large to have caused such havoc alone and either a second meteor collision at roughly the same time or/and a substantial volcanic lava flash flood (there was one contemporaneously in India) would be needed as contributory causes.

advanced creatures. It is clear that catastrophes can be random events that make it impossible to predict what the life expectancy of an advanced civilisation may be.

1. The time scale for the history of the Earth is very conveniently set down in a series of geological divisions which provide an internationally recognised framework. This has been explained in a little detail.

2. Next the process of the formation of the Earth and its general inner structure is considered. The three rather different regions of core, mantle and crust are recognised and the chemical associations with each are set down.

3. The internal thermal regime is treated in a way that brings out the importance of dimensionless numbers. The Rayleigh, Reynolds and Prandtl numbers are derived explicitly but there are a wide range of others not mentioned. These numbers are vital to physics and engineering because they allow the development of models of real physical situations through the principle of dynamic similarity.

4. Arguments of this type suggest that the passage of heat in the Earth is through free convection rather than simple conduction. Such a system of convection cells will have a profound influence on the surface and especially on the distribution of land.

5. One result has been a surface structure made up of a set of nearly a dozen large "plates" floating on the crustal material which move relative to one another. Mantle material swells up between them to make ocean floor and the subduction of one plate under another leads to a volcanic structure than makes continental material.

6. The distribution of land over the surface of the Earth and the time scale can now be found from radioactive dating (for the time scale) and the remanent magnetism of rocks for the relative distributions. These techniques allow the movement to be found and so the time scale for changes to occur on the surface. This turns out to be 2 or 3 cm. per year (coincidentally the amount that the Moon is moving away from the Earth in its orbit which is, incidentally, the rate at which our finger nails grow!).

7. It is in this way that the continents have been formed and rearranged on the plate system. This is structure of plate tectonics. The result has been a forming and reforming of the continental structure over the Earth's history. On occasions all the land has been together while on others it has been spread about. A partial list of the formation and disintegration of continents is given. The last total accumulation of land was Pangea

(Gr: all Earth) 180 Mya — this single land mass has since broken up to form the two land masses we recognise today.

8. Comments are made about the surface temperature and the way it depends on the orbit round the Sun. This leads to a consideration of the past climates. Information about the climates at different times in history is included.

9. The atmosphere is a most important part of the Earth and its composition and history are described. In particular the astronomical factors affecting the atmosphere are considered in a little detail.

10. Volcanoes have played an very important part in the history of the Earth and this is considered in the chapter.

11. Ice ages have provided a very traumatic time for life on Earth. The ice ages are considered and it is realised that we are still moving out of the last one.

12. Finally, the effect of volcanoes on the Earth is briefly considered and the random nature of the effects recognised.

Further Reading

1. Janson-Smith, D. and Cressey, G. *Earth's Restless Surface*, Natural History Museum Publications in the Earth Sciences.
2. Stewart, I., 2007, *Earth: The Power of a Planet*. (Five one hour films.)
3. A description of the Earth and all it contains as a single reacting entity keeping itself fit for survival is proposed by Lovelock, J., 1988, *The Ages of Gaia*, W. W. Norton, New York, with later books.
4. Ernst, W. G., 1990, *The Dynamic Planet*, Columbia University Press, New York.
5. Kious, W. J. and Tilling, R. I., 1996, *The Dynamic Earth*, Diane Publishing Co., PA, USA.

Chapter 3

Life in Water: The Precambrian

There is still no general agreement of how life started although there are hand-waving guesses: it might be that our present difficulties are due very largely to our not yet understanding the problem properly. It is not even clear at this stage whether life started on Earth or whether instead it came, at least in part, from outside (panspermia). If the latter were the case it does not make the problem of the origin irrelevant but simply shifts it from the Earth elsewhere and in the process makes it even more difficult. There would then be absolutely no indications of the physical conditions associated with its beginnings whereas an Earth origin would have been in conditions that we can duplicate at least approximately in the laboratory.

Our discussion of evolution in this chapter and the next will suppose a single thread of linked activities from the earliest times. It must be said now that a single evolutionary thread is an assumption. It is the simplest interpretation of the observations that we have collected so far and has led to no ambiguities but it may turn out later not actually to be true. It is possible, though on the present evidence not in any way substantiated, that life started more than once. This could have been the same process but at different places at essentially the same time (which is actually not unlikely) or it could have been at different times and different places. In these cases life at different locations could have been repeats of the same pattern or it could have involved different forms. As one example, actual life as we know it is universally left-handed (left chirality) but another, as it turned out, unsuccessful form could have been right-handed. As a matter of fact the Universe itself shows the same tendency to a left handed bias. We will neglect these alternative possibilities now because there is no evidence for them and so nothing to talk about but they should be kept in mind as possibilities for the future.

In this short chapter we consider the beginnings of life as far as we know it and its development up to about 530 Mya. This is a huge chunk of time

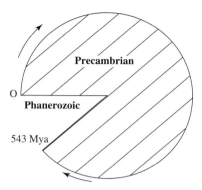

Fig. 3.1. A pie chart showing graphically the relative times that life was essentially microscopic and when it developed beyond that. The time up to 543 Mya occupies 318° while that after occupies the remaining 42° slice, a ratio of very nearly 8:1.

and it may seem strange not to break it down — it is about 88% of the current age of the Earth. The overwhelming scale of this period is made graphic in Fig. 3.1.

It must be recognised at once that, although this overall age division is quite firm, the intermediate ages to be quoted in what follows must, by their nature, be approximate to some degree. The relative occurrences, however, remain reliable. Most of the Precambrian period was dominated by bacterial life and this microscopic form would not have been generally visible to the unaided eye. Their remains now are very few and far between and can be found only by the most painstaking rock analysis. The end of this period saw the beginnings of macroscopic life of the most elementary form. This led (in the Cambrian) to a most amazing outpouring of new life which, though each individual was not very large by later standards, was certainly large enough to be immediately visible to the naked eye and especially with hard parts that form clear fossils in the future. In very general terms this is the first time true fossils unambiguously began to be formed which can be recognised without microscopes and it is this reason that marks the geological division of 530 Mya as an important dividing line for life. Fossils will become widespread and common from this time onwards. This outpouring of post-bacterial macroscopic life forms is usually called the Cambrian explosion and is an important, but in many ways mysterious, event in the history of living things. It marks a boundary in the animate calendar — the period before, where life had been microscopic or rudimentary, is called the Precambrian (because it was before the Cambrian). The period afterwards up to the present day, the period of fossils and clear living remains, is called the Phenerozoic.

Most of the Precambrian may seem uninteresting except that the development of life at the end could only have happened if considerable preparation had been made earlier on even though, without abundant fossils, the details will probably always remain hidden. The development of the wide range of life that fossils proclaim seems not to have been a simple thing to develop. Why nature might have undertaken this particular, obviously difficult, task will be explored in Chap. 6. It must be added that some authorities have questioned the reality of the sudden introduction of new life in the Cambrian explosion suggesting that some of the earliest fossils might have been wrongly dated and should be re-allocated to an earlier time. This would make the transition from microscopic to macroscopic life the more gradual and so more appealing. Certainly there is evidence of larger organisms before this but they were not vertebrates and have left little record of their being here. The technical details of dating cannot be entered into here but suffice it to say that a gradual transition presents many problems as attractive as it may be and a discontinuous transition appears the most acceptable on the basis of what observational evidence there is. This will be taken in what follows.

3.1.   Constructing the Very Early Times

Conditions on Earth at its formation could not be more different from conditions today. It was seen earlier that no surface material has yet been identified from this precise period but general evidence of these times is provided by the examples of the Moon, Mercury and certain of the satellites of the major planets which still have surfaces of this age. It appears that the initial surface was molten and calculations suggest that the initially liquid surface will have cooled sufficiently to become essentially solid after some 200 My, which a about 4,350 My ago. The surface would have still have felt very warm, however, with a temperature perhaps as high as $60\,^{\circ}$C.

We have seen before that the hot forming Earth also had an intrinsic atmosphere associated with its formation. It is generally accepted that this very first primeval atmosphere would have been composed very largely of hydrogen and helium but these very light gases would have been lost very quickly from a body of Earth mass. A second atmosphere will have quickly replaced the initial one formed from gases that were exuded from the molten Earth (that is out gassed — and especially from volcanoes) and the Earth is sufficiently massive to retain these heavier gases as a rudimentary covering to form an atmosphere which thickened increasingly with ongoing volcanic activity. Water was one of the components ejected but the major components were carbon dioxide, ammonia, sulphur and similar noxious molecules. There would not have been an oxygen component at this time. At first the water would have been largely in the form of steam which went into the atmosphere but with falling temperatures it will have rained onto the Earth's surface where it condensed as water to form a more or less complete thin covering for the surface. This would soon have increased in depth as the temperature fell further to provide a shallow coating with volcanic islands poking through in the form of rudimentary atolls. What rocks that still remain associated even remotely with this period are exclusively volcanic. Boiling lava regions will have been widespread with abundant boiling lava mud pools. These are found now in many places and evidence that they were present very early on is found now in fossilised form in rocks. Taking the known half lifes of the radioactive elements on Earth including short half period radioactive atoms such as U^{233} allows a simple calculation to be made which shows that the level of background radioactivity at the surface will have been more than twice the current level. The half life of U^{235} is closely the same as the present age of the Earth and this would have been supplemented by shorter life radioactive elements that have since essentially disappeared.

Evidence from the oldest rocks known so far (around 3.50 My in South Africa though rocks in the Isua Greenstone belt in Western Greenland are older at 3.8 MY) show fossilised vents (smokers) where hot fluid had streamed out, just like the modern smokers in the mid-ocean ridges (see Fig. 2.8). The richness of the minerals observed in these localities now suggests that the same situation would have applied then as now. The rich life forms observed around smokers depend ultimately on the presence of bacteria to start the food chain. Fossil forms of the bacteria Archaeospheroides barbertonis have been found in sedimentary rocks showing life had been established by this time though there is no indication of how widely spaced it might have been.

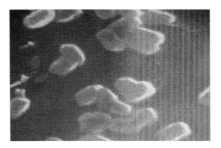

Fig. 3.2. Images of bacteria associated with sedimentary rocks in Barberton, South Africa. The rock is 3.2 By old. This is the earliest direct evidence yet of life on Earth and must be closely related to the very first living forms.

A date of 3.2 My has been assigned to them. It does, indeed, suggest the possibility of life having formed in the region of smokers. On reflection, it is a natural place for aquatic life to begin. Life remained confined to the seas for about the first 4,000 My of the Earth's history.

Several consequences of the presence of the substantial natural satellite of the Earth have been highlighted in this connection. The shared angular momentum between Earth and Moon will have several consequences. The shared gravitational interaction provided the tides (ocean, solid material and atmospheric) which have played an important part in the evolution of life. The constant value of the angular momentum of the Earth–Moon system has a particular consequence for both bodies. Friction forces are slowing the Earth's rotation. These are the result of the friction involved with the flow of the water tides on Earth driven by the Moon's gravity. A small amount of energy is transferred from the Earth to the Moon which increases the lunar angular momentum by a very small amount. There must, then, be a corresponding decrease in the angular momentum of the Earth to compensate and this is manifested by a very small decrease in the rate of rotation — the day is getting longer but by a few microseconds per year. The degree of slowing has been detected as an annual rate of about 10^{-7}s. A slowing Earth means a reducing angular momentum which must be complemented by an increasing lunar angular momentum to keep the total constant. It is known that at the present time the Moon in its orbit is moving away from the Earth by about 3.8 cm per year. This makes no noticeable difference now but it will over the lifetime of the Earth–Moon system. The Moon will have moved out from an initial orbit perhaps a third of the present value.

*

A: Single Cell Organisms

3.2.   Life Begins

The first noninanimate molecules probably appeared after a few hundred million years of the formation and were probably well established by 4,000 Mya, after about 550 My of the Earth's history. Certainly elementary bacteria have been found dated 3.2 By in one place as we saw above but indirect evidence exists of a more general appearance. Early rocks show a greater abundance of the isotope ^{12}C than expected and this isotope is found today to be particularly associated with recognized living materials. We can expect this association to have been evident from the beginning.

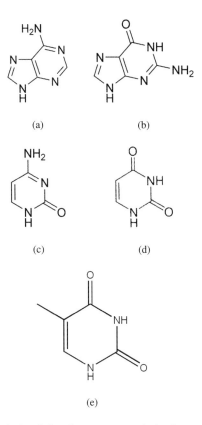

Fig. 3.3. The characteristic skeletal structures of the base components of RNA and DNA. (a) is the adenine base; (b) is the guanine base; (c) is the cytosine base; and (d) is the uracil base for RNA and (e) is the base thymine which replaces uracil in DNA. These nitrogenous bases are linked with a sugar — phosphate backbone. RNA usually forms a single strand and DNA usually a double strand but this is not always so. The components are all formed from C, H, N, O.

The very earliest life was probably self-replicating molecules with a structure very closely similar to simplified RNA molecules. The RNA molecule known today is the combination of the four base molecules adenine, guanine, cytosine and uracil. The form of these four components is shown in Fig. 3.3. They are joined to a backbone of ribose sugar and phosphate. The whole forms a single strand of nucleotides which can assume many complicated three-dimensional shapes. The emergence of RNA implies the prior presence of the four complicated base molecules adenine, guanine, cytosine and uracil. Here is a problem. It is the very early presence of this quartet that is at the centre of the discussion of an Earth or space origin — how could such complex molecules have formed so quickly and essentially from nothing on Earth? Once they are in situ it is then thought life could begin its evolutionary career by chance encounters. It is probably not irrelevant that the most elementary bacteria known so far are found inhabiting the black smokers in the mid-ocean ridges in the deep oceans. The simplest of these are, indeed, found to be chemoautoprophs which fix the carbon that is necessary to sustain life by breaking down chemical bonds to release energy. This synthesis of organic molecules obtains energy by the oxidation of electron donating molecules. For the present case the bacteria convert the hydrogen sulphide and ammonia issuing from the smoker gases. Although these bacteria are very primitive they represent a step forward in complexity from those that must have provided the original life. As yet the mid-ocean ridge structures have only been partially explored and virtually nothing is known about the evolution of life in these regions. For the land the first believed "fossils" are marks with a signature of the animate isotopic carbon ^{12}C and are found in the oldest sedimentary rocks which are about 3,800 My old. The period between 4,550 and 3,800 Mya forms a coherent unit and is called the Hadean Era. It formed a very fiery prologue of what was to come.

The first chemoautotroph organisms must have lived under shallow water or in fluid mud pools at the surface partly because this would have allowed them to develop and partly because there would have been very little solid land available for habitation. This differs from the present conditions around black smokers which are in very deep water (perhaps several kilometres below the surface). Away from black smokers the creatures could have produced the energy they required by using carbon dioxide from the atmosphere as a source of carbon and so oxidising inorganic materials for energy. The ability to reproduce distinguishes then from simple organic materials because it represents the most elementary form of genetic material. These organisms resembled prokaryotes in being only single celled, what genetic material there was, needed for reproduction by simple cell division, will not have been confined by a closed membrane within the volume of the organism. They will have been small, broadly in the range 10^{-7}–10^{-4} m. Life would not have been obvious to an observer without special optical aid. This was not to be the case a later on.

It was later in this era that prokaryotes appeared to have evolved the more efficient process of glycolysis for the production of energy. This is a set of chemical reactions that release energy from organic molecules such as glucose. The process is called the citrus cycle or the Krebs cycle but is complicated and we will not delve into it here. The process uses ATP in which one or two phosphate grouping(s) is/are lost to release energy. This process is still used universally today — nature hangs on to good ideas when it finds them. It is interesting that prokaryotes remain a dominant life form on Earth even now. During this era the Earth gave as close a vision of Dante's hell as it has managed to do so far but it was the world in which the first elementary life appeared and actually flourished. It seems that heat, water, mud and radioactivity all played a part in the initial formation of living things.

It was at this time, around 4,000 Mya, that DNA began to supplement RNA. This involved the replacement of the base uracil by thyrocine (see Fig. 3.3). The single short band strand structure also became a double helix structure that could be very long. The reason for the change was stability: RNA easily hydrolyses but DNA does not. This change placed living materials on a much more stable footing and would allow a more complex structure to develop. It is interesting to notice that even the modern bacterial deep sea life associated with black smokers is based on DNA and so, unfortunately, the bacteria cannot be exactly the same as those which started life.

3.3. Life Develops

The surfaces of the Moon, Mars and Venus give evidence of beginnings of very heavy bombardment by planetesimals over a wide range of sizes around the time of the formation of the Earth and rather later. This apparently lasted for a few hundred million years and subsided. There was, apparently, a late heavy plantesimal bombardment some 3,900 Mya which will have disturbed the now developing life. One might have thought that such an event would be harmful to the new life but this seems not to have been the case. Between 3,900 and 3,500 Mya a very significant thing began. It was during this period, when the Earth was some 700–900 My old, that the initial prokaryotes (single celled without a nucleus) began to separate into two closely similar forms called now bacteria and archaea. This meant that the initial life form (the universal ancestors) was lost gradually and very probably forever. Whereas the bacteria inhabit what may be called normal conditions, archaea moved also into extremes. This involved some changes in the internal materials and structures but still did not isolate a nucleus. The result was the ability to live under extreme pressures, either very high or very low, to live in extremes of hot or cold, to live in extremes of great salinity and so on. These extremophiles were essential for subsequent adaptive radiation.

Around 3,500 Mya bacteria develop a primitive form of photosynthesis. This allows the generation of ATP, an animate battery, by exploiting a proton gradient. This mechanism is still used by essentially all organisms. This early photosynthesis did not, however, produce oxygen as it does today (Figs. 3.4 and 3.5).

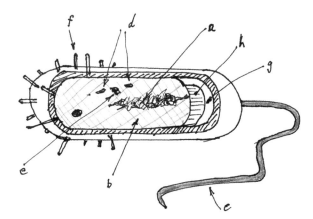

Fig. 3.4. A schematic section of a prokaryote. The nucleoid (a) is composed of circular DNA. There is a bacterial flagellum (c) for movement and pili (f) for attachment. Also present are (b) cytoplasm; (d) ribosomes (complexes of RNA and proteins); and (e) plasmid. This is all covered by a cell wall (g) and a plasma membrane (h). The whole is enclosed in a capsule. This is a complicated structure which must have resulted from a considerable evolutionary development.

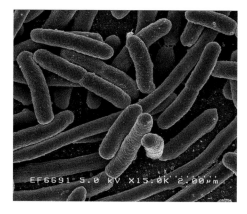

Fig. 3.5. Showing a group of modern bacteria. The early samples would have been not dissimilar.

3.4.   ATP: A Biological Battery

It is useful at this point to break our argument and trace this very important early development for the provision of energy in a cell. Living material needs an energy source as does everything else. About 3,900 Mya cells appeared which closely resemble prokaryotes. They needed an energy source for their various functioning and found this by extracting energy from the oxidation of inorganic materials. This very elementary process gave sufficient energy for elementary organisms but something better was needed if more ambitious evolutionary steps were to be taken. As prokaryotes evolved they developed the much more effective process of glycolysis which allows energy of organic molecules like glucose to be set free. This was a major development because this turned out to be an energy source to weather successfully the test of time and is still important today. Glycolysis employs the adenosine triphosphate molecules (ATP) as a short term energy currency. About 400 My later (some 3,500 Mya) bacteria developed an early form of photosynthesis but which did not, however, produce oxygen as a by product — that must wait for another 500 My. The ATP now was generated by exploiting a proton gradient across a membrane which is a process almost universally used today for the same purpose.

ATP has got the chemical formula $C_{10}H_{16}N_5O_{13}P_3$. It is, then, a fairly sophisticated molecule. The structural skeletal picture is shown in Fig. 3.6. It involves a ribose sugar with chemical formula H_4O_5. This is the same sugar that forms the basis of DNA (Figs. 3.7 and 3.8). Attached to one side of the sugar is a base, linked an adenine ring, a base with the chemical formula H_3N_5. Attached to this is a group of three phosphate groups. The links

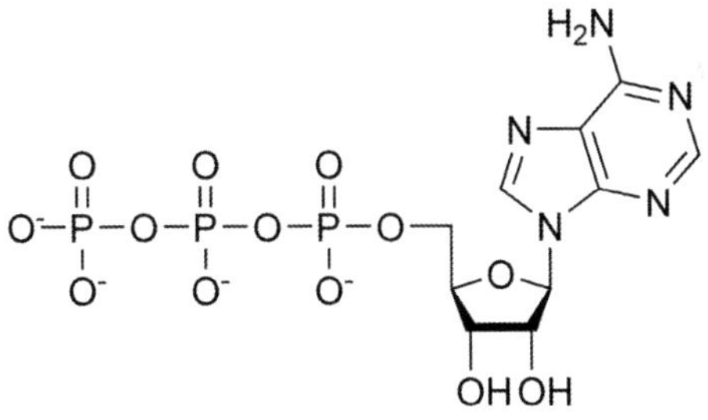

Fig. 3.6. The ATP molecule skeletal.

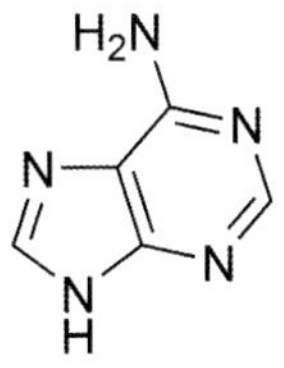

Fig. 3.7. The adenine ring skeletal.

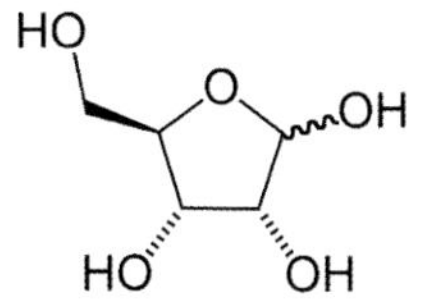

Fig. 3.8. The ribose ring skeletal.

between the phosphate groups are particularly significant for the present purposes because it is there that the energy is stored and is available for use. The link between the second and third phosphate group is especially significant. Although the bond energy is of normal value, energy is produced when water is allowed to react with each element of a fracture of the bond. The ATP changes to become hydrated (and so is ATD) and the inorganic phosphate is also hydrated. The reduced enthalpy at standard temperature and pressure (S.T.P) in the double process is $-20.5\,\mathrm{kJ/mol}$ with a change in the free energy of $3.4\,\mathrm{kJ/mol}$. The ATP molecule is built up when energy is spare to be released when an excess is required which puts the molecule in the category of a battery.

The chemical reaction for the provision of energy is

$$\mathrm{ATP} + \mathrm{H_2O} \rightarrow \mathrm{ADP} + \mathrm{P}_i \quad (+\ 3.4\,\mathrm{kJ/mol}).$$

Here, ADP is adenosine diphosphate with chemical formula $\mathrm{C_{10}H_{15}N_5O_{10}P_2}$ and P_i is the inorganic phosphate with chemical formula $\mathrm{HPO_3}$ (Fig. 3.9).

The conversion of ATP into ADP is the reaction for the immediate provision of energy under the instruction of an enzyme. This is a reversible process giving the means of replenishing the energy when the creature is resting

$$\mathrm{ADP} + \mathrm{P}_i\ (+3.4\,\mathrm{kJ/mol}) \rightarrow \mathrm{ATP} + \mathrm{H_2O}.$$

For greater energy usage two phosphate groupings can be shed leaving adenosine monophosphate (AMP). The energy for the reattachment of the phosphates comes from food or sunlight. More precisely the synthesis occurs in the cytosol by glycolysis, by cellular respiration in mitochondria and by photosynthesis in chloroplasts. To be more explicit, the conversion of ATP to ADP in humans is achieved by the aerobic respiration in the mitochondria while in plants the conversion uses photosynthetic means to convert and store sunlight.

It is interesting to note some of the various uses of ATP. First it is required for nearly all anabolic reactions. These will include: joining transfer RNAs to

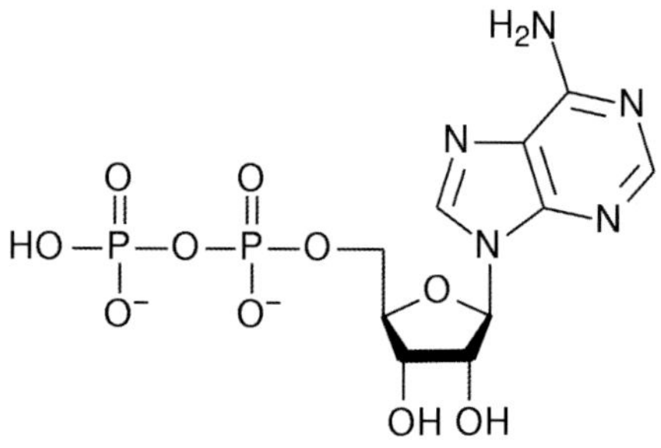

Fig. 3.9. The skeletal structure of the molecule ADP.

amino acids in the assembly of proteins; the synthesis of polysaccharides; the synthesis of fats; and very importantly the synthesis of nucleoside triphosphates for incorporation into DNA and RNA. It is separately central to the transport of molecules and ions. In this category it is central to the working of nerve impulses; in muscle contraction; in the beating of cilia and flagella such use in sperm; and for bioluminescence. In mammals its extra-cellular release indicates the shortage of oxygen in the blood and its release from the stretched walls of the urinary bladder indicates that the bladder needs emptying. These are some examples which explain why phosphates are such an important ingredient of the diets of all living things.

3.5. Life Expands

At about the same time cyano-bacteria (green algae) began to form layered structures as colonies. Those with dome shapes are known as stromatolites while the round ones are oncolites. They are essentially colonies of bacteria together with dust and gravel all held together by calcium carbonate secreted by the bacteria. This makes a solid mound which could grow up to a metre or more high off the bed of a shallow lake. These structures formed in colonies of a few dozen giving the appearance of a pin machine familiar in fun shops. A photograph of fossil stromatolites is shown in Fig. 3.10. The presence of microscopic life was now much more obvious. There were no

Fig. 3.10. Fossil stromatolites.

Fig. 3.11. Showing a bed of modern stromatolites in Sharks Bay, W. Australia. Much of the Earth would have had this appearance when life first took serious roots.

creatures to interfere with them at this stage of history so they had no predators. Although they had their hay day around these times there are some colonies even found today. The best known are probably those in Shark Bay, Western Australia, where conditions are distinctly saline so nothing else will grow there (see Fig. 3.11). This probably protects them from foraging and burrowing creatures. It must be admitted that columns of this kind can also arise from purely inorganic materials and the difference between the animate and inanimate forms is not easily recognised.

3.6.   Oxygen and Internally Differentiated Cells

This process develops over the next few hundred years by cyno-bacteria where water is used as a reducing agent. As for any "engine" there is a waste product and this now is oxygen. For the first time oxygen was being produced on Earth and the initial form is being changed by living things. This was a truly momentous development was one of the key factors in the move to advanced life. It took place about 2,500 Mya, when the Earth was already 2,000 My old. It affected the planet in a crucial way and, indeed, was perhaps the first time animate matter had affected the planet Earth. Initially there were small quantities of oxygen produced in this way which went into the seas but this slowly accumulated with several consequences. One involved other types of bacteria for which oxygen, now new to Earth, was actually a poison. These consequently began to play an inferior role. Another consequence was for the oxygen to combine with free iron in the seas to form iron oxide. This was laid down in the usual way and produced eventually the banded rocks of iron oxide that are found today. Once all the iron in the oceans was converted to oxide there was nothing for it but for the free oxygen to enter the atmosphere through evaporation. Thus began the steady accumulation of oxygen in the atmosphere so characteristic of the Earth today. The quantities were small at first but they have grown steadily until today oxygen accounts for about 27% of the whole. Although the effect was to poison some bacteria initially which could not assimilate oxygen some survived and a minority remain today living in isolation from any oxygen cycle.

Oxygen in an atmosphere is often taken as an indictor of life on the planet but this statement requires explanation. It may be an indicator of advanced life but it has been seen already that the first photosynthesis did not involve the production of oxygen. Once oxygen has appeared, however, it can be applied to improve the efficiency of energy production for the organism. More especially, it is now possible to extract energy from organic molecules such as glucose (sugars) and so make the production more robust. The citric acid cycles and oxidative phosphorylation developed for this purpose and produced methods which are still used by animate matter today quite universally.

3.7.   Complex Cells

When the Earth was about 2,400 My old the simple cells were supplemented by more complex ones. It seems that the archaea found a close relative in cells that developed enclosed specialised regions, called organelles, with various and diverse functions within the cell. These new cells are called eukaryotes. One example of organelles are mitochondria which derive energy from organic molecules. Many eukaryotes have got chloroplasts which are organelles deriving energy from light and use it to synthesise organic molecules. The cell had developed from a simple collection of essential items to an ordered collection of separate components acting as a single unit.

How this came about is the subject of much debate. The concentration of a set of particular activities into separate compartments each encased in a lipid boundary is very much equivalent to a stone age round house (where everything is achieved in a single space) developing into a Roman house (where there are compartments for specific activities). One possibility for differentiated activities is that it became more effective for individual activities to be defined in a particular space. Another is that a set of different objects came together for a more effective living. An intermediate possibility is that an existing cell was invaded and both the host and invader found it beneficial for the intruder to stay and contribute to the whole. This is certainly the generally accepted view of the mitochondria as an energy source. Some cells contain just one such energy unit while others might contain several. The idea of a symbiosis extends what is known already in other areas (for instance lichen is comprised of fungus and algae) but, as for other areas, it is difficult to see how it first came about. It is difficult to become more positive as far as early cells are concerned because so little evidence remains. One thing we can say: the move to organise single cells led the way to the formation of multicellular organisms rather later on.

Again, a process of photosynthesis has been developed capable of further development as a major energy provider. These developments were essential if life was to move to a more complex phase. They show a very interesting feature of evolution. Characteristics did not arise bespoke for the next move but rather the next move used characteristics that were already there. Was evolution just serendipity or the manifestation of an eternal plan? Or could it be both? Darwinian evolution is clear: the next move is always what is the most competitive for individuals, and one thing leads to another.

The Earth had existed at this stage for some 2,500 My and life is still restricted to single cells. The atmosphere still consists mainly of CO_2 but that was changing, not in a dramatic fashion but gradually.

3.8.   Sex Arrives — Genetic Diversity and Stability

The stage is being set for major steps forward in the distribution and complexity of life. Confined to the oceans, seas and lakes until now, some cynobacteria began to inhabit the damp soil on the edges of the waters. This was the first step towards a move of organisms onto the land. A development of this kind is possible when environmental circumstances are right to allow the organisms to make the necessary adaptations for land living. One requirement is the availability of sufficient oxygen to fuel more complex living. Another is complexity of life itself to use it effectively.

To this end the cells are becoming appropriately organised even though they stay separate and there is a good system for the production of energy. The restrictive thing now for further evolution is the limited genetic material for the reproduction of the cells. Reproduction is still by simple division which does not introduce any variation of the material. Some change occurs through mutation but this is an unimportant mechanism at this stage. The cork was taken out of the evolutionary bottle about 1,200 Mya by the introduction of the rule that two random cells must be involved in the reproduction process rather than only one. More simply this was the introduction of sexual reproduction. The more partners necessary for procreation the greater the possible variations but the more partners necessary for procreation the more difficult it is to achieve in practice. The difficulty of getting three or four creatures together at the one time could be severe. Two is the most practical choice which nature has made universally.

Living creatures were still confined generally to the seas, the oceans and the lakes. It is most likely, however, that some cyano-bacteria began to use the wet regions at the edge of the water as a habitat and so would move gradually to the damp regions. This was the first move of animate matter to the land. It had taken over 3,000 My for this to happen. The way was now clear for single cells to dominate the entire world and they have ever since. They are by far the most abundant and successful life form of today having expressed themselves in a very wide range of variants over every conceivable physical environment. They have lasted so long and there seems no mechanism which would be capable of bringing their reign to an end. They will surely be present at the end of the world.

*

B. Multicell life

3.9. Primitive Senses

Once sexual reproduction was established (during the Protozoic) the pace and security of evolution could increase significantly. This had taken a long time from the beginnings of life, indeed all of 2,500 My. The first new development was multicellular life. This seems to have resulted from colonial algae in the first instance but it soon developed into an elementary seaweed. Sponges (order Porifera) soon appeared. These are the most primitive animals with partially differentiated tissues but without muscles. They have no nerves or internal organs and have no means of locomotion other than being carried by the currents in the sea. Closely similar but with means of locomotion are jelly fish (order Cnidaria). These are certainly multicell creatures but they have no precise outer profile and no outer protection. They are generally successful poisonous predators and they rely on enemies keeping away from them for fear of being stung.

After very nearly 4,000 My with microscopic cellular life being the only life on Earth evolution began to move, slowly at first but with growing speed, towards more complex forms. About 600 Mya there was the first radiation of animals in the sea. This began with Porifers (sponges), Cnidaria (jellyfish) and Platyhelminthes (flat worms) along with other multicellular creatures. These were invertebrates without a clear backbone. The flatworm was a significant newcomer to the ocean some 600 Mya. It later showed a great radiation and is still very much with us today. It might seem an insignificant player now but then it was the first creature with any sort of brain and with a very rudimentary nervous system. This early example also had light sensitive cells which were the precursors of eyes. Although this was extremely primitive and generally ineffective by later standards it was, nevertheless, cutting edge at the time. At least the principle had been established of a central controlling mechanism and so was available for the later development. Like surveyance instruments, the senses are of little value if the information they offer cannot be interpreted and acted upon. One important control developed at this time was of light sensitive cells (eyes). Eyes are of limited use under water but hearing is of vital importance. It is almost certain that techniques of hearing will have begun to be developed at this time

3.9.1.   *The amazing Cambrian explosion*

Up until now the various creatures had been generally small and without horny parts (invertebrates). Fossil remains have been no more than marks on the ocean floor where the creatures had slithered and moved. Soon after this, between 565 and 525 Mya a large range of creatures quite suddenly appeared with hard parts which have left fossils. It seems that the evolving of hard material was the trigger for this expansion which is called the Cambrian explosion. It initially involved life in the water but soon moved onto the land. It introduced all the phyla of modern animals.

The beginning of the Eon (the first Era, the Palaeozoic 4,007 My–4,302 My) has proved to be one of the most extraordinary and controversial eras of them all. It witnessed an enormous explosion of types of organisms apparently with little or no previous roots. So great and abrupt was the transition from simple microbia that some workers have even doubted whether the fossil record for the whole period has been interpreted correctly. To add to the confusion, some fossil species of the period have not been catalogued as being related to any known species existing before or after. A particular example is the quite extensive group of fossils found at Mistaken Point, Newfoundland, along the southern coast of the Avalon Peninsular. These are imprints in mud of soft bodied organisms most of which had not been found anywhere else and defy definition in terms of known living organism types. Typical of many are large leafy forms some with stalks pinned to the ground (hold fasts), others are like a small cabbage and others still like a large bush. Still others have the appearance of a branching entity like a small tree. Others have a spindle shape, pointed at each end while still others are disc shaped. The fossils at Mistaken Point can be accurately dated because they are covered by layers of volcanic ash which can be readily dated accurately. The date is found to be 565 Mya and is the only very accurately dated layer of this period. The nature of the surrounding sedimentary rocks shows that the fossils were laid down under deep water, well below the level where either solar radiation or surface wave profiles could penetrate.

Not all the Newfoundland fossils are unique: some are, in fact, similar and some even identical, to fossils found at Charnwood Forest, England. It can be remembered that Newfoundland and England were very close together at that time since the Atlantic Ocean had not opened up. The fossil fields of these two localities will have been very close together then and it is not surprising that they will show similar fossils even though the fossils themselves provide mysteries.

3.9.2. *Questions about the Cambrian explosion*

It is a remarkable occurrence that an explosion of species seems to have taken place over a short period thought until recently to have been about 40 My during the period 565–525 My ago. More recent studies appear to have narrowed this down even further to as few as 5 My, which is instantaneous from a geological point of view. During this period evolutionary changes occurred which created all the essential phyla of modern animals. They also introduced elementary eyes and hearing. The nature of these events have long posed problems for theories of evolution (Charles Darwin was particularly worried about these events) since it suggests that a wide range of new species appeared without the required period of testing for fitness. Certainly, the changes that appear to have occurred during this period are dramatic and the nature of life changed forever. The profundity of these changes can be seen from a few examples.

The first example is trace fossils which are marks left by organisms as they move over a two-dimensional sedimentary surface. At the end of the Proterozoic these were simple surface marks left by arthropods such as trilobites, resting, or furrowing, or emerging, or walking or even striding. There was no apparent attempt to move from the simple two-dimensional geometry. With the coming of the Cambrian the geometry became three-dimensional with the introduction of burrowing creatures. These trace fossils are particularly important because they show the presence of creatures that may not have hard parts of their bodies and so will not form true fossils.

There is some evidence that creatures with hard parts had, in fact, appeared before the Cambrian. These include a stalked structure named *Namacalathus* and a cone structure called *Cloudina* both of which seem to have become extinct before the Cambrian era arrived. After these a series of small skeleton creatures appeared including early molluscs and several sponge spicules. Following these there followed a range of small shell-type creatures including early brachiopods. It was not until a little later that true shell fossils appeared that can be widely accepted as trace fossils. This move from elementary creatures to more complex structures across the Cambrian boundary is extraordinary if it occurrence over as short a period as 5 My without any ancestor trials.

We have seen that the Cambrian is marked by a very large number of exceptionally preserved fauna. The most important are the Burgess Shale fauna of British Colombia, Canada (Middle Cambrian), Chengijang China (Lower Cambrian), Sirius Passet, Greenland and Orsten Sweden (Upper Cambrian). The Burgess Shale has been particularly prolific in fossils: it was discovered in 1909 and continues to be the source of many fossils still. Indications are that it was a region of deep water about 590 Mya subject to land slides. These appear to have covered the living creatures on the bottom at one time providing a detailed picture of life on the sea bottom at one instant of time. The fossils are dominated by arthropods although sponges and *echinoderms* are also represented. Many taxa have been particularly interesting because they do not fit into any modern catalogue of taxonomy. In particular, many of the fossils from the Burgess Shale do not fit into the current arthropod classes such as insects, crustaceans and chelicerates. The Chinese and Greenland specimens also include specimens that cannot be included in modern catalogues. During this period the first primitive plants, probably evolved from algae, moved from the edges of lakes onto dry land. It is very likely that the elementary plants and algae had a symbiotic relationship, the first that there is any real evidence for. Once the elementary plants had invaded the land other organisms could hardly be far behind.

Many questions have been raised about this multiplication of the rate of evolution and the almost step change in the form of life on Earth. Several comments must be made.

(a) There has been great difficulty in achieving a common and consistent time scale for different parts of the Earth. Modern work is aimed at improving on this situation but it remains possible that some fossils belong to an earlier period than is apparent at the present time. If this is proved to be the case it would at least alleviate the appearance of a rapid evolutionary change at this time. This aspect is still the subject of active research although this could raise even more difficulties. Indeed, very modern work is suggesting that it is not 40 My but nearer to 5 My. If the very modern work proves true and the time period of the explosive radiation was about 5 My the period becomes even more peculiar in the absence of earlier living creatures.

It has been suggested that microscopic creatures of the type *Trichoplax adhaerens* were present in the Precambrian seas. These are the most elementary animals known and are the only known representatives of Placozoan phylum. They are a flat plate like animal with typically a diameter 0.05 mm. They have three layers, the middle one being fibrous tissue. They have no component parts — no head, visible gut or nervous system. They move using a myriad of tiny legs though their physiology is as yet unknown. They are believed to be related to *Dickinsonia* (see Fig. 3.13). Analysis using DNA suggests they have very ancient origins and could well have been one of the earliest animal forms. There have been claims to have found them in Precambrian rocks but this would require microscopic aids and finding one would be a very chance affair. A cartilage back bone could have allowed them to evolve quickly into early macroscopic creatures. Such creatures could provide the link between the Precambrian and Cambrian creatures that would be required by Darwinian evolution.

(b) If the Explosion did, in fact, occur as seems to be the case at the present time, the questions arise of why it happened at all and why it happened when it did. These may well be related questions. In order for an evolutionary step to take place it is necessary that the biological structure shall be correct and also that the environment shall allow it. It also requires the previous animate landscape to provide a niche for new radiation, which usually means an extinction of a few or many of the pre-existing species. Things are simpler if there is no already existing species to get in the way. A most important biological requirement for the effective radiation of new species is probably the availability of sexual reproduction. Now more than one organism is involved and change is no longer dependent on chance mutations. A second important requirement is the availability of an energy source, through ATP, that could support substantial animate objects. This requires appropriate levels of oxygen in the atmosphere. With the biological scene set and the material requirements made clear, the question then is whether the environment can offer the chemical components necessary to carry through the changes. A crucial factor could well have been the availability now of an elementary

cartilage to give strength to larger organisms. In these terms the Explosion
must have come about because the environment could now support the new
species. The environment is constantly changing and can change very quickly
so changes could be expected to occur quickly once the external conditions
were right.

3.10.   Some Images from the Middle Cambrian Burgess Shale

The richness of the images from the Burgess Shale is really quite remarkable.
Up to the present time some quarter of a million fossils have been unearthed
and the tally continues apparently unabated. A complete ecosystem appears
to have been encapsulated by a mud slide underwater. We give here a few
images of the fossils that have been found. They represent a very strange
world (Figs. 3.12–3.17).

Fig. 3.12.   An image of an early *trilobite*. This was an early success story and the species
lasted for some 300 My in various forms that evolved from one location to another. It
was armoured on the top but had a soft underbelly and lived on the sea bottom. The
size varied widely from species to species: from a few centimetres to about 1 m. All were
segmented although some were rather thinner than others. They generally lived on the
sea floor and the end segments probably acted as flaps for locomotion which could have
involved swimming. They seem to have had many predators and many of the enormous
number of fossil shells that are found are just the remaining armoured top. They died out
quite suddenly, probably due to "over fishing" by sharks some 300 Mya.

Fig. 3.13. *Dickinsonia costata* presents rather a mystery because it has an unknown affinity since no one has worked out what it was to everybody's satisfaction. They were between 4 and 1,400 mm long. There is no obvious head for the ovoid shape with one end thicker than the other. There are symmetrical ridges with an axis of symmetry. There have been suggestions that it resembles leaves, or lichens or fungi. It has been compared to Charmia which was a Pre-Cambrian life form with a line symmetry marking its length with a ribbed structure similar to a long leaf. It appeared to have been fixed to the bottom of a deep ocean — presumably it did not use photosynthesis.

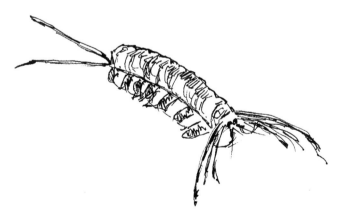

Fig. 3.14. *Kerygmachela kierkeaadi* was fast moving blind organism which moved in the water column. It had 11 pairs of lateral lobes for agile and manoeuvrable swimming, together with 11 pairs of small legs located at the base of the lobes. It had front and rear cerci to act as feelers. It had a small mouth which presumably would limit the size of prey it could devour. It was typically centimetres long.

Fig. 3.15. *Anomalocaris* (strange shrimp) was closely related to anthropods with its segments. It is believed to have been a carnivorous predator of great speed and agility. It moved by flexing its body and side lobes in an undulating sinusoidal motion of its whole body to rear up and sink down. It had a large head with a pair of large eyes which could move out on storks — the eye was probably compound. Its circular mouth was under the head and consisted of 32 circular plates, 4 large and 28 small. The creature was large, specimens up to 2 m in length have been found in China. This very formidable animal died out at the end of the Cambrian.

Fig. 3.16. A schematic representation of *Opabinia regalis*, a segmented animal with an un-mineralised exo-skeleton and a length in the range 4–7 cm. Each segment carried a set of gills and a flap appendage for locomotion. A prominent characteristic is a substantial proboscis with grasping pins at the end. It had a mouth under the head near the base of the proboscis and presumably used the proboscis much as a modern elephant uses its trunk. It appears to have lived on the sea bottom using its "trunk" to search the sands for food. The precise genus is unknown — it could have been an arthropod or even a trilobite. It seems to have represented a dead end in evolution. Its head carried five eyes on stalks giving it virtually a 360° field of view. The head was not formed from the segments. It must have appeared very strange in real life but was probably a very dangerous predator.

Fig. 3.17. *Halkieria* lived very widely and fossils are found on essentially every continent. Its length was typically around 10 cm and consisted of a "chain mail" calcerous scales type body with a more solid shell at each end. The body was flat and was a rubbery and slug like under the armoured top. It is difficult to describe and place in modern terms but was probably an early mollusc.

3.11. Summary

In this chapter we have considered the development of life during the first 4,000 My of the Earth's history. It is useful here to summarise the main points of the discussion so far and to collect a list of the important development points.

1. 4,565 Mya: the formation of the Earth as a thermal process in a molten form. The surface is part of a deep convective motion.
2. 4,300 Mya: the Earth now has a solid surface but at a high temperature with many surface volcanoes
3. 4,000 Mya: very elementary life had certainly formed. This may well have been self-replicating molecular structures of the form of very simple amino acids. Presumably RNA/DNA must have been formed about this time.
4. 3,900 Mya: the first cells appear. Now the self-replicating structures can maintain an environment different from that of the environment. These elementary cells were very probably similar to modern prokaryotes without a cell nucleus or

other organelles bound by a membrane. Oxygen began to be ejected as a waste product.

5. 2,500 Mya: the first organisms arise that can employ oxygen in the atmosphere. The amount increases and those cells that cannot adapt to use it either become marginalised or do not survive. The appearance of oxygen has the most profound implications for energy provision later.

6. 2,100 Mya: eukaryote cells arise with an enclosed central DNA and specific regions for different organelles in contrast to the "one room" prokaryote cells. This allows more stability during movement. The eukaryote cells are associated now with advanced life.

7. 1,200 Mya: sexual reproduction appears. This allows for faster evolutionary processes.

8. 900 Mya: choanoflagellats appear with the possibility of more rapid mobility promoted by a flagella ("tail"). These are often regarded as the ancestor of all animals.

9. 600 Mya: elementary multi-cell creatures arise — probably a type of sponge.

10. 560 Mya: the first animals arise with nerves and muscles and a defined form and shape. These are cnidarians with a radial symmetry.

11. 550 Mya: the most elementary form of flatworm appeared. It had bilateral symmetry (long and thin) and is distinguished in having an elementary brain and light sensitive cells in its head (elementary eyes). This was a most significant development because it meant that a control system had a beginning.

12. 540 Mya: acorn worms appeared distinguished in having a circulatory system based on a heat which also acted as a kidney. They also had a gill-like structure for breathing.

13. 530 Mya: *Pikaia gracilens* is the first animal known to have a notochord which acted rather like a backbone. This primitive creature was about 4 cm long and was similar to the modern lancelets.

It will be realised from Table 2.1 that this list just overlaps the Cambrian period which begins at 543 Mya but it is appropriate for our purposes to include the introduction of the circulatory system and the notochord. It will be realised that at this point essentially every feature for the development of more advanced life had been introduced at least in embryo — all the future phyla were now possible.

Could the development of these various life forms have come about virtually instantaneously as might be thought to have been the case? Each life form must be given appropriate material for its construction and be viable in its living. These creatures may be the experimental trials for this but it is still a remarkable event that these creatures appeared at all. Favourable environmental conditions would allow the further development of these already established components into the most advanced forms. It should be remembered that all these developments took place under water. Life had not yet invaded the land. This was to come soon and then evolution could really take off when the environmental conditions were right. It might be thought that evolution had been slow during this period but it was probably the most difficult period of all. The basics for further development had been laid down which means the evolution had been "bottomed". This is not to say that further evolution did not have its problems but they would be more of a secondary nature.

Although the Cambrian Explosion may have seemed dramatic from the point of view of the appearance of new and unexpected living forms it takes on a less startling appearance when viewed in terms of symmetry properties. These can be described in a simple way by using algebraic topology. This is concerned with spatial properties that are preserved under a continuous deformation (called a homeomorphism) such as stretching without cutting or sticking together. The most elementary life has the topology of a sphere and this is also the most elementary geometrical structure. This can be transformed into a flat plate, a rod, a rugby ball, a wavy surface, a potted surface or even conceptually compressed to a point by simple rearrangements of the shape such as by squeezing or pulling (the different examples are said to show homotopy equivalence). Again the shape can be static as in bacteria or mobile as in amoeba. The internal contents can be free (as in a prokaryote) or bounded by membranes (as in a eukaryote). In practice the stability of features with a spherical geometry requires the living volume to remain relatively small. The new life after about 543 Mya had a different topology — it took the form of a sphere but with a hole through the middle to form a doughnut. This is topologically different from a sphere because a sphere and a doughnut are not homotopically equivalent. When pulled out a doughnut cannot become a sphere but can form a thick hollow pipe which, biologically, is the structure of all advanced life.[1] It allows for a stable, mobile creature

---

[1]It is homotopically equivalent to a mug with a handle.

with a mouth, gut, anus and bounded internal regions specifically allocated to particular organs like brain, kidney and so on. Arms, legs and reproductive functions arise naturally from deformations of the basic shape. This structure can offer great flexibility for creatures of large size and strength. There is no difficulty in spheres and doughnuts existing together so long as they can each survive. Physically the time became right then for these different latent possibilities of symmetry to be available for exploitation 543 Mya. Such general forms would be available for the evolution of exo-life elsewhere, should it exist. This indicates how a geometric approach can offer a different perspective to biological problems.

Further Reading

Much of our definite information about life during this Eon has been discovered in the last two or three decades and has not yet been fully evaluated in book form. Nevertheless we can quote the following:

1. Schopf, J. W., 1999, *Cradle of Life: Discovery of Earth's Earliest Fossils*, Princeton University Press, New York.
2. Oparin, A. I., 1961, *Life: Its Nature, Origin and Development*, Oliver and Boyd, London (translated from the Russian by Ann Synge).
3. A classic book is Oparin, A. L., 1968, *Genesis and Evolutionary Developments of Life*, Academic Press, New York (translated from the Russian).
4. A book by one of the leading modern explorers of the Burgess Shale is: Conway Morris, S. 1998, *Crucible of Creation: The Burgess Shale and the Rise of Animals*, Oxford University Press, Oxford.
5. An earlier discussion of life on a planet is in Maynard Smith, J. and Szathmáry, 1999, *Origins of Life*, Oxford University Press.

Chapter 4

Life Develops in the Phanerozoic

In the last chapter we followed the fortunes of evolution during the very extended Pre-Cambrian time and saw very little direct activity as far as macro-organisms were concerned. For most of the time life was bacterial and only towards the end did creatures emerge that were capable of leaving fossils. This must not be interpreted as little significant activity in the Precambrian — far from it. In fact the ground was being laid for the great developments of the Phanerozoic. Let us recount.

Separately from the beginnings of life the blueprint was established for the form of life to come. RNA and DNA were established and cells had been developed. These developed to have an internal structure with bespoke organelles. The first oxygen producing bacteria appeared and the quantity ejected as waste became large enough to affect the composition of the atmosphere. This was so much so that much of the living organisms of the time, though not all, used oxygen as an essential ingredient for life. Later in the period sexual reproduction came about to provide stable diversity for evolution. Some 900 Mya micro-organisms developed a means of propulsion through flagella and elementary multi-cellular creatures appeared. These developed into creatures with a firm shape and form and internal segregation which involved nerves and muscles. Even light sensitive cells appeared which quickly led to elementary eyes. A circulatory system was developed and even a rudimentary kidney appeared. Finally the first steps were taken to develop a backbone in the notochord. All this is a pretty impressive list even though it covered a large period of history. This cannot be regarded in any way as little activity. One can see the shape of future evolution and all the basics were set for it. Indeed, the so-called Cambrian explosion is evidence that much had, in fact, been achieved. Of the greatest importance has been the evidence of the Burgess Shale fossils which between them show a snapshot of preserved ecology of the time. The lack of hard fossils before about 530 Mya could be said to show evolution proceeding under wraps but from that time onwards it showed itself in a tantalising collection of extraordinary remains becoming the more extraordinary the more advanced the creatures became. Only when the many components were ready and environmental conditions became friendly could new creatures come to life. These were all small in the first instance (generally of the order of a few centimetres) as one might expect to begin an epic experiment. It is interesting to notice than many of these early creatures had some form of optical capability showing that evolution can act rapidly if predators and prey are each to survive.

Not all the subsequent forms of life have actually made it through: indeed, 99.9% of the living creatures that arose suffered extinction at some stage. This would be regarded today as a great slaughter — evolution can be very cruel. Extinctions can, however, offer enormous possibilities for evolution so one must be philosophical. We will consider the extinctions later and look at the possible causes for them. The important thing has been the overall continuity of the evolutionary development through time. Some species have hardly changed over vast periods of time but the complexity present seems never to have decreased over the evolutionary process. Old forms have persisted continually in some cases but new forms have also been introduced. It is this constant introduction of new forms to meet new environmental conditions and the adaptation by some already there to accept the new conditions that has given the evolution process the firm appearance of a steady progress. Whether this is chance or design is another question.

The first new complex creatures to appear were, like their forebears, strictly marine being confined entirely to the seas and rivers. This divided life into fresh water and sea water because the seas would already have been salt since they were receiving minerals in the rain draining off from the land. The land was entirely devoid of any life at this time having an appearance rather like the surface of Mars has now. The land was invaded for the first time around 530 Mya and this is essentially the starting point for the arguments of the present chapter. The invasion was very slow at first but gradually gained momentum. It took a little time for creatures to appear that were not directly associated with the water. The pace slowly quickened and our studies here will be taken up to the point when the first creatures appeared that provided the direct springboard for the development of hominoid. In the meantime evolution showed experiments with a wide range of creatures including some of the largest and most fearsome that have ever lived on the land or in the seas. Many creatures had enormous protective armour which meant that predators had to develop great strength to be able to attack them. Side by side some of the smallest creatures co-existed with them to begin what was to be the line leading to mammals and so to Homo sapiens. It was about 60 Mya when these direct precursors to modern mankind first appeared and this point will end the present chapter. The period beyond that, from about 55 Mya to the present time, is for the next chapter.

4.1. Invasion of the Land

The predominant life rather more than 530 Mya at the end of the Proterozoic eon was almost certainly marine worms (*metazoans*) living on the sea floor. They left tracks as they moved around and it is astonishing that some of these tracks have been preserved in the sedimentary rocks to this day. The paths are often very clear although they are really virtual fossils because the creatures that made them are not evident. An example is shown in Fig. 4.1. This was, for the worms, a two-dimensional world but things suddenly changed around the beginning of the Cambrian. Now the creatures began to burrow into the sediment of the seas rather than simply move across the surface — the elementary creatures had discovered the third dimension.[1] Presumably this could have been either to gain mineral food or prey or to avoid predators. Suddenly the marine *metazoans* developed a mineralised skeleton, essentially of calcium carbonate. Incidentally, this was the first time that carbon dioxide was withdrawn from the atmosphere by living things and the second

Fig. 4.1. A fossil track of a creature, believed to have been a *trilobite*, that moved in the mud of the deep ocean.

[1] It took many millions of years for mankind to recognise the importance of the fourth dimension!

interference with the Earth by living creatures, the first being the production of atmospheric oxygen. This meant that for the first time hard fossils were formed when they died and so left a very direct record of their form and life. These creatures included sponges, moss animals and sea mosses (*bryozoans*), sea urchins (*echinoderms*), corals, molluscs, clams (*brachiopods*), and elementary crabs (*arthropods*). The *trilobita*, so evocative in the popular picture of palaeo-life in shallow water habitats, first appeared very early during the Cambrian period some 540 Mya. They were well established by the end of the Cambrian (488 Mya) in a great diversity of variations and lasted into the Devonian (416–359 Mya) having existed for some 200 My (Fig. 4.2).

The word *trilobite* refers to the fact that these creatures were composed of three parts. The parts were longitudinal giving a right, left and middle: the left and right were essentially instruments of movement while the middle was the head, thorax and tail. They had a protective shell but this covered the top only. The mouth was underneath showing them to be bottom dwelling creatures. The head contained also feelers that could have acted as taste areas or sound areas or both. These creatures had eyes except one species that lived

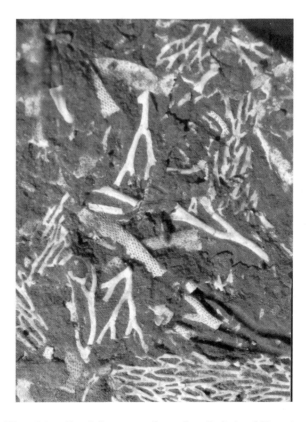

Fig. 4.2. Fossil *bryozoans* from the oil shale of Estonia.

in very deep water where vision was of very limited value. The eyes were made of calcium carbonate and of compound structure with many prism-like components — perhaps several thousand. We will consider these later. The trilobites were armoured on the top surface only and malted their armour periodically. The creatures themselves and their discarded armour will all fossilise: although the trilobites were actually widespread throughout the world as a most successful animal they were not as prolific as the many fossils that are found now might suggest because the discarded armour will also have fossilised. Their reign effectively ended 245 Mya. The precise reason is not yet clear but it did coincide with the rise of the early sharks and it could be that they were "fished" into extinction. We shall say more of this later. Living had begun to take on a modern appearance. This being so, it was inevitable that life should also move to the land and it did at about this point in time.

The pressure to move onto the land became overwhelming during the Ordovician around 475 My ago so that the entire surface of the Earth became the domain of life. The life that emerged from the water was elementary and small. No life in the water or on the land was much like the life we are familiar with today but neither was the environment. The first life to venture out of water was primitive plants which began to populate the moist land surrounding water. These plants seem to have developed from green algae and were accompanied by fungi in what could well be a symbiotic relationship. This move to the land coincided with a buildup of oxygen in the atmosphere well beyond the 1% level. Once oxygen had collected in the atmosphere in significant quantities the ultraviolet light from the Sun formed a useful ozone layer (the oxygen isotope O_3). This offered a growing protection for land creatures against the higher frequencies of the solar radiation to some extent comparable to the protection offered by the sea for marine creatures. Such protection on the land is vital if advanced life is to form there as will be seen in Chap. 2. Once oxygen is available in the atmosphere it is also available separately to provide a source of energy capable of meeting the requirement for the activity of complex creatures (should they form later — which, of course, they duly did).

Land creatures are subject to three new problems not met with by creatures whose habitat is the water. One is the possibility of the loss of water (fluids) by the creature due to evaporation from the skin and other atmospheric forces. A second is the absence of a broadly constant temperature environment. Water is a very good thermal blanket but the combination of the Sun and the atmosphere is not. The third is a new need, the need for support against the constant force of gravity and especially the hazard of falling and of having objects fall on the creature. The solution to these problems lay in the development of some form of hard shell covering at least the most

Fig. 4.3. Fossil lilies.

vital parts — enter the arthropods. It is also necessary for land creatures to breath air directly through elementary lungs and not through gills in the water. These evolutionary changes took place over the next 100 My. This period was only some 2% of the then age of the Earth and broke the microbial form of life which had covered the previous 98%. Evolution was now moving fast. Among the first creatures to invade the land (during the Devonian) were "millipedes" and spiders (*Myriapoda*) followed a little later by scorpions. Insects evolved on land as in the water. They moved with many small appendages underneath their bodies able to propel them at some speed. The earliest land fossil tracks yet found had got 23 pairs of appendages, or legs, for movement. Judging from the spaces between the "foot" marks the creature must have been about 1 m long. This is surprisingly large for this period of evolution (Fig. 4.4).

The waters had also seen new things during this time. Jawless fish appeared. Being jawless does not mean they have not got a mouth. This was a round orifice used for sucking up nutrients from the sea bed or drawing in small creatures passing by. At this time *trilobites* began to radiate and prosper though to reach a peak of diversity millions of years later. These creatures will provide a very rich range of fossils for the next 100 My or more. Some

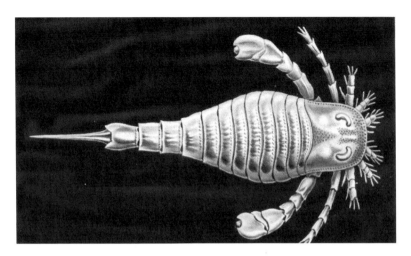

Fig. 4.4. *Haeckel eurypterid* (or Sea Scorpion) which lived from the Cambrian to the Permian (510–248 Mya). The average length was about 20 cm although there is evidence (from a fossil claw) that other examples could grow to 2 m or perhaps more. They lived in fresh water and later in salt. They pre-date the earliest fishes. They are related to the modern horseshoe crab.

fish did now develop jaws — a very significant event. It is believed that this came about by the development of the first gill arch. About this time the first land predators appear (among the *Arachnids*). About 400 Mya (again in the Devonian) some fresh water lobed fish, *Sarcopterygil*, developed four legs and even elementary fingers and toes. This was the first development of tetrapods (Figs. 4.5–4.7).

The legs were to make movement easier among the various weeds and mosses that were appearing in the water and were certainly not associated with walking on land. The possibility was there, however, for stable locomotion on land and this was taken up quite quickly afterwards. In due course this consequence was a more determined colonisation of the land later even though legs had not originally been developed for that purpose.

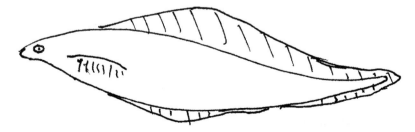

Fig. 4.5. A jawless *Haikouichthys*, probably a primitive agnathan related to the Lampreys, which lived during the period of the Cambrian explosion. It was about 2.5 cm long.

Fig. 4.6. A jawless *Cephalaspido morphi* which lived during the Silurian to the early Devonian. It had an armoured head as protection against jawed predators. It had a size which varied between those of a goldfish and a trout.

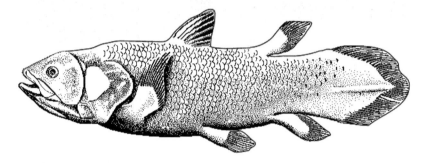

Fig. 4.7. The *coelacanth* which first appeared in the mid-Devonian some 410 Mya. It was related to the *Scarcopterygil* and so to lungfish and tetrapods. It was thought to have become extinct during the Cretaceous but a living example was found in 1938 off South Africa and since then many have been found in deep water in the Indian Ocean Rim. The living examples are very little different from the Devonian. They move their fins in same way that we move our legs and arms when walking. The average weight is 80 kg and the average length 2 m. They are said to be poor eating and do not sell well at the fish markets.

An evolutionary note: five toes because the norm for a range of creatures but this was a choice from experience. Creatures with up to eight toes are known through fossils but five seem to have been selected eventually.

The first members of more familiar faces appeared around 400 Mya in the Silurian (443–417 Mya). This saw a radiation of fish of all sorts including the first sharks and lung fish. In the open oceans other predators appear including the nautiloids with substantial tentacles (see Fig. 4.8). They were the forerunners of squids. These marine forms mark the Silurian period. The first *ammonites* appeared around 400 Mya in the Devonian which were destined to provide a very prolific source of fossils. They appear to have formed from *nautiloids*. Later they became very large and very ferocious with extended tentacles at the mouth and thick armour. At their peak the largest formed a hard coil a few metres in diameter. The Devonian was to see a radiation of many fish types. Of particular note is a fish with an armoured head. Predator and prey produced a fine balance for life in the Devonian.

(a)   (b)

Fig. 4.8. (a) An *ammonite* fossil. The size can be judged by comparison with the person carrying it. The largest are known to have grown to nearly 10 m in diameter. They were very vigorous hunters. They are believed to ancestor to the modern squid. They inhabited widely. They had two primitive "box camera" eyes. (b) *Nautilus_Ocean_Thailand* a modern version but virtually the same as the earliest forms. Now it lives in the Eastern Indian Ocean. The tentacles are clearly evident of which there may be as many as ninety. The nostril is also visible through which it ejects a water jet for mobility. The eyes remain primitive.

The earliest land plants appeared in the late Silurian, around 420 Mya. The earliest plants were nonvascular, meaning they lacked roots, stems or leaves. The movement of water and nutrient minerals was performed by special mechanisms developed by each species. Examples found today are algae, mosses and liverwort. In the late Silurian these differing mechanisms were put on a more sound footing with the appearance of elementary vascular plants. Now there are lignified tissues for conducting water, minerals and the products of photosynthesis around the plant. These were mainly stick-like "plants" but without leaves or roots called now *psilophytes*. Other examples are ferns and club mosses. There was no animal life at this stage to forage on such plants but this was about to change as important things happened on the land.

4.2. The Seed and the Amniotic Egg

Perhaps one of the most significant things that occurred at the Devonian/Carboniferous boundary about 350 Mya was the appearance of seeds.

This development allowed the plant embryos to be protected when spread around the soil and led to the rapid spread of plants over the land. The Carboniferous was a time of prolific plant, bush and fern growth. Vast forests arose of *lycopods* (clubmosses), horsetails and tree ferns. The fossilized form of these plants and trees forms the coal we use today which is around 300 My old. It contains the solar energy collected by photosynthesis all those years ago and the extreme compression it suffered has made is a very high density energy source. It is, of course, a finite source although it is still a very substantial one throughout the world. From this time onwards the continental land surfaces would be covered by a wide range of foliage outside desert regions.

The development of the seed proved a crucial thing for flora: there was a corresponding development for reptiles. Until about 300 Mya the essential place for reproduction was in water but about that time the egg with the shell, the amniotic egg, appeared. The amniotic egg is quite sophisticated and proved a most significant development. The egg consists of a collection of closed fluid regions providing all the nutrients necessary for the development of the embryo including the processing of waste products, until the new creature is able to survive in the outside world. All these components are enclosed in a shell which may be hard or leathery. The shell has the important feature that it allows air to enter and carbon dioxide to leave. It can be thought of as providing marine conditions on the land. The embryo develops to the stage where it can exist outside when the shell shatters to reveal the brand new "chick". Some can survive alone but most often the parents must take care of it until it learns to look after itself. This applies to a wide range of creatures, the oviparous vertebrates. The whole process is a little masterpiece of design and engineering and yet eggs are produced in very large quantities. Some are squandered by accidents every year and also the eggs of many species are used as food by many other species. Time is required for incubation and is a period of great vulnerability. Nature is a prolific waster of embryos to maintain the healthy continuity of a species. A chicken's egg is an example both of the amniotic egg and of its use as a food, in this case including us. The development of the egg allowed a wide range of reptiles to evolve safely on land and allowed evolution to take a new direction.

The amniotic egg later led to the important development of mammalia where it was developed to become a cavity retained within the female in the form of a womb (as an internal egg). Nutrients could now be added by the mother as required, oxygen taken in and carbon dioxide exhumed by the mother. This makes the early development of the embryo very much less hazardous and, more importantly, can accommodate a more complex embryo than can reasonably be accommodated by an egg structure. It is also possible to rear fewer embryos to maintain a given density of population because put crudely there is less waste. The move would seem ultimately to be towards the possibility of propagating ever more complex creatures with fewer offspring per pregnancy. The Devonian period ended with a major mass extinction some 354 Mya which gave space for evolution in other directions. This led, about 300 Mya, to the appearance of the *Amniota* which are reptiles that can now reproduce on land through the laying of eggs.

4.3. Creatures Come and Go: Some Fly Away

At about the same time older inhabitants were disappearing. For instance, trilobites, once so common, became rare and were soon to disappear altogether. Wide forests of conifers filled the land together with club mosses (*lycopods*) and horsetails. These will eventually form coal and the wide extent of coal fields today. There was a great expansion of *gymnosperms* (from Greek naked seed) in which the seeds were not enclosed in a "shell". An example is the conifer where the seeds are contained in a cone. There were no flowering plants at this stage: the value of flowers in evolution took another 170 My before the environment was ready to accept it.

Among the foliage a most extraordinary phenomenon could be seen — winged insects. Large dragonflies were very common, some with a wingspan of up to 6 ins (≈ 15 cm). They appeared to be essentially the same as the (much smaller) dragonflies (*Odonata*) with which we are very familiar today, including the thin colourful double wings. Around 250 Mya (in the Permian) the lizards (*archosaur*) separated as a group from other reptiles. This led ultimately to a radiation which included crocodiles, alligators, dinosaurs and the earliest flying reptiles (*pteranodon*). In the seas the *teleosts* evolve from the ray-finned fish to become the dominant group. The essential feature was an extended jaw which allowed for more effective eating.

The Triassic (248–199 Mya) saw some significant developments on land and sea. In the sea reptiles returned to the sea from an as yet unidentified land reptile. One example was of the order *icthyosaur* (Fig. 4.9) which thrived between 230 and 90 Mya. To look at it was apparently a cross between fish and a dolphin but with large flippers for fast locomotion. It is believed capable of speeds up to 40 km/h. It had been a Triassic oxygen breathing land animal but reverted to the sea. It was large: 2–4 m long and the largest could weight up to 1 tonne. It was capable of diving very deep. Having been a land animal it is not surprising it was vivipaurus, with the embryo developing inside the body of the female. Birth of the young took place in surface waters. A distinctive feature was its eyes. They were the largest of any vertebrate being some 260 mm across. This is known because most vertebrate eyes have got a sclerotic ring under the eye covering the eye region. The eye is a very distinctive feature in Fig. 4.9.

Another carnivorous reptile was the marine lizard *plesiosaur* (see Fig. 4.10). It was probably a slow swimmer but had exceptional maneuverability with a very long neck to allow the head to swivel very effectively in any direction. It was cold blooded and breathed air. It is a modern relative of this class that is said by some to inhabit Loch Ness in Scotland and others in Lake Khaiyr in Eastern Siberia. There is considerable doubt as to whether this can be true. Isolated creatures do not form a viable life form: a limited number of creatures is necessary to form a stable population so more than one should be visible but only single members are quoted. Yet

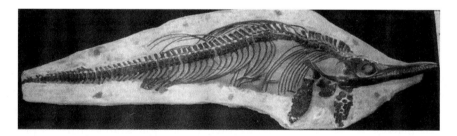

Fig. 4.9. A mounted *ichthyosaur*. Its long snout and very large eye are both very clear. The two front flappers are also clear but those at the rear are not.

Fig. 4.10. A drawing of a reconstruction of a *plesiosaur* or marine lizard.

Fig. 4.11. A sketch of *mosasaur* — how it probably looked in life.

another marine lizard was the *mosasaur* (see Fig. 4.11). This was a very ferocious predator and seems to have the snake as its closest living relative. It lived between 85 and 65 Mya. It was a very powerful swimmer and lived in warm shallow seas. It breathed air and reached sizes between 3 and 17 m. It could weigh up to 2 tonnes. It had no known predators but would attack almost anything else. Its most distinguishing features were a hinged jaw and a flexible skull very much like a modern snake though its head could be 10 m long. It had very strong nutcracker teeth. So much for the sea.

The land masses were joined in Gondwana 250 Mya which included Antarctica, India and South Africa. The dominant *archosaur* which covered the whole region was the plant eating *Lystrosaurus* (Fig. 4.12): it radiated into six known species over the ten or so million years of their existence. They were about the size of a modern pig and were distinguished by two large horns covering the mouth. It is thought that these were used for digging up plant material. This genus died out when the environment became too dry and water became scarce.

4.3.1. *Dinosaurs appear*

On the land conditions became hot and dry some 200 Mya favouring flora and fauna best adapted to these desert conditions. The fauna will have been determined by the flora on which it lived. The main floras were gymnosperms

Fig. 4.12. A sketch of *Lystrosaurus* the pig-sized vegetarian showing also its horns covering its mouth.

which offered a very sparse diet for creatures relying on them for food. The favoured fauna were *archosaurs*. These were, and still are, creatures which developed two holes on each side of their skulls. They include a wide range of creatures including birds. To gain sufficient nutriment a vegetarian creature (a herbivore) must eat a lot of food: this requires a large gut and so a substantial frame to support it. Carnivorous predators would consequently need to be large to gain fresh meat from such a physically large meat source which meant we enter the era of the giant lizards. One immediately thinks of dinosaurs but not all large carnivorous archosaurs are dinosaurs. Dinosaurs are a characteristic type of terrestrial reptile that ruled the planetary surface during the Mesozoic (250 Mya–65 Mya). They did not invade the waters — there were certainly large sea creatures at that time but they were not dinosaurs. They are archosaurs with limbs including feet that were held erect below the body. In excess of 500 genera have been identified but it is confidently believed that three or four times this number remain yet to be discovered. The land creatures ranged in size from the very small (smaller than a cat) to the very large (*Terasaurus Rex* and larger). They all had a substantial tail for balance. Among the largest ones were fearful carnivore predators but the smallest were themselves preyed on. Some, perhaps the majority, were bipedal in walking or running but others were quadrapedal (for instance *Ammosaurus* and *Iguanodon*, Fig. 4.13). Still others could walk equally well on two legs or four. Although many were carnivorous others were herbivorous. Some had crests or horns on their heads presumably to provide balance for a bipedal gate. They developed during their evolution — the earliest examples were small bipedal predators. There is strong evidence that some later members had features, such as the theropod with feathers *Shruvuuic deserti* whose skeleton was found in China recently. The feathers

Fig. 4.13. *Iguanodon* was a member of the group Sauropodomorpha of herbivorous dinosaurs that included the largest animals that have ever inhabited the Earth. This member could grow to a height of 7 m at the shoulders with a length of more than 10 m to the tip of its tail.

seem already to have been well formed showing they must have developed rather earlier. Apart from the appearance of the fossil, the existence of beta keratin is a major protein in feathers. It is a corollary that these members were warm blooded because feathers would surely be needed to retain body heat. It would seem that feathers were not initially connected with flight, another example of earlier development of a feature which later was a basic requirement for something else.

4.3.2. *Flight*

There has yet to be discovered the actual origin of flying creatures and their ability to fly — no one is quite sure. It did seem to happen during the Jurassic which was the hey day of the dinosaurs. It is an obvious thought, supported by a range of fossil examples, for everything to have started with dinosaurs. There are two very opposite possibilities for the origin of flight. One is that tree living creatures learned to glide from tree to tree or from tree to land as a first step and developed from there. The other is that land hopping creatures learned to extend their time in the air. There are, in fact, serious objections to either hypothetical origin. Certainly two very early, but unrelated, examples of birds are known and there may be others yet to be discovered. The use of the air to fly appears to have been an achievement of the Cretaceous although the first steps seem actually to have been taken in the Jurassic. In evolutionary terms the different initial appearances could be yet another example of convergent evolution.

There is considerable confusion from fossil evidence but it is clear that flight was not a simple event. The fossil evidence itself is not easy to interpret because, such is the delicate nature of things that fly, complete fossilised creatures are difficult to find. Typically a set of shoulder bones or leg bones or something like that has to be the basis for reconstruction. There is a basis for advance, fortunately, because some complete fossils have been found which are for animals that could not have flown while others are for animals that certainly could have. The interesting thing is that they did not appear in a chronological sequence — bird-like creatures appeared after true birds had already evolved. It appears that the appearance of feathers was not of itself associated with flight. Certainly there were feathered dinosaurs and the feathers could well be tailored for flight but may alternatively be a way of preserving heat in the creature. The situation will hopefully be clarified in the future as more fossils become available for study. As an example of creatures that could not have flown are: the large *Gigantoraptor*, the fearsome *Velociraptor* shown in Fig. 4.14 and the most birdlike found so far *Unenlagia comahuensis* in Fig. 4.15. *Gigantoraptor* lived some 85 Mya in the late Cretaceous. It was typically 8 m long and probably weighed about 1.4 tonnes. It had a beak but no teeth. Its large claws suggest it was a carnivore. A partial fossil was found in Inner Mongolia in 2005. *Velociraptor* lived in the late Cretaceous between 83–70 Mya. An adult would typically be a little more than 2 m in length and 0.5 m in height. It would probably weigh about 15 kg. It had a jaw with 28 widely space teeth on each side. Its "sissor" claws suggest a carnivore. Its body was covered by feathers. *Unenlagia com.* was some 3 m in length and lived in the mid-Cretaceous some 90 Mya. It was certainly too large to fly. The main pieces of 20 fossilised bones were found in 1996 in the Patagonian region of Argentina from which the present profile has been constructed.

(a)   (b)

Fig. 4.14. (a) An image of how *Gigantoraptor* is believed to have looked like and (b) how *Velociraptor* probably appeared. It had three long vicious claws on its arms. Each creature was too large to fly with arms too weak to support flight although each had at least a partial covering of feathers. They each had bird-like feet with three toes.

Fig. 4.15. How *Unenlagia comahuensis* is believed to have appeared. The arms carried feathers and would be strong enough for flight. The toes were those of a bird and there was a substantial feathered tail. The creature was about 3 m long and lived in the mid-Cretaceous some 90 Mya. It is the most bird-like nonbird yet found.

A creature that surely could fly was the reconstructed *Jeholornis* (*Shenzhouraptor*) shown in Fig. 4.16. It lived in the early Crutaceous in China where its fossil was found in 2001. It had feathers somewhat like a chicken or a goshawk and it is presumed to have flown in a somewhat similar manner.

The earliest evidence found so far of a genuine flying creature that is universally accepted is *Archaeopteryx* which lived during the period 155–150 Mya (at the end of the Jurassic). It appears to have inhabited what is now South Germany but it is possible that other fossil remains will be found elsewhere though none have appeared yet. It had a mean size of about

Fig. 4.16. The believed image of *Jeholornis*. This was a true bird about the size of a large chicken.

0.5 m or 1.6 ft which is about the size and weight of a European magpie. It had a pronounced do-do-like head and a wide tail. It had a beak with teeth. It was covered with flight feathers and each feather was asymmetrical appropriate to flight. Although it had wings it had much in common with theropod dinosaurs and not a great deal in common with modern birds. It would appear to have been too weak for substantial sustained flight and could well have used gliding with some wing assistance for its locomotion. X-ray analysis of its brain cage suggests that it had a fully developed neural system necessary for flight. How the flight came about is not clear: there were no high trees in its environment, only low shrubs, so it can hardly have been through initial gliding exercises.

Flight took on a new meaning in the Cretaceous with the arrival of the more robust *pterosaurs* (see Fig. 4.17), often popularly known incorrectly as the *pterodactyl*. These creatures appeared about 150 Mya and survived in various forms until 65 Mya. Although there was some doubt after the initial discovery of the fossil remains whether these creatures could have flown in the normal sense it is realised now that they did in fact fly but could also use air currents very effectively to glide as large birds do today — for instance the albatross and the condor or even the humble sea gull. Indeed it seems that they travelled enormous distances over land and sea. Fossilised eggs have been crushed but not shattered which suggests they had a flexible leathery outer skin rather than a fragile shell. The wings were composed of stretched fabric and muscle between the arms and the thorax. The "hand"

Fig. 4.17. Sketch of what a *Pteranodon* might have looked like in flight. This is the classic popular image of birds as dinosaurs but *pteranodon* was a separate reptile. It had a beak without teeth and a wing span that could be in excess of 9 m. They lived around 70 Mya. There was doubt for some time about the ability to fly making it a glider. Since then it has been realised that its bone and muscle structure would have allowed it to fly over very considerable distances, probably like a modern albatross, part gliding on wind currents over the ocean surface. Its food was mainly fish.

had a greatly extended fore finger to accommodate a wide wing stand which could have extended to 3 m (about 10 ft). They were truly giants of the air. They were able to move on land though perhaps not comfortably where they were bipedal. The bones were hollow to reduce weight but the tubular form still left them very strong to crushing. This is fully acceptable for flight where the stress is transverse and a hollow tube provides sufficient strength. For walking the effect on the bone is a longitudinal crushing and there it is necessary to have a filled bone. The overall design of the *Pteranodon* was accepted as suitable for scaled repeats. In this connection we can mention *quetzalcoatlus*, the very largest flying creature that has been found so far. This was a toothless pterosaurus characterised by a large stiff neck and great size. The wings span of fossils are 5.5 m (about 18 ft) and the overall length of the bird about 4 m (about 12 ft). These experiments with size were not repeated but nature still models a shape by differing sizes. There are indications that these experiments did not give a continuous line to modern birds. Indeed, it seems that modern birds have had a different origin about 200 Mya but the details are still very obscure. The view now is that birds generally represent the evolved form of the dinosaurs and on this view the dinosaurs are still with us. It is possible that forms with two sources are present now as another example of convergent evolution.

4.3.3. *Mammals appear*

Very significant changes happened about 220 Mya. Evolutionary changes in *synapsid* reptiles, the dominant animals in the late Permian, moved to produce the first primitive mammals. Thus began the group *Eucynodontia* which includes mammals and mammal-like creatures. The *Synopsid* reptiles were really a cross between the reptiles that had been and the mammals that were yet to come. *Archaeothyris* is one of the earliest known reptiles which lived about 320 Mya. It was lizard-like about 50 cm head to end of tail and is distinguished by having wide opening jaws. Another typical example of the early members was *Cynognathus*, a wolf-sized dog-like creature widely distributed around the world. It had a double palate which meant it could eat and breathe at the same time. It was able to both tear and to chew meat. There were some creatures even more closely mammalian. Thus *Adelobasileus* (living between 239 and 208 Mya) was very close to the common ancestor of all mammals judged from the rather scant fossil evidence from their head. The fossil remains of *Sinocobasileus* with an age of 208 Mya showed mammalian features that were even closer. At the end of the Cretaceous the three modern types of mammals were already established: these are monotermes that lay eggs, marsupials where the female has a pouch to rear her young and placentals with a womb. The earliest fossilized remains of a true marsupial were found in China. The creature died 125 Mya. It was about 15 cm long and probably weighed no more than 30 gms. Close by there is evidence of soft skin and fur. It seems the animal lived in the trees. The population of marsupial continued to grow making the period of their largest numbers between 90 and 65 Mya. Mammals with warm bodies filled a niche at night when the dominant dinosaurs were unable to operate. The nocturnal mammalia lived primarily on insects which also would be active at night (Table 4.1).

Table 4.1. A comparison of features of reptiles and mammals.

Characteristic	Reptile	Mammal
Hair	None	Hair or fir
Teeth	More or less uniform teeth: conical shape	Vary in shape and size: front teeth for tearing, rear flat for chewing
Lower jaw	Made up of several different bones	Single dentary bone
Ear	One bone only (stapes)	Three middle ear bones (hammer, anvil and stirrup)
Skull attachment	To spine by single point of contact	Double faced occipital condyle allowing strong maneuverability
Legs and arms attachment	To the sides of the body	Below the body
"Third eye"	Small hole in skull through which pineal body extends	Not known in mammals
Breast bone	None	A feature of mammals
Body temperature	Variable from species to species	Held constant
Mammary glands	None	Always

4.3.4. Developments apace

There are claims from imperfections in fossil leaves that here is the first evidence that plant viruses abounded. Viruses remain enigmatic to some extent and it is not clear when they first formed. They have not got the means for self-reproduction and generally rely on normal cells for this so if this has always been true they could hardly have appeared before normal life itself. Nevertheless, their structure is certainly more elementary in some fundamental ways.

About 200 Mya some modern amphibians evolved to make things rather more homely. In particular the *Lissamphibia* appeared — the humble frog (*Anura*) and the salamander (*Urodela*). In the realm of the flora the *angiosperm* plants evolved, that is plants with flowers of various kinds. Presumably perfumes also became common at this time. This certainly made the world more colourful and attractive but there was a purpose. The flowers were designed to attract the growing number of insects and other creatures to visit them. In the process they would brush pollen within the flower and from flower to flower. This allowed the plants both to propagate more easily but also to mix genes more widely and so to evolve the more strongly. It is interesting to note that some flowers are entirely male, others entirely female while others still are mixed (said to be complete). The appearance of *angiosperm* plants appears to have been accompanied by a major acceleration of animal evolution. It is also true that the species of dinosaur doubled during the period 95–65 Mya (which was the last 30 y of the Mesozoic). Obviously flowered propagation can only happen if appropriate insects and flies have evolved.[2]

Around 125 Mya there arose *Eomaia scansoria*, a eutherian mammal with the appearance of a modern dormouse which climbed small shrubs, whose fossil remains have been found at Liaoning in China. This little creature was to be a source of modern placental mammals. It turned out that mammals could thrive during periods of great adversity. 65 Mya is a date that just about everyone knows because it saw the end of the reign of the dinosaurs during what has become known as the Cretaceous–Tertiary extinction. The essential effect which caused the extinction appears to have been a strong cooling of the climate though there is still controversy as to why this should have occurred. It is known that two events can be dated to this time. One is an impact with a giant meteorite as is evidenced by a thin global layer in the rock record of titanium. This is a rare element on Earth but this not so for some meteorites. The other event of the same date is a massive lava flood in Asia which ejected thousands of tons of lava on to the surface over an extended period of time. Either event will have sent great quantities of dust and obnoxious chemicals into the high atmosphere affecting adversely the solar radiation reaching the surface. Either event would have blocked out the Sun for a considerable period but the two together would have been

---

[2]In the modern world bees play a major role in fertilising plants.

totally devastating. Lack of sunlight would have played havoc with the flora. The chemicals in the atmosphere (especially those from the extended volcanic eruption) would have produced acid rain for a considerable period thus attacking both the food of the animals and their very skins. The youngest and oldest animals would have been the most seriously hit which would surely have affected the stability of a species. There was not a simple cut-off time for the demise of the dinosaurs but rather a short time period when they perished as a group of reptiles. Their demise left a major void in the fauna pattern since about half of the animal species alive at that time were eliminated.

A major gap in the spread of fauna is open for exploitation which mammals began to fill. Small creatures would live underground and be immune from the chaos above to an extent. If they lived on insects they could well survive because insects are very robust and could survive many situations where larger creatures could not. In any event, there was continuity because the earliest true primates, called *euprimates*, appeared 55 Mya, 10 My after the significant extinction. The details of the evolution during this period are not yet established but new fossils are continually being unearthed so it might be expected that the situation will be clarified before too long. They appeared widely in Europe, Asia and North America. One example is the fossil remains found at Clarks Fork Bain in Wyoming. This is *Carpolestes simpsoni* which had grasping hands and feet with nails rather than claws. Most significantly it has forwards facing eyes allowing it a bifocal vision allowing the estimation of distances (Fig. 4.18). Another fossil example is *Teihardina asiatica* found at Hunan in China. This was a diurnal creature with small eyes and similar to *Marmoset*, the New World monkey. It thrived

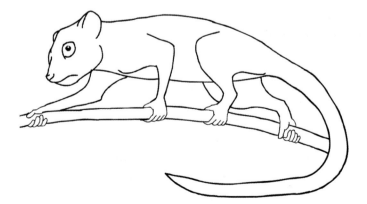

Fig. 4.18. A presumed image of *plesiadapsis*, a member of the *Carpolestes simpsoni* family. It had hands which would have allowed it to climb trees with great agility.

between 56 and 47 Mya and seems to have developed differently in different places (called polyphyletic).

An interesting terrestrial creature arose and lived during the period 90–40 Mya. This was a wolf sized carnivore *Mesonychids* that appeared first in Asia but the genus eventually spread much more widely. The genus *Dissacus* spread to Europe and North America as a jackal sized carnivore but the genus which appeared in Mexico was more bear sized. It had distinguishing features in its feet. The early creatures had five toes but later members had four. Each toes ended in a small hoof (unlike a wolf) which appear to have allowed the animal to move very rapidly. The importance of this creature from the point of view of evolution is that it seems to be a central link in two different developments. First it seems to have hoofed creatures in the family tree. The separate small hoofs concentrated together to form hoof feet and so can be regarded as the pre-cursors to horses and similar creatures. The second thing is rather astonishing: there are strong structural similarities between the *Mesonychids* (which lived on land) and modern *Cetaceans*, namely whales, porpoises and dolphins. Certainly these sea creatures are mammals that give birth to active young, that breath air but have a tetropod bone structure with the arms and legs adapted as fins. The argument is that these once land creatures found a better future in the sea than on the land and so transposed their lives that way.

About 40 Mya the *euprimates* were already showing considerable diversification. One was of a primitive monkey (*Prosimians*). Creatures existing today are lemurs and tarsiers. The tarsiers are small diurnal creatures with large front facing eyes. The move to *Homo sapiens* was set here, in retrospect, and this is a convenient point to draw this part of our story of evolution to a close.

4.4. The Role of Extinctions

The story developed so far gives a rather smooth transition from one era to another but in fact this is an extreme over simplification. The conditions of life on Earth are very much controlled by the state of the environment and this has changed radically from time to time and especially early on in the Earth's history. Again, more recently, continents have combined and then separated again. Meteors have struck the surface, some small but some large; volcanoes spewed forth molten lava, some in small quantities but others in enormous quantities. Again, the orbit of the Earth round the Sun changes with a quasi-period of 23,000 y due to the gravitational influences of the planets (and particularly of Jupiter and Saturn) on the Earth and Sun.

The consequence is that the Earth itself shows a precession as a spinning top. All these various effects change the Earth environment and so influence the life on it. At another level the effect of new predators could be decisive on occasion: for instance, some workers associate the elimination of trilobites with the development of sharks with a taste for them. Similar eliminations are familiar in recent history due to the several actions of mankind.

There are more personal reasons why a species may be lost. The temperature of the environment may fluctuate too wildly leaving no time for a species to adapt to a mean; there may be a lack of oxygen in the seas due to temperature or chemical effects; the genetic structure may be such as to limit necessary variations. There is also the effect of small numbers in breeding stability and possible effects of inbreeding. All or any of these things can lead to the loss of a species and have had this effect over geological time. A loss need not be an extinction and we need to consider this term carefully.

An individual extinction is said to have occurred when the members of a particular species show a substantial irreversible fall in number over a very short period of time. The substantial fall may, in the severest form, represent a complete obliteration of the species. A significant extinction describes the case when several or many individual extinctions occur essentially simultaneously. When a species disappears a hole is left which can be filled by another species. This will always tend to be a new one and in this way evolution has taken place. Individual extinctions are found throughout palaeo-time but there have been five occasions when significant mass extinctions have taken place. It must be remembered that these can be recognised only from fossil evidence and microbial life leaves essentially no such evidence. Extinctions can, therefore, be assigned only after creatures made a firm skeleton that would be preserved. In numerical terms, extinctions can only be recognised with certainty over the last half a billion years. Before that the detailed history of microscopic life must remain unknown perhaps forever. The five major extinctions will be reviewed briefly.

4.4.1. *The Cambrian extinctions*

A large extent of the world was covered by relatively shallow salt water seas, sometimes called epeiric seas (the North Sea is a modern example). All life was marine. There were at least four extinction events in the Cambrian which occurred at irregular intervals. The largest event was the last at about 500 Mya towards the end of the epoch. The earliest ones saw the demise of the oldest group of trilobites (the *olnellids*) but later ones saw the perish of the reef-building *archaeocythids*, the clam-like marine vertebrates the *brachiopods* and the microscopic eel-like *conodonts*. Altogether some 50% of the marine life perished. The cause of such a widespread loss of creatures has not yet been agreed upon. There is evidence from South America of glaciation during the Cambrian so it would be natural for creatures requiring the higher temperatures to be unable to survive. Again, the formation of ice will decrease the quantity of liquid water available as a habitat especially in the continental shelf regions. Life normally there would have nowhere to go. The effect of cooling the sea water will be to drive oxygen out of it. Reduced oxygen in the seas would again suffocate life there. The Cambrian is known for its extinctions but also for the Cambrian explosion which produced essentially all the phyla (body parts) of animals existing today. It seems — no extinction, no evolution.

4.4.2. *The Devonian extinction*

A mass extinction occurred towards the end of the Devonian epoch. It affected marine life severely although it seems to have shown little effect on terrestrial flora. The effect was mainly on the primitive sponges that secrete calcareous skeletons (the *stromatoporoids*) together with the ragose and tabulate corals. From this time onwards coral reef building activity appears to have been rare until modern times. The marine invertebrates (*brachiopods*, *trilobites*, conodonts and the micro *acritarchs*) together with all jawless fish and the heavily armoured jawed fish (*placoderms*) all disappeared from the records. Indeed, 70% of marine invertebrate taxa did not survive. Glaciation (with evidence from Brazil) has been blamed as has a meteor impact but the argument goes on.

4.4.3. *The Permian extinction*

The environmental situation 248 Mya was affected by the formation of the giant continent Pangea. This was formed by the accumulation of all the land mass available at the time. It was surrounded by the extensive Panthassa Sea very much like a very extended Pacific Ocean. This arrangement of land from constituent pieces must lead to a serious loss of coastline and so of important warmer continental shelf waters. This alone would cause the demise of a range of smaller marine creatures. The extensive land mass will also have its effect. It will be damp close to the sea but throughout its inner area will be hot and dry. The conditions inland would favour creatures which thrive on heat and dry and this was the lizard family. There was, in fact, a very substantial development of land vertebrate fauna after this time.

The extinction at this time was the most extensive yet known which has led to it being called the Great Dying. The details are not entirely clear but it appears to have involved two extinction pulses separated by about 5 My. The extinction is marked by the demise of 96% of all marine specimens and 70% of all terrestrial vertebrate specimens. It is the only known case where insects were also involved in the mass extinction. Particularly vulnerable were marine organisms which produced calcareous hard parts, for instance from $CaCO_3$, that had low metabolic rates or weak respiratory systems. Added together, the loss amounted to some 75% of all living creatures perishing but more positively 25% survived. Such a massive loss of living creatures slowed the evolutionary pattern substantially so recovery took a long time by the standards of the then recent past.

What could possibly have caused such chaos? There seems not to have been a substantial meteor impact at that time but there were massive volcanic eruptions lasting tens of thousands of years. One was at Emeishan in China but the really substantial one was tat of the Siberian Traps currently dated at the very end of the Permian and so coincident with the extinctions. The lava flow covered $200,000 \, \text{km}^2$ half of which was on a coal-forming seam and so rich in carbon. The outflow lasted for several tens of thousands of years. An important characteristic was that some 20% of the outflow was pyroclastic this explosive behaviour throwing ash and debris high into the atmosphere. There is firm evidence that the value of the carbon isotope ratio $\frac{C^{13}}{C^{12}}$ suffered a 10% fall world wide in the end of Permian fossils and rocks. The production of C^{12} is traditionally associated with decaying vegetation.

To account for these effects requires a very strong atmospheric warming component and methane is an obvious such source. This could reasonably come from methane clathrates, or methane hydrates. This is methane molecules trapped in water molecules and occurs when the surface temperature is greater than 25°C. This will require high levels of CO_2 and other greenhouse gases in the high atmosphere. The ozone layer would be at most highly depleted and more likely destroyed by these events. This would allow very high energy photons to strike the surface with the effect of causing cell mutation damage. Associated with these conditions is the state where the oceans are completely depleted on oxygen throughout the volume except in the surface region — this is called an anoxic event. It has occurred many times in the past. It is seen that here is a rather complicated event.

The sequence of events leading to the extinction could have been in two stages as follows. First there were the volcanic eruptions and especially that of the Siberian Traps. Siberia was located in the region of the equator at that time so an explosive eruption would scatter ash and chemicals high in the atmosphere. Ash cuts out sunlight and causes the surface temperature to decrease. Cold-blooded creatures will perish if this continues for any length of time. In any case acid rain will be a consequence and this will destroy flora on which many creatures live. Acid rain, even though even weakly acid, will also act adversely on the skin of fauna. This could also include insects. The ash will eventually fall to Earth but the chemicals, and especially CO_2, will remain in the atmosphere. The ash having gone, the Sun can radiate through to the surface. In this second phase the greenhouse gases in the atmosphere will raise the temperature of the near surface layers. This could easily reach values in excess of $25°C$ leading to anoxic conditions in the seas. With no oxygen in the water plankton and other marine creatures die. Rotting flora and fauna in the water give a range of chemical elements there and especially sulphur. The major volume of the oceans would be poisonous to oxygen breathing living creatures. It will take a length of time for the temperatures to fall again to low values. The canvas of life will be nearly blank. It is interesting that life after this Great Death was very different from that which ruled and lived before it.

4.4.4. *The end of Triassic extinction*

This extinction is dated about 200 million yeas ago although the precise date is subject to debate. It coincides with very substantial lava floods in the central Atlantic magmatic province. This apparently triggered the opening of the Atlantic Ocean. The marine loss has been estimated at about 22% of marine families. The loss of invertebrates is still controversial.

4.4.5. *The Quaternary–Tertiary extinction*

This is sometimes called the K–T boundary, the K is from the German Kreidezeit for Quaternary and the T is, of course, for Tertiary. Everyone knows this event — it was 65 Mya and saw the demise of the dinosaurs. There is clear evidence through the world wide layer of iridium in the rock sequence at the same point in the time scale that a very large meteor struck the Earth then. Iridium is rare on Earth though more common in the cosmos because it is a siderophilic element and so has an affinity with iron. It will have sunk to the core region with the iron leaving very little in the surface regions. Indeed, the culprit seems to have been found: it is the oval buried crater at Chicxulub on the Yucatan peninsular in Mexico. It has an average diameter of 180 km. Its falling will have produced a large tsunami which has been traced in the rocks. It landed in a gypsum (calcium sulphate) rich region and so would have produced sulphur dioxide as an aerosol. There is some theoretical evidence for believing that the meteor that struck Chicxulub was one piece from the asteroid 298 Baptistina which broke up 160 Mya and a second piece struck the Moon 108 Mya to create the crater Tycho. The break up of an asteroid size body is not unknown — such an event was observed in 1994 when the comet Shoemaker-Levy broke up in Jupiter's strong gravitational field. The general view is that this is the end of the story but unfortunately it is not.

Many workers find it impossible to accept that the effects of the fall of a single meteor are sufficient to satisfy the observations. Certainly there were other meteor falls at about the same time although they were smaller. Thus there is the Ukrainian Boltysh crater about 65 Mya with a crater size of 20 km; and there is the Silverpit crater in the North Sea with a diameter of 20 km and an age between 60 and 65 Mya. There was also very substantial volcanic activity in the Deccan Traps at about the same time. This combined attack on the Earth's environment would certainly be sufficient to cause great havoc on the flora and fauna. This is reminiscent of the Permian Great Death. A range of species were eliminated. All nonavian dinosaurs became extinct. No remains of dinosaurs have been found this side of the K–T boundary line.[3] In fact the dinosaurs had shown a decline in diversity before the final extinction. Other groups were also lost. *Mosasaurs, plesiosaurs* and *pterosaurs* also disappeared from the record. Several species of plants and vertebrates also disappeared. On the other hand, mammalian and various bird groups did not become extinct but were able to pass through the boundary effectively unaffected. With new gaps in the evolutionary scene those species that survived had clear niches to allow development.

4.5. Evolution of Eyes

The evolution of eyes has proved in the past a major stumbling block for evolution theory. Charles Darwin admitted that it was a major problem for him and his concept of natural selection because — in the crudest terms — a half an eye is not much use.[4] He indicated how he suspected the matter would be resolved and his prediction has, in fact, been realized. Modern research

---

[3]This is not quite true — a few fossils have been found but they can be explained by later rearrangement of the sedimentary sequences.

[4]This may be true for the possessors of complex eyes but is not true otherwise especially early on in the evolutionary sequence.

has allowed the evolutionary sequence of the eye to be followed through its various stages and can be summarized by half an eye is strategically better than no eye at all. The story that has unfolded both from fossil evidence and from the study of living creatures providing a sequence of the development of ever more complex optical forms from the simplest beginnings. The evolutionary sequence appears to have taken a very short period of time some 540 Mya. The fossil record also shows evidence of parallel evolution (convergence) of optical abilities (see the later figures) though more recent work links the ability to see with the occurrence of a particular gene, TOX6. In this way sight was invented once only and convergence represents different applications of a single principle.

The principle of sight is simple to set down but complicated to achieve. The basic light processing unit is the photoreceptor whose job is to collect photons from the environment. The photoreceptor consists of two molecules contained in a membrane. One is the light sensitive protein opsin. This can be enclosed by a pigment called the chromophore which allows colours to be distinguished one from another. The absorption of a photon by the chromophore initiates a chemical reaction which transduces the energy into electrical energy. This is then relayed to the nervous system. The photoreceptor cells are part of the collection of cells called the retina which transmits the visual information to the brain for analysis and recognition. The most simple "eye" is sketched in Fig. 4.19. It is clear that for this process to work the brain must have more than a certain minimum capacity. Marine eyes, which were the first to develop, operate in the "visible" frequency range $(3.7 \times 10^{14} - 7.5 \times 10^{14}\,\text{Hz})$ of the electromagnetic spectrum because these frequencies are best transmitted by sea water with minimum attenuation. It is thought by some workers that this initial feature of eyes has been carried through to the variants ever since even for land creatures but with some

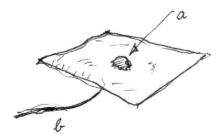

Fig. 4.19. The simplest eye was a light sensitive spot on the outer skin consisting of an opsin molecule enclosed by the chromophore (a). This is linked directly to the brain by an optic nerve (b). The light sensitive patch is open to the vagaries of the environment.

local variations. For instance, some creatures can detect ultra-violet light and some infra-red light.

The first photo sensitive patches appeared during the Cambrian, about 540 Mya. They were extremely elementary (see Fig. 4.19) and gave no more information than whether there was light there or not. This actually can be very valuable information because it would tell a creature whether it was night or day or during the day whether it was under cover or not. There could be no indication of the direction of the light nor any details of the nature or form of the light source. There would not be a need for colour in the very early forms.

As direction became important a sense of directionality was achieved by sinking the photo-detector system in a well (Fig. 4.20). The open mouth of the well was quite wide at first relative to the depth but a detailed picture became possible as the width of the mouth was reduced to the form of a small circular aperture. This essentially produced what later became known as the pin hole camera. This arrangement gives the more detail the smaller the aperture but at the same time the illumination in the eye becomes less the smaller the aperture. There must be a compromise which is determined by the environmental problems of the owner. The cavity within is open to the sea or the air and so to extraneous matter getting into it and fouling the receptor and separately to attack by parasites. The obvious answer is to close the aperture with transparent tissue. To keep the tissue thin the inner chamber can be filled with a transparent fluid forming a region called the humor. The eye now has got a fully fledged directional capability giving a complete picture as the region containing the eye is rotated. Distance awareness is still missing but this is provided by placing a lens on the orifice

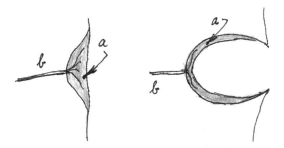

Fig. 4.20. Showing the development of the sunken eye to give some directional properties. In each case a is the photoreceptor and b the optic nerve to the brain. The depth of the well can increase and the aperture decrease to improve the operating characteristics of the eye. The photosensitive layer in the sunken eye can be more extensive that in the flat layer.

which can also be reduced in size. The lens will give focusing possibilities and also increase light collection. The whole needs protection and strength which is achieved by enclosing the front of the eye by transparent tissue called the cornea. Finally flexibility of focus can be achieved either by lengthening or shortening the whole eye or, more conveniently, by introducing a lens which can be made thicker or thinner. Muscles are required to achieve this. The most robust way is to manipulate the lens. Finally, the quantity of light entering the eye can be controlled by a shutter that can be made larger or smaller — this is the iris. The final addition is the aqueous humor in front of the iris. In many designs the eye has a blank spot where the optic nerve leaves the eye but it is possible to avoid this. *Homo sapiens* has got a blind spot in each eye but this does not cause problems in everyday life. The eye scans quickly to accommodate for the spot where no signal is received (Figs. 4.21 and 4.22).

The requirement of 360° vision has been met by a compound eye positioned towards the side of the head which is the position in many birds now that might sit exposed and will need to see an attacker in time to escape. An alternative approach is to place the eyes on stalks and this is the solution developed by, for instance, the common crab. Other creatures, and especially predators, will need to be able to judge distance with some precision. This can be done effectively by using parallax. For this two eyes must be placed with a separation distance between them and the head will need to be held steady while running at speed, perhaps over rough ground (Fig. 4.23).

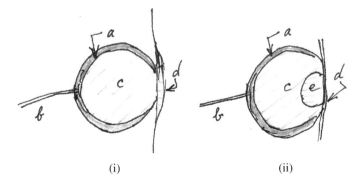

(i)   (ii)

Fig. 4.21. More advanced forms of eyes. The outer orifice is sealed by a transparent membrane against dust and parasites forming an inner cavity. The cavity is filled with a transparent liquid partly to maintain the shape of the eye and partly to aid the optical properties. Focusing is possible in (i) to a limited extent without a lens by changing the shape of the eye; the lens in (ii) allows for more careful focusing and adjustment. In the figure — *a* is the photoreceptor, *b* is the optic nerve to the brain, *c* is the transparent humor, *d* is the transparent front cover and *e* is the lens.

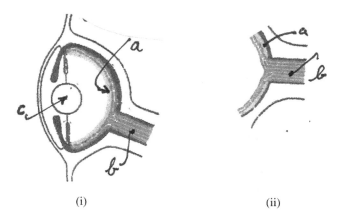

(i)   (ii)

Fig. 4.22. An example of convergence but with a difference. (i) is the representation of the eye of an octopus. There is no blind spot. (ii) shows a small modification in invertebrate eyes — the nerve system is outside the photosensitive layer leading to a blind spot. The eye is the same in every other regard. In each diagram a is the photoreceptor, b is the optic nerve system leading to the brain and c is the lens.

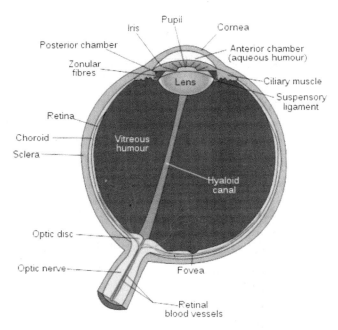

Fig. 4.23. A cross-section of the human eye showing its complicated structure. It is most suitable for hominid living but has not got the capabilities of some creatures where night vision is important (for example, cats), or where fine detail is crucial (for example, birds of prey), or where viewing in water is important (for example, squids). Will there ultimately be modifications for space?

Although we have presented the development of the eye as a single progression of design in fact this is not really the case. The eye has been "re-invented" a number of times (Richard Dawkins says 40 times and that is probably not far wrong) in different circumstances and represents a good example of parallel evolutions. There are indications from observations that the evolution of eyes took place very quickly rather before the Permian explosion taking no more than a few tens of million years to be achieved. Computer studies have confirmed the possibility of these speculations from observation. Making changes in steps in which there is a 1% improvement of vision, and neglecting negative improvements as would happen in nature, it seems that as few as 400,000 generations will move from an elementary eye to an advanced type. This may sound a lot but if the average life time of creature is one or two years this means that the whole evolution can be accomplished in a half a million years or so and certainly less than a million years. This is instantaneous on an evolutionary time scale. Some workers have suggested that the explosion of life at that time of the late Cambrian owes much to the evolutionary development of the various optical capabilities.

4.6. Brief Comment on Climate

Climate is the mean conditions locally against which we can measure the effects of change. Deviations from the local mean have been seen to be a controlling element in evolution. The effects of the mean climate itself are also significant. Gaining an insight into early climatic conditions is very problematic because the oldest rocks provide a very sketchy record of the past. Information becomes more reliable once fossils and older rocks can be studied, which becomes possible after about 550 Mya. The result is interesting.

The general ambiance has remained very similar throughout the history of the Earth. There is clear evidence from rock samples that liquid water has been on the surface since the earliest times. This is interesting because the radiation reaching the Earth from the Sun has generally increased although the Earth's orbit has remained much the same throughout. The early conditions of volcanic eruption will have filled the atmosphere with dust and reducing chemical gases and these would have tended to cut out the sunlight and so depress the temperature. The widespread distributions of heat from volcanoes will, however, have tended to keep the surface warm. This will not have been the case as the intensity of the volcanism decreased. Fewer larger volcanoes can lead to cold conditions and there is evidence in the record that the Earth has frozen over for long periods on more than one occasion before 550 Mya. The slowly accumulating land will have frozen over mush of the globe usually with the exception of the equatorial region. Both the distribution of land and the presence of ice will have profound effects on climate.

The local weather has remained remarkably constant over the last 500 My. The local temperature has hardly varied 10°C above or below the present temperature. There have been ice ages in which regions of the north and south polar regions and beyond have been covered with ice and glaciers have been a fixed feature of the planet. However the equatorial regions have been totally clear of ice and snow. Indeed there have been periods, such as in the Jurassic, when the poles were entirely free of ice. This comparative constant temperature environment has been of great assistance to the evolutionary processes.

4.7.   Summary

We list here the steps that would seem to have been essential for the later development of Homo sapiens.

14. 505 Mya: Appearance of 1st vertebrates — jawless fish related to *lampreys* and *hagfish* with cartilaginous internal skeletons but without paired fins. Precursor of modern fish with gills.

15. 480 Mya: The first jawed fish appear, the jaw forming from the first of the gill arches. The head and thorax only were protected by strong articulated armour plates.

16. 400 Mya: Thee first *coelocanth* appears — an elementary fish still found today almost unaltered. Its four fins move as the later tetra-animal legs/arms.

17. 375 Mya: Late Devonian lobe finned fish (genus *sarcopterygian*) has tetrapod-like structural features.

18. 365 Mya: Legs developed by some lobe finned fresh water fish. This gives rise to *tetrapods* Front limbs bent backwards at the elbow and back legs that bent forwards at the knee. Early tetrapods had a flattened skull containing a two-lobed brain. There was a wide mouth and short snout. Upward facing eyes showed they were bottom dwellers.

19. 315 Mya: First amphibian with recognisable limbs but lacks wrists. *Ichthyostega* is tetrapod with arms, legs and finger bones. Used for crawling through weeds and mud in shallow water.

Not a land dweller. It may have spent brief periods out of water. At this time lungs first appeared in *amphibia*. Modern *amphibia* retain features of these very early members.

20. 300 Mya: Earliest known reptile — *hylonomus* — probably similar to a modern 8 in lizard. Had sharp teeth and probably ate millipedes and insects which had also transferred to the land. Is the precursor of amniotes and mammal-like reptiles. Development of the amniotic egg: reptiles now reproduced on land and could conquer regions away from water. Reptiles developed an advanced nervous system unlike amphibians.

21. 256 Mya: Mammal-like features appear about this time. Quickly after the appearance of the first reptiles they split into branches the *synapsida* and the *diapsida*. The former had enhanced jaw muscles associated with two holes in the skull behind the eyes (the temporal *fenestra*). The mammal-like reptiles called *therapsida* developed from *synapsida* to become the direct ancestors of mammals. The later forms of *therapside* developed a second palate — they could now eat and breath at the same time. This might be associated with warm blooded creatures.

22. 220 Mya: The first mammals came from *eucynodonts* — small, shrew-like animals feeding on insects. It is likely they had a constant body temperature and mammary glands. The brain developed a unique neocortex region. This region is involved with the higher functions such as sensory perception, spatial reasoning, motor commands: in humans language and conscious thought.

23. 125 Mya: Modern placental mammals arose from *eomaia scansooria*, a small door mouse-like creature

24. 100 Mya: Common genetic ancestor of mice and humans appeared at about this time.

Note

Very recent fossils finds have increased dramatically providing firm evidence for the links between species. One example found in Northern Canada is the marine *Tiktaalik roseae* which shows both fish and land tetrapod features. This creature lived some 350 Mya. Another found in the Liaoning Province of China is *Microraptor gui* which gives evidence of early flight. Living between 120 and 110 Mya it was a 0.8 m long dinosaur with pennaceous feathers that gave a wing-like surface to the arms *and* the legs and the tail. This gave it the appearance of having two pairs of wings. Although it would not have been able to fly in the sense of a bird it could certainly have glided, swooping in the phugoid fashion familiar in modern aeroplanes with its feathered appendages displayed in biplane fashion. This suggests at least some later avions descended from glider forebears. Generally missing links in the evolutionary train are being found expanding the details of a continuous evolutionary process.

Further Reading

1. Fortey, R., 2000, *Trilobite! Eye Witness to Evolution*, Harper Collins, London.
2. Fortey, R., 2002, *Fossils, The Key to the Past*, Harper Collins, London.
3. Fortey, R., 2004, *Life: An Unauthorised Biography*, Harper Collins, London.
4. *British Palaeozoic Fossils*, 1975, The Natural History Museum, London.
5. *British Mesozoic Fossils*, 1983, The Natural History Museum, London.

Chapter 5

Hominids — Homo Sapiens

Some 60 Mya a branch of evolution turned towards the appearance of upright biped creatures in contrast to both the semi-biped and quadrupeds that had so successfully ruled the continental Earth in the past. The new bipeds showed crucial advantages over previous creatures in that the front two of the four limbs became available for uses other than for stability, transportation and simple tearing. Actually, there had been experiments with biped movement before. The largest dinosaurs, for instance *Tyrranosaurus Rex* which existed between 68 and 65 Mya, had only small but powerful forelimbs with two primary digits (basically claws) with probably a third vestigial digit developed for tearing and this was the standard dinosaurs design. There seems as yet still no agreed reason for the elementary forelimbs (other than heavier ones would have upset the creature's balance) and they were not used for further evolutionary developments of *Tyrranosaurus*. Millions of years later this freedom for forearms was reintroduced but under very different circumstances with a very different type of creature. The forearms now were associated with fingers. In these new circumstances new possibilities were associated with the development of extended brain functions. This led ultimately to a brain power with capabilities far in access of the requirements of daily living and which could be applied to a range of what might be called intellectual pursuits. One of the greatest new developments was in speech; another was in artistic perception including the graphic representation of events and the environment taking these creatures to new heights. Again, there developed the curiosity to observe and describe the environment around them and to make functional links between apparently different phenomena — this is the beginning of science and technology. We are describing the appearance of modern mankind. But it all had very humble beginnings.

Mankind (Homo sapiens sapiens) appears so very different from all the other animals in so many ways that earlier wisdom had presumed it to have

had a different origin.[1] There seemed to be no evolutionary background to
mankind which only left the possibility of an appearance after a separate
and more recent creation. This did, indeed, appear to be the full story until
quite recently. Some 60 ya, however, evidence of earlier human-like creatures
began to appear in the fossil record (especially in Africa) and it began to
indicate, contrary to the earlier beliefs, that mankind had indeed been subject to the rules of evolution just like the other living things on Earth. Over
the years more and more discoveries have led to the recognition of such an
evolutionary tree. The story is still far from complete but an evolutionary
sequence is now well established. This discovery may set mankind at a pinnacle of the evolutionary pattern at the present time but takes away any
special place in the order of things. It raises questions such as whether the
whole evolutionary pattern was structured to create mankind and what the
future of the evolutionary tree might be, among many others. This whole
pattern of new discoveries has offended people in the past because the earliest
roots of Homo sapiens are linked to primates such as gorillas, chimpanzees
and orang-outans. The different forms branched off several million years ago
from common ancestors, not from each other, and the distinction between
the different species is now very clear.

5.1.   Nomenclature

It is necessary to make clear the nomenclature of our subject that we will
use here more especially because popular ideas, and those of science fiction,
are often not correct. This confusion has arisen because the classification has
changed in the past and the present style was introduced only in 1980.

The super-family of primates is called *Hominoidea* and has two divisions. One is the family *Hylobatidea* which are essentially the *gibbons* and are
sometimes called the lesser apes. The other is the family *Hominidae* which
includes *orangutans, chimpanzees, gorillas* and *humans*. It is now substantiated that the common ancestry of humans and apes was broken at a branch
point after which they took different evolutionary paths. The human family
is included in the genus *Homo sapiens* and mankind itself is named *Homo
sapiens sapiens*. We are concerned now with the line *hominid* from the time
that it split from apes. The story is still far from complete. There is difficulty
in being sure that a given set of bones do indeed belong to a biped ancestor
of modern humans. A complete skeleton is never found and the dating of the

---

[1]Charles Darwin knew better but it has only been relatively recently that his conjectures have
been confirmed.

remains that are found is sometimes far from precise. Nevertheless, the fossil evidence is now forming a clear and consistent story of the development of humans according to the laws of natural selection. We will now present a summary of the evidence for the evolution of hominids, starting where the last chapter ended at about 44 Mya.

It is interesting to notice that all apes are agile tree climbers except *gorillas* and *humans*. They are all omnivorous in their diet. They are native to Africa and Asia although humans have now spread throughout the world.

5.2.   Development of More Modern Forms

The general picture that is emerging is as follows. True primates appeared about 40 Mya and other changes to the environment also came about. Of great importance to later humans is the development of elementary grasses from the angiosperms. The *Anthroids* developed as man-like in form only and a most significant development some 30 Mya was the split between the *Anthropoid* into infra orders *new World monkeys* (*Platyrrhini*) with prehensile tails and old World primates (*Catarrhini*). As the African and South American continents drifted further apart the *Platyrrhini* moved with South America while the *Catarrhini* moved with Africa. It is an interesting fact that the males of the New World moneys then were colour blind but gained colour vision during the next 25 My. At about that time the *old World monkeys* themselves split into the two sub-families of *old World monkeys* (*Cercopithecoidea*) and *Hominoids*. The *old World monkey* has not got a prehensile tail and one example is the baboon. None of the *Hominoids* have got visible tails (although there may be vestiges on the skeleton — and this included modern humans). Examples here are lesser and great apes and hominids. *Gibbons* are lesser apes: *orangutans* and *chimpanzees* are great apes. It appears that our human ancestors separated from the *orangutan* about 13 Mya.

This became the time of apes with as many as 100 different types of ape moving about in the forests and especially in the trees. The apes dominated over a period of several million years but between 10 and 6 Mya the climate changed and this upset their empire. It became dryer and cooler. Forests gave way to grasslands and savannas and the whole environment was changed. With trees in decline the habitat of the apes disappeared so the big days of the planet of the apes are now over unless they can adapt to the new environment. Many did not with the result that apes became more rare. Monkeys, on the other hand, proliferated and so did hominids. The open spaces make this also the heyday of horses, much smaller than our modern breeds, but able to move freely over the savannas of the northern hemisphere.

The horses are set to decline, however, due to the competition with the *artiodactyls*.

It is worth noting that chimpanzees and humans today share 98% the same DNA structures and their haemoglobin molecules differ only by a single amino acid. These similarities hide profound differences, however. *Chimpanzees* cannot speak but interestingly they seem incapable of contracting aids. The *chimpanzees* show great genetic diversity, overwhelmingly greater than humans. In spite of this there are very interesting similarities of behaviour between the two groups. Both form friends and acquaintances and use these to influence their peers. They have enemies among themselves. They form alliances and can well distinguish friend from foe. They know good behaviour from bad and are adept at group living with their own personalities. They each ultimately find a mate and move on to establish their own family unit. There is one extremely significant similarity. The larynx is repositioned in both *chimpanzees* and *humans* during the first two years of life moving down to between the pharynx (which lies immediately behind the mouth and is central to breathing and eating) and the lungs much lower down. This move is essential for speech in humans as they get beyond the age of two. It is interesting that the facility for speech appears to have been available in humans in rudimentary form before the divergence long before speech is believed to have actually appeared in humans. Presumably the 2% difference in DNA contains the vital components for speech in humans.

5.3.   Hominids Diversify

A diversity of hominid forms has been recognised over recent years and the complexity seems still to grow. This is exactly what is to be expected if *Homo sapiens* followed a line of evolution controlled in general terms by Darwin's vision of the survival of the fittest. As on so many other occasions, the initial expectations proved to be quite wrong. It was supposed at first that the thing that distinguished *Homo sapiens* from the rest was brain complexity and this implied brain size. This identification was supported, or even induced, by the fact that human brain sizes are rather more than double those of *chimpanzees* and *apes*. The evolutionary sequence would, according to these ideas, be increasing brain size which will lead along the way to upright walking and the use of the foremost "legs" as arms and hands with the prehensile fingers and thumb now converted from tree climbing to more imaginative daily tasks. This is, of course, very logical but is not exactly how nature works. Time and again we have seen the development of a facility for one purpose being expanded to allow much more diverse use later on — for instance legs, arms and digits designed for marine use later expanded to allow free movement on land. The present case was in that category but rather different. Before some 5 or 6 Mya the land was covered by trees and life abounded in them in the form of apes. And then the apes suffered a severe extinction and became much rarer although one member, the hominids, survived and flourished. What ever caused all this? The hominid story is still largely undiscovered but many clues have been unearthed during the last decade and more are being discovered all the time. The evidence is usually very fragmentary being one or two bones from which much is constructed as we describe.

Geological and palaeo-biological evidence is that the climate changed and trees suffered a substantial extinction as a result. Forest areas were cut back drastically giving way to savannahs and grasslands as the whole ecosystem changed. Creatures that lived and fed in trees lost their habitat and consequently died out. Creatures that could benefit from the new environment survived and prospered. This environment was ideal for herd animals roaming the grasslands and it proved an ideal place for the bipedal hominids. There was a wide diversification of species some of which we are now finding out about. The discoveries are random so they are not being made in an ordered sequence. We will present the main features of the work so far but will place them in a time order of developments rather than a time order of discovery.

5.3.1.   *7–4.5 Mya*

The period between 6 and 7 Mya is a crucial period for us because it was then that it is believed that our human ancestors separated from the *chimpanzees*. The precise division has not yet been found with any certainty but the last common ancestor must have been close to *Sahnelanthropus tchadensis* found in 2001 in Chad, Central Africa and named in July 2002. It is known through an almost complete cranium and given the nickname *Tomnai*. It has got brow ridges and small canine teeth similar to hominids. On the other hand it has got a very small brain size of about $350\,cm^3$ which is more like primitive apes. It has not been agreed whether this creature actually was bipedal or not because not enough of the skeleton has yet been found. More especially there is not yet any indication of where on the skull the neck joined the head. This is crucial because it determines the balance of the head on the body. For bipedal creatures the neck is joined under the head.

Fossils of several creatures were unearthed in Western Kenya in July 2001 and named *Orrorin tugenensis*. Fragments of arms, thigh bones, lower jaws and teeth were catalogued. These remains were of creatures about the size of an adult female chimpanzee. Some workers believe there is evidence for bipedalism and for an ability to climb trees but others still need to be convinced. Nevertheless, the remains belong to the transition period between 6 and 7 Mya and must represent some form of intermediate creatures. A species now dated at about 5.8 Mya is *Ardipithecus ramidus* named in September 1994. The remains are essentially skull fragments. There is indirect evidence that is was about 1.2 m (that is 4 ft) tall and probably was a forest dweller. If this is, in fact, substantiated it will have repercussions on the present theory that links the move to the savannah with bipedalism. A number of fossil fragments were discovered between 1997 and 2001 and were dated between 5.8 and 5.2 Mya. These have been assigned to a new species *Ardipithecus kadabba* separate from *ramidus* although at first it was thought to be a sub-species. None of these fossils are in any way complete and there is really as yet insufficient information to be definite about them and their lives. It is not known yet what was the size of the brains of these creatures. It is clear, however, here was the beginning of the new world of the hominids.

5.3.2.   *4.5–3.0 Mya*

Many of the fossils falling within this range of ages had a small brain. One was named in 1995 by Leakey working in Kenya as *Australopithecus anamensis*. The material forming the basis of this identification was: 9 fossils primarily from Kanapoi, Kenya; and 12 fossils (mostly teeth) from Allia Bay, Kenya. This species seems to have lived between 4.2 and 3.9 Mya. It had a mixture of physical properties. The teeth and jaws are reminiscent of fossil apes. On the other hand, a lower humerus is very like a human form while a partial tibia suggests strongly that the owner was bipedal. It must be admitted that there is no positive confirmation that these various fragments are actually from the same species although the circumstances of their discovery very strongly suggests this.

At the same time as *anamensis* a different species was thriving leaving much more authenticated fossils. This was named as *Austalopithecus afarensis* which existed between 3.9 and 3.0 Mya. There was an ape-like face with essentially no chin; the jaws protruded and the back teeth were large. The other teeth were more like human teeth although rather more pointed. The shape of the jaw was between the square shape characteristic of apes and the parabolic shape of humans. The shapes of the pelvis and leg bones leave little doubt that these creatures were bipolar although the gate was for walking and not for running. There is also no doubt that these creatures were physically very strong by modern human standards. Their height was between 1.07 cm and 1.52 m and the males were substantially larger than the females, a condition called sexual dimorphism. This is a condition that is discernable still in mean terms in humans today. The toe and finger bones were longer than modern humans. This is taken as evidence by many workers that these creatures were still adapted for climbing and movement in trees. Others regard this as evolutionary baggage that had still to be lost. One find included a body with 40% of the skeleton bones present. So unusually complete a find had to be accompanied by celebrations and one song of the time was the Beatles song "Lucy in the sky with diamonds". Hearing and singing such a song made it obvious that the skeleton be nicknamed Lucy and the name seems to have stuck. It is not known really whether it is actually male or female.

Two other species are associated with this time interval although the evidence about them is less than certain and especially whether they were truly bipedal. One was *Australopithecus bahrelyhazali*. This fossil was discovered in 1993 in the Bahr el Ghazal valley close to Koro Toro in Chad. The catalogue number is KT — 12/H1 but has been given the nickname Abel. The found material was: a mandibular fragment; a lower second incisor; both lower canines; and all four premolars still fixed within the dental alveoli. The remains are thought to be 3.9–3.5 My old. The denture is, in fact, similar to that of *afarensis* and could fall within the variation of this species. More specimens need to be unearthed before any firm conclusions can be drawn but it is certainly unique in being the only Australopithecine found in Central Africa. If the find is authenticated as a new species it will give the third African region associated with early hominid.

The second species was *Kenyanthropus platyops* (flat faced man of Kenya) which has been dated between 3.5 and 3.2 My old. It was discovered by a member of the Leakey team in the region of Lake Turkana, Kenya, in 1999. More than 30 skull and tooth fragments have been found, the teeth being intermediate between ape and hominid. The reconstructed head shows a flat face (hence the name), a small brain but high cheek bones similar to a hominid. It seems to have had a small ear hole, like a chimpanzee, but this could be the result of the distortion of the head bones as they lay shattered. A toe bone indicates that it walked upright and so was not chimpanzee. Dr Meave Leakey of the Kenya Museum has suggested that this is a new genus of ancestors separate from Lucy. She believes it is separate from the *Australopitheus* species which might have been a dead end as far as the evolution of humans is concerned. There is some disagreement among anthropologists as to whether *platyops* was, indeed, a new species rather than a variant of *afarensis* or *africanus* which have got similar cranial sizes. The importance of these issues is the nature and extent of the adaptive radiation of hominids at this time. The one clear conclusion is that humans did arise from such adaptations rather than a single creation.

5.3.3.   *3.0–2.0 Mya*

A creature known more fully is *Australopithecus africanus* which existed 3–2 Mya. It was quite slim, as was the earlier *afarensis*, but the body size was slightly greater than the older species. It is clear that *africanus* was bipedal. Its brain size was in the range 420–500 cm^3 which is also slightly greater than that for *afarensis* but only about 40% of a modern human brain size.[2] The *africanus* brain would seem to be too small to allow for speech. The shape of the jaw is now fully parabolic like modern humans but although the teeth were much larger than human teeth their shape is generally similar. The size of the canine teeth is smaller than those of *afarensis*. In general terms *africanus* appears to be an evolved form of *afarensis*. One species known only though a partial skull is *Australopithecus garhi* named in 1999. The skull differs from other skulls of *Australopithecus* in an elementary morphology. For instance, its teeth are remarkably large and especially the rear ones presumably used for heavy chewing. Some other remains were found nearby with a ratio of humerus to femur of human magnitude but yet a ratio of lower to upper arm more like that of an ape. These may be unrelated to *garhi*, however. Here is insufficient evidence to develop more information about this species.

An interesting find, which could possibly be a direct descendent of *afarensis*, was made in Ethiopia in 1935 and again in 1985 in West Turkana, Kenya. The catalogue (field) number is KNM WT 17000 and the find has been named *Paranthropus aethiopicus*. The lower jaw and tooth fragments found in Ethiopia were later supplemented by the finding of a skull in 1985. The skull has unusually high levels of manganese (presumably from the soil it was buried in) which has turned it very dark — hence it has become known as Black Skull. The brain capacity is small, about 410 cm^3. The physical size of the whole creature is not known but it could have been about that of *afarensis*, which is between 3.5 and 5 ft tall. It is known to have lived in mixed woodland and savannah. The skull is the earliest example of a robust hominoid in the Pliocene. The skull dates from 2.5 Mya and it is likely that the species lived between 2.7–2.5 Mya. The primitive features are very reminiscent of *afarensis*. A most interesting feature of Black Skull is a pronounced

---

[2]cm^3 is used to mean cubic centimetres.

sagittal crest on the temporalis mosele and zygomatric arch. This formation, which comes with adulthood, is not uncommon even in mammals and acts as an anchor for the muscles of very strong jaws and jaw bones. It is believed by some anthropologists that *aethiopicus* will turn out to be the direct descendent of *P. boisei* and *P. robustus*, to be described later, since there was a strong similarity with jaw shape and size. It seems, however, that *aethiopicus* — together with other *Australocines* — was on a different evolutionary hominid branch from that leading to Homo sapiens. We need to know more about this hominid.

A hominid which lived about 2 Mya was *Homo rudolfensis* mentioned above. Fossil remains of this species were discovered in 1972 at Koobi Fora on the east side of Lake Turkana, Kenya. The field number is KNM ER 1470. The fossil has been assigned an estimated age of 1.9 My. The find was a series of fragments of a fossil skull which were pieced together after their discovery. The brain size was then given as $752\,\text{cm}^3$. Since then more has been learned about mammal fossils and especially a rule has been discovered giving a precise relationship between the mammal's eye, mouth and ears. A new reconstruction of the skull on the basis of these relationships has reduced the brain size to $526\,\text{cm}^3$ which is more in keeping with its rather primitive features. There is insufficient evidence to draw significant conclusions about this species.

5.3.4. *2.0–1.0 Mya*

The hominids now begin to become more sophisticated. *Australopithecus robustus* may have been one of these. All the fossils have been found at several sites in South Africa. At a cave at Swartkrans the remains of 130 individuals were found. A study of the teeth suggested that *robusus* rarely lived more than 17 y. It thrived during the period 2.0–1.2 Mya. It appears to have had a body similar to that of *africanus.* Its skull with its lower jaw and its grinding teeth seem to have been rather larger than those of africanus. Its brain size was small probably in the range 410–530 cm^3. It had a large flat face with essentially no forehead but very pronounced brow ridges. Most of the specimens found have got a sagittal crest which is in keeping with the large jaw and chewing teeth. They seem suited to food in a dry environment. It is likely that the male stood about 4 ft (about 1.3 m) high with a mass of about 55 kg: the female was smaller with a likely height of only 1 m and a mass of perhaps 40 kg. This is a rather large sexual dimorphism. It is interesting that robustus skeletons have been found associated with animal bones which might suggest that this species used bones as digging tools.

Autralopithecus boisei (now renamed *Paranthropus boisei*) lived between 2.6 and 1.2 Mya overlapping with *robustus*. These two species were generally similar except that the face and cheek bones of *boisei* were rather more massive and the teeth correspondingly larger. Indeed, *boisei* is described as the largest of the *Parannthropus* species. The brain capacity is quoted as 530 cm^3 which again is quite small. The species showed sexual dimorphism, the male being larger than the female. The first specimen was unearthed in 1959 at Olduval Gorge, Tanzania by Mary Leakey the well known anthropologist. The find was a well preserved cranium numbered OH 5 and nicknamed Nutcracker Man on account of his very large molar teeth. The fossil was dated as 1.7 My old. A second skull was unearthed in 1969 at Koobi Fora in the Lake Turkana region by Richard Leakey. He believed he had evidence of *boisei* being the first hominin to use stone tools. A partial skull was found at Kronso, Ethiopia, in 1993. It was designated KGA10-525 and assigned an age of 1.4 My. The youngest specimen was found at Olduval Gorge with an age of 1.2 My while the oldest yet found was discovered Omo, Ethiopia, with an age of 2.3 My. Some anthropologists consider *robustus* and *boisei* to be simple variants of a single species. More information is needed before *boisei* can be properly understood. Neither *robustus* nor *boisei* have been regarded by anthropologists as in the direct line for *Homo sapiens*.

5.4. The Line Homo Sapiens

The genus Homo encompasses modern humans and their close ancestors. This means that all close relatives are included in the group (Genus) which is estimated to be about 2.5 My old. This is the time of the first stone tools (the Oldowan industry) and defines the start of the Lower Palaeolithic era. It must be admitted that there is no universally accepted species from which the taxa Homo have radiated. Continuing excavations yield new fossils and this situation may well be corrected before long.

It might be of interest to rehearse the linguistic origins of the word Homo. The word homo is the Latin word for man in the sense of mankind, a human being or a person. Human is derived from the Latin humanus an adjective cognate to homo and derived from the original Proto-Indo-European language dh g ham meaning earth. The Hebrew adam means human and is cognate to adamah meaning ground. The Latin equivalent is humus meaning soil. These various words and their cognates are thought by some linguists to have the super-root ad–ham. It is clear that the root Homo has very deep meanings rooted in earth. Modern *Homo sapiens* first appeared in archaic form perhaps as much as 600,000 ya but certainly by

200,000 ya. This includes two main members and a third has been discovered recently.

Some hominids seem to have been handy men and this seems to have been the case for *Homo habilitis* whose remains have been dated in the range 2.4–1.5 My. There was strong evidence of tools with his fossilised remains. He had many primitive features but had many similarities with *australopithecies*. In particular his face was flatter than that of *africanus*. His brain size was larger being between 500 and 800 cm^3 which is nearer the human value and the shape of the brain is also more human like. One important area is Broca's area which is the centre of speech. This area in one sample shows a distinct bulge suggesting the creature was able to take part in at least some sort of elementary speech. *Habilitis* has been a rather controversial species and many anthropologists believe the present classification contains more than one species. The present range might well be too wide to contain the variations of a single species.

Up to this point we report bones of the hominids found exclusively in Africa but with *Homo erectus* we have remains found in Africa, Asia and Europe. This species existed between 1.8 Mya and even as late as about 0.3 Mya. The first skeleton was found near the Solo river in central Java as long ago as the 1890s. Many of the early and spectacular finds were made in China. It seems to have migrated from Africa around 2 Mya and spread through much of the southern Asian countries. It is generally accepted as being the first hominid to migrate from Africa. Most of the original fossil finds were destroyed during the Second World War but accurate casts had been made (and which survived) which now are accepted as sufficiently accurate for academic study. These are lodged partly in the American Museum of Natural History in New York and partly in The Institute of Vertebrate Palaeo-ontology and Palaeo-anthropology in Beijing.

The males were quite tall with an average height of 1.79 m. The females were rather smaller, perhaps by 20% or 30%. This is a dimorphism rather larger than modern *Homo sapiens*. The good height was matched by a larger brain size varying between 750 cm^3 for former members and 1225 cm^3 for later ones. This gave the possibility of evolutionary skills not known in hominids before.

Perhaps the most significant thing about *erectus* was its use of stone tools of some sophistication. One could be forgiven for thinking that here was the first "hunter–gatherer". There is evidence of the controlled use of fire and this could suggest some form of communal living and certainly of direct communication. There is even evidence from Olduval Gorge of the eating of mammoth. *Erectus* remains the most successful of the *Homo* species and gave rise to a number of variants towards the appearance of man. *Erectus* was probably the first hominid to have some control over its environment.

A species (or some say simply a sub-species of *Homo erectus*) which lived between 1.9 and 1.4 Mya was *Homo ergaster* or the working man. Remains have been found widely in Africa, in Tanzania, Ethiopia, Kenya and South Africa. There are several variations compared with *erectus* such as thinner skull bones, a very reduced dimorphism and a smaller straight jawed face. Its brain size was large, somewhere between 700 and 850 cm^3. It is estimated that the male *ergaster* may well have stood nearly 2 m tall, remarkably tall for hominid up until then. It is distinguished as the first known hominid to have the same proportion of longer legs and shorter arms as modern humans. The most complete skeleton was discovered in 1984 at Lake Turkana and was found to be 1.6 My old. It was nicknamed *Turkana Boy*. The terminology "working man" was given because various hand axes and cleavers were found near the bodies. There is also evidence from charred animal bones that this species used fire and made camps (the beginnings of a cuisine?). These are all characteristics of *Homo sapiens.*

A few fossils have been discovered in the Republic of Georgia which are believed to belong to new species that lived about 1.8 Mya. It appears to have had a small brain capacity of about 600 cm^3. It has been named *Homo georgicus*. Its status is still under investigation.

5.4.1.   *1.0 Mya to the present*

Homo erectus survived well into this period but other primates also appeared. The one we would like to know more about is *Homo antecessor* which lived possibly between about 900,000 and 600,000 ya. The species was defined when over 80 bone fragments were found in the Ataperca Hills, near Burgos, Spain in 1994–1995. About 200 stone tools were also found together with some 300 animal bones. The remains were dates at 780,000 y old. The important finds were a upper jaw bone (maxilla) and a frontal bone of a child of about 10 y of age. Many other remains have also been found subsequently in the same region. It seems that this has been occupied continuously by homonia up to the present day. There has been no *antercessor* skeletons found in Africa to this date.

The males were probably on average between 1.7 and 1.8 m tall and weighed in the region of 91 kg. The brain size was between 1,000 and 1,150 cm^3 which is large but not as large a modern *Homo sapiens*. There is just an indication that *antecessor* might have been right-handed. If this is so it distinguishes him entirely from apes which have no handedness. Skulls show a bulge at the rear (occipitsl bun associated with consciousness and running), a low forehead and a vanishingly small chin. It has been suggested by some anthropologists that *antecessor* could reason and had a symbolic language. Markings on one also suggest that he may have been a cannibal.

Further members may well be added to this group. One tentative member could be *Homo cepranensis* following the finding of a scull cap south of Rome in Italy. It is believed to be between 0.9 and 0.8 My old. It is estimated to have had a brain size of about 1,000 cm^3 but little else is known. It could be associated with *antecessor* with the same brain size but there is no direct proof to date. Another possible candidate is the owner of a few fossils found in Zambia. It has been named *Homo rhodesiensis* and appears to have lived between 300,000 and 125,000 ya. It seems to have had a brain capacity of about 1,100 cm^3. It is tentatively linked with a possible species involved with the fossil *Rhodesian Man*. Most experts now link *Rhodesian Man* with the group *heidelbergensis*. The waters are mudded.

To add more confusion three craniums found in Ethiopia are between 160,000 and 15,000 y old and have been associated with a new species *Homo sapiens idaltu*. The brain sizes are quite large at approximately 1,450 cm^3. Little is yet known about this group.

It is generally accepted that the extinct *Homo heidelbergencis* (*Heidelberg Man*) was the precursor of *Homo sapiens*. This species lived during the period about 600,000 and 200,000 ya. It covers a very diverse set of fossil skulls which are reminiscent of both *erectus* and modern man. Its brain size had an average volume of 1,200 cm^3, greater than *erectus* but smaller than humans. The teeth seem to have been less strong than *erectus* but more strong than humans. The average height was 1.6 m and there was some sexual dimorphism. Receding foreheads and chins and substantial brow ridges were evident. There is no clear dividing line between *erectus* and archaic *sapiens*.

We have a tangible link with *heidelbergensis*. At Roccamonfina volcano, Southern Italy, three *heidelbergensis* (two adults and a child) have left the earliest *Homo* footprints in the soft volcanic ash before it became hard. Volcanic ash can be dated rather accurately — the stride of the adults can be interpreted to provide a height for them and also an estimated age for the child. The depth of the prints could in principle give an estimate of the weight of the people. This is a remarkable find that brings us affectionately close to an ancient family.

Homo sapiens neanderthalensis, or *Homo neanderthal*,[3] is a recent hominid which appear to have existed between about 350,000 and 30,000 ya.

---

[3]They are named after the Neander valley in Germany where their remains were first discovered.

These creatures are very vivid in the general imagination as brutish inarticulate savages but this is being found to be a very misleading description. Certainly they were very strong but they had a very hard life. They were very stocky to preserve heat and appear to have been the first hominid to live outside temperate zones. They lived in small communities with a clear social life. It appears from the condition of the bones in some relics that they took care of the sick and injured. Their average life span was about 30 y and there is evidence of members living beyond the time when they could fend for themselves. This implies some care for the older people in the community. It seems they buried their dead according to some ritual which suggests thinking, imagination and a sense of the past with some awareness of a spiritual aspect of life coupled with an acceptance of elementary hygiene. They hunted with stone tools and spears though the spear was probably used more for stabbing than for throwing. The later members learned to make jewellery, for example in the form of strings of shell beads. *Neanderthal* had a large brain volume with a mean value of $1{,}450 \, \text{cm}^3$ which is some 7% greater than the modern human size. The brain cage was rather longer and lower than in modern humans and there was a distinct bulge at the rear end which might suggest a capacity for language. This possibility is confirmed by the finding of a hyoid bone in a fossil in Kebara cave, Israel, dated 60,000 ya. This bone is essential for speech and could well be a positive indication that *neanderthal* could speak. It appears that the gene responsible for speech (FOXP2) also first appeared in hominids at about this time. If they could speak their stocky frame with a broad chest would provide a powerful source of sound. Their larger skulls and broader sinuses would have given them a deep and resonant voice rather similar to modern opera singers. Communication between people is vital for successful communal living and hunting. They certainly appear to have been very determined hunters of all the fauna of the time and had a full diet of meat and vegetables. It is a small step further to suppose they had a language for communication (a very exciting possibility) but there is no direct evidence for that or for any writing capability.

In spite of their large brains *neanderthal* showed no evidence of improving with time. Although they lived over a period of some 200,000 y the designs of their stone tools and jewellery remained essentially the same. They may have inherited the initial thought from *erectus* but they seemed unable to develop what they had. As a consequence it is not easily possible to say which artefact comes from their early period or which from their later. This, presumably, gives a possible clue to why they declined: they were unable to change with changing circumstances. This may well have been the key ingredient that allowed *sapiens* to survive where *neanderthal* failed but there is another possibility.

Some *neanderthal* DNA has been recovered from several sites such as from Moula-Guercy in France (which provides evidence of cannibalism), Feldhofer cave in Germany, Mezmaiskaya cave in Russia but most particularly from Vindija cave in Croatia. Any analysis presents extreme difficulties due to pollution from any or all of the modern humans that have handled the bones together with mice, rats and other vermin that chewed flesh from it; and fungus that might have lived on it. The last two can be recognised and probably allowed for but the human forms are more difficult. It has been suggested that the extraction of genuine *neanderthal* DNA presents similar problems to that of extracting one unknown book from five or six that have been shredded and mixed together. As an example, the relic from Vindija is a single bone that has been in a laboratory draw for the last 20 y having been identified (wrongly) as an animal bone. It has become now the basis of a quest to construct the genome for *neanderthals*. This is a massive undertaking but the work so far has already provided some surprises. First, the gene MCIR has been isolated which suggests that at least this specimen was a red head. The gene is not precisely the same as that in Modern Humans so there need not be a link. This is not the case for the next find — the gene FOXP2 which is responsible for speech. It is the same gene form as in Modern Humans which is the major surprise. One might have found this in either *neanderthals* or in Modern Humans but not in both if they were separate species. The suggestion is that it is possible that we obtained our speech from the *neanderthals* through interbreeding. This idea is very controversial at the present time because previous indications were that this was not possible. If it is true it would answer the unanswered and difficult problem of why *neanderthals* died out — they did not. Certainly, the two types of Hominid are known to have existed together for at least 10,000 y. It would also answer another very worrying question: how did the different races of Modern Humans evolve over the short period of 40,000 y that they have been out of Africa? The *neanderthals* were around for some 200,000 y which is certainly long enough to adapt to local circumstances. Interbreeding would pass this conditioning on the Modern Humans in a natural way. We must await further evidence from the genome project to gain harder knowledge of the relation between *neanderthals* and Modern Humans.

Homo floresiensis (nicknamed The Hobbit) was a recent miniature Hominid discovered in 2003, to everyone's total amazement, on the Indonesian island of Flores. The most complete find is of a female (nicknamed

The Little Lady of Flores or Flo), possibly about 1.09 m tall, with a brain capacity of 417 cm^3 and was entirely bipedal. This was dated 18,000 ya. Other remains of several individuals have also been found which indicate that the 1.09 m height was a general characteristic of this group. Stone tools have been found with dates spread from 94,000 to 13,000 ya. Several special features have been recognised which seem to characterise *floresiensis*. One is a wrist bone which is indistinguishable from that of *australopithensis* and of chimpanzees. Another is essentially no chin.

It seems that this species was basically a dwarf form of *erectus* that used stone tools and had learned to control fire. There is a controversial possibility that it also had a verbal language. All this is exciting but it did not go by unchallenged. There was a body of anthropologists that maintained that Hobbit was not a new species at all but was a member of a pygmy group who had a cranial disease such as Laron syndrome. It appears now to be agreed generally, however, that the retrieved skulls show no signs of microcepholy but on the contrary appear to have been fully healthy. It seems that there really was a species of hominids that existed side by side with humans up to about 13,000 ya though the two types never met. What is more amazing is that these Hobbit people hunted dwarf elephants whose remains have also been found. This may sound fantastic but it is not uncommon for a species of dwarf mammals to evolve on off-shore islands and here, it seems, is an almost modern version. The possibility that other hominid variants exist now in remote corners of the world cannot be ruled out.

Homo sapiens sapiens (often shortened to *Homo sapiens*). The modern forms of *sapiens* (Modern Humans) are about 195,000 y old. Characteristic features are a prominent chin, very weak eyebrow ridges and a sharply rising forehead. The skeleton is very gracile (even though obesity seems to be on the rise but the ability to over eat is a very modern option for some). The average height is 1.8 m (but this is variable with region on the Earth). The brain is divided rather differently from neanderthal and has an average size 1350 cm^3 which is 93% of that of *neanderthal*. There is a sexual dimorphism but it is not always very marked.

The first anatomically modern humans appeared in Africa at some time before 100,000 ya. They migrated to Asia by way of the Middle East taking the route by the coast. Skin colours were formed by the need to protect the skin from solar ultra-violet light. Physical characteristics were changed by the local weather. In this way the different "races" began to be formed. Obviously the use of the term race is wrong in this connection — the differences are, in fact, cultural but the wrong word seems to have become standard. Be that as it may, conditions were to change. Some 70,000 ya saw the last ice age which would have been a testing time for Homo in the north. In the south conditions were different.

Humans inhabited Blombos cave in South Africa and left a very remarkable artefact. They made tools from bones and among the relics found in the cave were several small blocks of red ochre. The sides of the ochre had marks scratched on them to form a pattern in the ochre. The initial ochre was not found locally but had to be brought in from a distance away. The markings have been studied very carefully and it is quite universally agreed that these marks are not random but form a pattern deliberately set down. This is taken as evidence that *sapiens* at that time had the ability for symbolic thinking and at least an elementary linguistic facility. The red ochre is well known locally in modern times as a cosmetic for ceremonial tribal functions so it must be presumed that these ceremonies were performed 70,000 ya. This shows an established social order at that time.

Although humans were well established in Africa and had moved into Asia some moved eastwards to Australia while others moved into Europe moving especially along the coast where the penetration was easier and faster. They were very successful and seem to have eventually replaced *neanderthals* wherever the two species met. By 40,000 ya they had become widely established and their skills developed. They were fearsome hunters and began to deplete the stocks of wild animals that they chose as food. Of especial note are caves where they lived and painted scenes with the greatest artistic skill. There are about 100 such caves known at the present time dated generally some 40,000 ya. Perhaps the one best known is at Lascaux cave or perhaps the cave near Les Eyzies in the valley of the Dordogne. The prototype was a cave near Cro-Magnon in S.W. France which was discovered in 1868. It contained a range of artefacts which showed a clear culture described as characterising the Cro-Magnon peoples. Since that time many such sites have been found all telling much the same story and it is generally not possible to differentiate one group from another. For this reason it has become customary to draw them all together under the collective name early modern humans (EMH). These show many of the features of modern humans. There are two distinguishing differences. One is that EMH were much stronger than modern humans (MH) emphasising their very hard and rugged life. The other is modern technology. EMH were only at the beginners' stage but would undoubtedly have matched our performances had they obtained our technology. In support of this it has been found that DNA samples from EMH lies within the variations of MH whereas that for *neanderthal* lie outside it. It is clear that EMH were really "one of us".

EMH lived during the period 40,000–10,000 ya. Among their artefacts that have been found are huts, carvings, antler-tipped spears, bone jewellery, evidence that they wove cloth and cured animal skins for clothing and other domestic uses but above all the knowledge to use magnesium and iron oxides to make paint. This takes us back to their great skill in adorning the walls of their caves in magnificent style using different coloured paints. The cave paintings are quite extraordinary and it is an unforgettable experience to have seen them. The caves are caverns like cathedrals and, although adequately lighted now, are nevertheless full of shadows and quite eerie. Presumably they would have been lit by blazing torches when they were in use. The subjects are animals of the field in full cry and in many cases being chased by hunters. The level of drawing and colouring is of the highest order and compares well with the very best of Renaissance art. There are also patterns of waves and dots that seem at first glance to have no meaning. And there is often the image of a hand painted over to leave a no-painted impression on the rock in the midst of colour all round. There are no symbols that could be interpreted as a written word. These images are not easy to understand. Are the hunting pictures celebrating hunting prowess? Or are they instructions to the young men — or information for young women about the activities of young men in the field?

Some light can be shed on these matters by essentially modern paintings in caves and on walls. Native societies in South Africa and in Australia have also produced very similar displays including the waves and dot and are available now to be asked about them. The patterns of dots and waves are, apparently, associated with conditions of trance. This could be entered into to combat illness or misfortune or to help success in particular connections. The "unpainted" hand represents a way of passing though the rock to enter the "other world". These interpretations of the cave paintings, if valid for EMH, give a picture of a very sophisticated society fronting a clear spiritual background. Here there is a very rounded set of people easily recognised today even though they lived in the Upper Palaeolithic.

Modern humans expanded over Europe and Asia and some 30,000 ya crossed from Siberia into North America across the Bering Straits several times in antiquity before it entered from Europe much later on. It was as late as 10,000 ya that humans reached the very tip of South America, Tierra del Fuego. All the world is now inhabited except the wastes of Antarctica which we hope will not become a friendly habitat. Before this, about 15,000 ya at the end of the last ice age, a group of humans in the Middle East branched out into a new and highly significant adventure. They developed sustainable and static agriculture rather than remain wanderers. Instead of temporary accommodation that could be picked up and taken away they built more permanent accumulations of dwellings and so invented cities. Animals were domesticated and used for food and the amenities of the towns and cities multiplied. The central region for these developments was what is now Iraq. It was here that the first of these modern concepts was first developed. The idea was good. They have now spread throughout the world and look set to cover much of the Earth's surface. Presumably they will ultimately be exported to the Moon and beyond.

5.5. The Future of Homo Sapiens Sapiens?

Where does sapiens go from here? That is an important and vital question. Surely sapiens cannot eliminate the Darwinian concept of the survival of the fittest. Might the concept need to change? Is sapiens that powerful? Why are we concerned with questions such as these?

Up to the present the changes identified as those of evolution have not involved the "interference" by the creatures themselves. This is not now the case since modern humans have been able to gain knowledge of evolutionary processes at least at a superficial level. Modern humans have discovered the microscopic, atomic and sub-atomic, world and many of the linkages between that world and the macroscopic world of everyday life. Many of the macroscopic forces have been related to the corresponding microscopic forces and the different rules applying to the macroscopic world and the microscopic world. Out of this has come the recognition that biological processes are controlled by atomic forces and the intricacies of DNA and RNA have been elucidated. The DNA structure of individual species is open for investigation leading to the comparison of the very basic structures of one species and another. We have witnessed one example of this in the comparison of the DNA of humans and of chimpanzees but it has gone beyond that. The first steps have been made to identify the genetic basis of disease with the aim of eradicating it. It is a simple step to give humans new attributes that have not arrived naturally by Darwinian methods — or has it? Could it be that the next step to the intelligence sapiens acquired from ordinary evolution would be the discovery of something of the nature of the Universe itself? One of the central features of evolution is the development of more complex species with more advanced features. It is interesting that the noticeable development of modern humans over the last 100,000 y has been towards a physically much weaker creature with smaller teeth but with ever developing ideas and abilities. Presumably this could be the pattern provided the changes can be accommodated by the facilities offered by the Earth. The ability to enter space is perhaps the most important new initiative. Humans can now live in the neighbourhood of the Earth (witness the International Space Station) and can send robotic vehicles throughout the Solar System. These may be the beginnings of a move to populate the Solar System itself.

The present developments are hardly sustainable if an equilibrium is to be established. The rate of growth of the population is staggering. In 1990 the world population was about 4.8 thousand million. By 2008 this number has been raised to 6.5 thousand million, an increase of 35% in the intervening 18 y or on average of 1.9% per year. This is a rate of doubling the population every 36 y or half the life expectancy of an individual member. This is a very rapid rate of increase. It is not only the people involved. They require various facilities such as food and heat and this means a corresponding increase of pets and farmyard animals. All living creatures emit methane which is a greenhouse gas. One consequence will be a rise in the local surface temperature. Too great a rise will put humans at risk because there are upper and lower limits to the temperature that they can stand. There will also be limits to the facilities that can be provided by the Earth itself. All these make for a difficult future to predict. Add to this the individual feeling and consciences of the members to provide a volatile background. It is true that there are species that behave in a collective way such as bees and ants but these differ in that the individual members of such colonies have not got free will. What indications there are for humans tend to see the individual free will there as destructive of the community rather than constructive. Clearly the future for humans is far from clear. The species will need to be fit in order to survive. But what the description "fitness" will need to mean is another story. It is hard to accept that Modern Humans are purely computerised machines because they have an inventive brain and a vision outside simple logic. Hopefully this will introduce new possibilities in the future.

5.6. Summary

The Hominid grouping has developed over several million years. A number of evolutionary invariants have been established and there may well be more yet to be found. It is clear that MH did not appear on the Earth as a "finished product" but underwent the same evolutionary development as all the other creatures.

Table 5.1. A list of known species of *Homo* together with their dates of living, the places where the fossils have been found, how many found and the volume of the brain. These values are not of final accuracy so the dates may not be entirely reliable. New members may be added in the future. The findings have covered a wide spread of Afro–Asia. NonHomo finds have not been included.

Species Homo	Approx. dates (My)	Place Found	Brain volume $(\text{cm})^3$	No. of Fossils
habilis	2.4–1.5	Africa	660	Large number
erectus	1.8–0.03	Africa, Java, China, Caucasus	850–1,100 (late)	Large number
rudolfensis	1.9–?	Kenya	?	One skull
georgicus	1.8–?	Rep. Georgia	600	Very few
ergaster	1.9–1.4	E. & S. Africa	700–850	Many
antecessor	1.0–0.6	Italy	1,000	Two sites
cepranensis	0.9–?	Africa, Europe, China	1,000	One skull cap
heidelbergensis	0.6–0.25	Africa, Europe, China	1,100–1,400	Large number
neanderthalensis	0.35–0.03	Europe, W. Asia	1,200–1,700	Large number
rhodesiensis	0.3–0.12	Zambia	1,300	Few
sapiens sapiens (EMH)	0.2–now (0.04–0.01)	worldwide (Europe?)	1,000–1,850	Everywhere
sapiens idaltu	0.16–0.15	Ethiopia	1,450	Three craniums
floresiensis	0.1–0.012	Indonesia	≈ 400	Seven individuals

More evolutionary development is surely to come in the future. What will future palaeo-searchers conclude about us?

Note

On 19 May 2009 the announcement was made in PLoS ONE of the preliminary analyses, by an international team of experts led by Dr. Jøra Harum, of a fossil lemur-like transitional form between *Strepsirrhini* and *Haplorrhini* and called *Darwinius masillae* or IDA for short. It was claimed that this is an example of a "missing link" involving a distant ancestor of humans. The fossil is complete and in an unbelievably fine state of preservation. It is judged to be 47 million years old (mid-Eocene) as measured by the $\frac{^{40}\text{Ar}}{^{39}\text{Ar}}$ dating methods. It is of a young female some 9 months old. It is the presence of the ankle bone *talus* that links the fossil specifically with humans. The furry creature has a body length of 24 cm and a tail 34 cm long. It had teeth (not a toothcomb) and nails hand and foot (rather than a grooming claw) which is typical of primates. It lived on fruit and leaves in the rainforest around Messel in Germany and was found in the Messel pit. The fossil is undergoing very comprehensive examination still because any links found between primates and other creatures are much sought after. Indeed, the lack of such evidence was the cause of considerable anxiety to Darwin in claiming an evolutionary history for *Homo sapiens* in common with all other creatures. More is sure to be heard of this work although there are sceptics who challenge the importance of *Darwinius* as a transitional form.

Further Reading

1. *Man's Place in Evolution*, Natural History Museum Publications, London.
2. Johanson, D. and Edgar, B., 1996, *From Lucy to Langauge*, Simon & Schuster Editions, New York.
3. To realise that *Homo sapiens* suffers the same genetic problems as other creatures see: Leroi, A. M., *Mutants: On the Form, Varieties and Errors of the Human Body*.
4. An interesting account especially of the origins of language is in

   Maynard Smith, J. and Szathmáry, E., 1999, *Origins of Life*, Oxford University Press.

5. Language is the unique characteristic of *Homo sapiens*. An account of the methods for discovering the multitude of ancient languages is given in

   Cleator, P. E., 1973 (reprinted), *Lost Languages*, Robert Hale & Co., London.

   Particularly significant contributions to the understanding of language have been made by Noam Chomsky. Two of many references are

   Chomsky, N., 1972, *Studies in the Theory of Generative Grammar*, Mouton, Paris; 1975, *Reflections on Language*, Pantheon Books, New York.

6. A very useful reference is

   Leakey, R. E., 1981, *The Making of Mankind*, M. Joseph, Ltd., London,

   which provides a very good summary of the way the discoveries were made. Although lacking any reference to the very recent work, it is written by a member of the family which has been a central player in the search for evidence of *Homo sapiens* in Africa. Quite exceptional illustrations accompany the detailed account.

7. Our interest in this Chapter has been with *Homo sapiens* but the evolution of other animals has also gone ahead. A reference to these activities is given in *British Caenozoic Fossils*, 1975, The Natural History Museum, London.

Chapter 6

A Universe of Exo-Life?

In the previous chapters and in the appendices, we have explored various
aspects of the Universe that can be expected to have at least some relevance
to the presence of life in the Solar System and beyond. We have established
that the Universe forms a single entity and have found that the various
aspects of it, animate as well as inanimate, have resulted from a single vast
evolutionary process. Now we are going to put all these pieces together and
see what conclusions we can draw about the likelihood of exo-life actually
existing outside the Solar System and what form it might possibly take.
Repetitions of previous conclusions will be accepted to allow the arguments
to follow uninterrupted.

6.1. Preliminary Information

There is every indication that the Universe is in a phase which began about
13.7×10^9 ya. It seems that it all started with what can be called an explosion
which expanded to form what we find today. It is one possible logical extension of the Hubble expansion projected to the past. This is the theory of
the hot Big Bang. It is still just a theory, but there is no serious alternative
save the refinement by Hoyle of the earlier Steady State theory of Bondi,
Hoyle and Gold. This theory does not involve an origin to the Universe, but
involves phases from an infinite past. The big bang theory does not explain
everything, however, and may need considerable modification in the future as
our observations grow, but that is how it seems at the present time. What
came before that is quite unknown, but presumably if the Universe is to
be "born" something must have been there and some mechanism must have
operated to start it off. Model universes can accept a dynamical state before
what we know, but cannot give any indication of the physical condition of

matter then.[1] Any indication of what this might have been might ultimately be given by observations, but it was a long time ago and the chances of success are not great. Certainly special features would need to be involved such as gravitational waves although they have not yet been detected and so proved to be more than a theoretical conjecture.

It is known that the present galaxies are moving apart at a speed proportional to their distance apart. Everywhere space is expanding, but there is no centre for the expansion. Wherever the observer is in the Universe, the expansion would be measured according to same rules as it would be everywhere else. This extraordinary fact, discovered by Edwin Hubble is the late 1920s, shows that the line of sight velocity, v, of a distant galaxy is proportional to its distance, D, from us. We summarise the data leading to this conclusion by writing $v = H_0 D$ where H_0 is a constant (the Hubble constant) whose value is to be found from observation. Theory subsequently showed that H_0 is really a parameter which changes with the epoch and is not actually a constant, but it can be taken to be a constant now. It has the dimensions of inverse time. The time scale for the expansion at any epoch is generally expressed by the inverse of the Hubble constant, $\tau \approx \frac{1}{H_0}$. Obtaining a reliable numerical value of H_0 has proved very difficult but, using observations from modern equipment, a reliable value seems ultimately to have been found. In astronomical units, this value is $H_0 = 72$ km per megaparsec. A megaparsec is a measure of distance: 1 megaparsec $= 3.0856 \times 10^{22}$ m. This gives H_0 the value $\frac{7.2 \times 10^4}{3.0856 \times 10^{22}} = 2.33 \times 10^{-18}\,\text{s}^{-1}$. The inverse of this is the time $\tau = 4.29 \times 10^{17}$ s which is taken to be the length of time the present epoch has existed — it actually all began this number of seconds ago. There are 3.15×10^7 s in a year so that in terms of years we can write $\tau = 1.36 \times 10^{10}$ y. This is very close to the age of the present epoch of the Universe given by more precise calculations. There is no evidence for material entities that have entered or left the Universe during this time so the initial state of the Universe contained everything necessary for the future developments. Put another way, everything that we see around us has developed during this period of time including life itself and all of this from the prescription set at the very beginning of the epoch. Present theories predict that the earliest chemical elements were hydrogen with about 23% helium together with a very small proportion of the very light elements, especially lithium and some beryllium. There were no heavier elements present. These

---

[1]It is not impossible that dark matter pre-dates the present Universe, but we cannot tell at the present time.

are essential for the formation of planetary bodies and also for the evolution of animate materials.

6.2.   A Stellar Time Scale

It appears that the very first stars must have been composed entirely of hydrogen and helium with zero metallicity. These are Population III stars although none have yet been observed. Their precise size is held within two tight constraints. One is a lower mass: the central temperature must be high enough to allow the proton – proton cycle to function (see Appendix 2). The other is an upper mass. The temperature of the condensing mass will rise prodigiously and so will suffer internal radiation, but this must not be so great as to prevent the gas cloud condensing. The theoretical result is a mass less than 100 solar masses, but greater than 10 solar masses. Such high masses will be associated with a short life time, probably no greater than 10^6 y. The death of these stars will have produced heavier elements up to the mass of iron which will have spread into the surrounding space to mix with interstellar gas and dust. This will condense to form the next generation of stars (Population II stars) which consequently will have a small though nonzero metallicity. The death of one generation of stars heralds the birth of another. The small amount of heavier elements in the surrounding gas cloud will not be very auspicious for the production of planets. It is the death of the population II stars that will project the full range of chemical elements into the surrounding space and so provide the materials for the next generation of stars, the Population I stars. The Sun is a population I star with about 2.6% of heavier elements. It will be realised that the proportion of heavier elements, the metallicity, of a given star can be used as a measure of its age. Theory suggests that the generation of stars beyond that of the Sun has is yet to be born although stars are known with a higher metallicity than the Sun. There is a great research effort to observe Population III stars. Their discovery will validate the theoretical conclusions that we just set down. The search will have to involve the very distant stars with very large red shift which will be very difficult to study. Perhaps the next generation of telescopes, available beyond 2012, will be powerful enough to do this.

The time scale for these things is of interest. Population III stars have an insignificant life. Population II stars can be expected to have about a solar mass and so about the corresponding life time which is about 10^{10} y. Our Sun is about 5×10^9 y old giving the present age since the beginning of our epoch of about 1.5×10^{10} y. This is surprisingly close to the age obtained from the Hubble constant mentioned earlier in this chapter. There was the beginnings of the heavier chemical elements during the early stages of the Population I era which would put the time at perhaps 10^{10} y after the beginning of the present era or perhaps 5×10^9 ya. There will not have been a clear division between the appearance and demise of the different star classes, but this timing will probably be not too far wrong. This would suggest, if the argument is correct, that the present grouping of solar type stars is the first to have full planetary companions and so the first able to harbour any form of advanced life. Interestingly,elementary life cannot trace its history back much further. It must have required trace amounts of the heavier chemical elements, such as oxygen, nitrogen, phosphorus, iron and others and in sufficient concentration and temperature, to allow the required chemical reactions to take place. Certainly life cannot be older than the first appearance of these various chemical components. Even though very elementary life could have lived in space without a firm planetary platform to live on its age cannot be much greater than the materials that can provide such a base. These arguments put the appearance of life as a later event in the development of the Universe rather than an earlier one.

6.3.   Abiogenisis — How Did Life Form?

If the Universe has got a finite age then life must have had an origin at a finite time in the past. There might have been just one formation in space with continual reproduction over the development of the Universe to provide the most elementary life (microbial) as an integral feature. Alternatively, the precursors of life might collect together at each location and form if the conditions were right. There has been interest in the beginnings of life on Earth from the earliest times. Aristotle (384–322 BC) taught that decaying organic matter generated life (though he did not explain where the organic matter first came from). It was not until the latter half of the 17th century that the Dutch scientist Anton van Leeuwenhoek (1632–1723) first discovered minute organisms (for instance in water) and showed the existence of a vast subworld that cannot be seen by eyes alone. Louis Pasteur (1822–1895) showed conclusively that bacteria and fungi cannot appear of their own accord.[2]

It was probably Charles Darwin who stated the modern view that life probably started from some form of organic "soup". He pointed out that such a soup would be unlikely to exist now because the initial conditions

---

[2]This led to the conviction that life must start from an "egg" — omne vivum ex ovo.

depended on the absence of pre-existing life. This means that any investigations now must be conducted under strictly controlled conditions in the laboratory. This approach was developed particularly by the Russian Ivanovich Oparin (1894–1980) who maintained that life started in an atmosphere similar to that of the planet Jupiter and mainly composed of hydrogen, methane and ammonia in the absence of oxygen. This is thought to be the composition of the very first primitive atmosphere which was quickly lost. The implication is that life could not begin now and that it would have to be a once for all process on Earth. Oparin wrote a very influential book entitled *Origin of Life* which was the starting point for later experiments by others. This approach was endorsed by J. B. S. Haldane (1892–1964) a British biometrist. We can mention here especially the experimental work of the American Stanley Miller (1930–2007) who showed that, starting with such components as Oparin's hydrogen, methane and ammonia processed in a laboratory flask, it is possible to produce amino acids as a product. It became clear that the basic bio-molecules of life can be formed as the result of elementary physical processes that could have taken place on Earth. It also became clear that a vital component for the formation of life is the presence of spherical droplets of assorted fat soluable molecules (coacervates). Although Miller did not produce the actual components of DNA he had shown that it is possible in principle for abiogenesis to have taken place on an early Earth.

The recognition that life could have begun on Earth leads to the question whether it might also have formed elsewhere in the Solar System. Mars and Venus are the two obvious places before conditions became what they are now. Any evidence that there might have been on the surface of Venus has probably been lost forever due to the most hostile conditions there, but any development of life on Mars might have left traces which are still discernable. The principal satellites of the major planets were shown by Cole (1928) not to have been frozen early on in their histories so there is the possibility of life having started there. Although the surfaces are icy now there will be internal heating to allow liquid water in their interior regions and life could have formed there in a fashion reminiscent of that which surrounds the smokers on Earth. Firmer conclusions could follow if it found that life has also arisen elsewhere in the Solar System. On reflection, it may need only one initial life source to fill a system of planets eventually and this would argue against the likelihood of finding life elsewhere in our System. Nevertheless, the emergence of technologically capable life (TCL) on Earth gives every confidence that it could also have arisen at other sites in the Universe and representatives may be active now. Past supernova explosions could well have destroyed living colonies which arose in the past.

To investigate the evolutionary development further it is necessary to distinguish between elementary (microbial) life, advanced life and TCL. Put another way, it is necessary to specify how far the evolutionary process has proceeded along the complexity axis. The first step will be simple microbial life presumable of the type that pre-dated that associated with stromalites or the bacterial accompaniments of underwater smokers. Why it might have stopped in many cases is clear from the history of life on Earth — environmental factors provided a primary control. These can be expected to be equally severe throughout the Universe although the toughness of microbial life might suggest that this type of living organism is very common throughout the Universe.

6.4.   Where Did Life Form — In situ or Panspermia?

Panspermia asserts that space is full of elementary living spores that reign down on any surface and can develop if the environment of that surface is friendly. In situ asserts that elementary life forms spontaneously on each body separately. A modification is that life from one body is placed on another through a transfer for instance embedded in meteorite material.

We have considered the in situ possibilities in Sec. 6.3. There is some direct support for an in situ origin for the Earth. This could have occurred any time between 4.4×10^9 ya (when liquid water will have been able to lie on the surface) and 2.7×10^9 ya when there is clear evidence of life from the isotopic ratios of the chemical elements carbon and iron (that is $\frac{^{12}C}{^{13}C}$, $\frac{^{56}Fe}{^{57}Fe}$ and $\frac{^{56}Fe}{^{58}Fe}$) and sulphur (through the isotopes ^{32}S, ^{33}S, ^{34}S and ^{35}S). These elements appear in sediments and minerals are generally accepted as relating to a presence of the remains of living materials. There are, however, also indications of an excess of the isotope ^{13}C in the earliest rocks supporting the presence of living materials at the earliest time on Earth. Collections of what appear to be elementary bacteria have been found in rocks dated some 3.5 By old (see Fig. 2.2). Living spores deep inside a rock would be well able to withstand the heat of entry through an atmosphere. There would, of course, be a single infection with living material at one site where as in situ the appearance would be expected to have occurred at many sites simultaneously to be effective.

Whilst assumption of an in situ origin has got very old roots, the concept of panspermia also dates back a long way. Anaxagorus (c. 599–428 BC) spoke of the possibility and there may have been others before him.[3] However, the idea lay essentially dormant until the 19th century when it was revived by such notable people as Jöns Berzelius (1779–1848) and Lord Kelvin (1824–1907). Hermann von Helmholtz (1821–1894) and Svante Arrhenius (1859–1927) also supported the concept. It was championed in the 20th century by a number of people, but particularly by Fred Hoyle and Chandra Wickramasinghe (1939) on the basis of the study of infra-red radiation from space.

In the Solar System, there are many bodies that could have received the injection of life. This will obviously be one test — if life also started on another planet other than the Earth. There is yet no evidence that that was so, but this makes the search for the beginnings of life on other planets and satellites all the more important. The only body other than Earth that has been tested so far is the Moon and that has not proved positive — a mark against panspermia. Mars is currently being tested in a rudimentary way, but it may require human examination in situ to make a thorough search. There is hope in the hearts of some observers of finding elementary life under the surface ice of satellites of the major planets (and especially of Europa, Enceladus and Triton), but there is no real possibility of making explorations there in the more immediate future. It is known that small samples of rock have been thrown explosively from both Moon and Mars to land on Earth. Since such interplanetary linkages are known there is no reason why spores should not have been transferred in the rock samples protected from the interplanetary cosmic particles. Indeed, one example of microbial life from Mars was alleged to have been found in a meteorite shot from the planet, but this is most likely to have been only apparent and to have arisen from the spraying of gold film on an inorganic specimen in preparation for viewing in an electron microscope.

Microbial matter is certainly most resilient to its environment. Extremophiles are found over the full range of conditions from extreme heat (in volcanoes and in ocean smokers under extreme pressure) to extreme cold (deep under the Antarctic ice); from extreme drought to extreme wetness; from rich oxygen to none (endolithic bacteria using chemosynthesis have been found in rocks and subterranean lakes); and under extreme pressure to vacuum where there is none at all. Bacteria can lie dormant for large periods of time perhaps many millions of years. Dormant bacteria have been isolated and restored from insects after tens of millions of years. One classic example

---

[3]A healthy characteristic of Greek thinking was the appearance at one and the same time of thinkers proposing both an idea and its antithesis. This led to debates which, alas, were not often concluded.

came from the unmanned lunar lander Surveyor 3 set down near Ocean Pro-
cellarum in April 1967 as part of the American Apollo Moon Program. Its
onboard television camera was recovered by Apollo 12 astronauts in November 1969. Examination in the laboratory back on Earth showed that a small
quantity of specimens of *Streptococcus mitis* had, inadvertently, been allowed
to enter the camera before the flight — presumably one of the engineers had
a cold then. It was found that they had survived after 31 months on the
Moon's surface, through the cold vacuum there and the considerable radiation level that will have bombarded them during that time. This is a small
measure of the resilience of bacteria.

One feature of panspermia is the time factor. Microbes will be entering
the planetary bodies continually and not just at the beginning of history.
This can have very important consequences. It could, for example, explain
why a particular chirality is associated with animate matter if the source
of all is common in space. Presumably the continual arrival of alien micro-
matter would do no more than supplement the stock already present. The
idea can be developed further. For instance, Hoyle and Wickramasinghe
have even suggested that bacterial infections on Earth now can be related to
the arrival of alien bacteria from space. Wickramasinghe has claimed to have
detected organically important molecules high in the Earth's atmosphere.
There is no physical evidence that panspermia has taken place, but the idea
does stress a novel feature should it actually be true. A continual renewing
in the Universe has been an attractive feature of Hoyle's work. It must be
admitted, however, that the formation of bacterial matter in space simply
shifts the location of its origin from Earth into space says nothing about the
actual formation process of living material itself.

6.5.   Where Can Life Live?

This depends very much upon which kind of life it is. Microbial life can
be expected to be ubiquitous. We know it can live almost anywhere not
requiring a firm base and exist under almost any conditions. If it can form
in space, it can live there. The Universe could well be alive with life, but
an observer would need to take a powerful microscope to know it was there.
Adverse environmental pressures would be likely to prohibit any significant
evolutionary development. Microbial life would be in equilibrium with the
Universe, but would have no discernable influence on it.

More advanced life would be more demanding. It would need a solid
planetary surface holding a permanent water sea complex with flora and
fauna appropriate to its level of development. Technologically advanced life

would require more. Apart from appropriate flora and fauna for living, it would require the presence of mineral resources for power and material for construction. This would mean a planet of broadly Earth composition orbiting in the zone around the central star where liquid water can lie freely on the surface, the so-called habitable zone. An important feature for TCL is local environmental stability. There are many aspects of this. Some are concerned with the planet itself, like its volcanic structure and the movement of its continents but others are more astronomical. There are two very basic requirements that can be isolated immediately.

One requirement is approximately constant, equitable conditions for a sufficient period of time for TCL to evolve and prosper. There are two requirements here. One is a central star of mass such that on the one hand it offers a haven to life of at least 10^{10} y (assuming the experience on Earth is not untypical) and on the other hand which does not emit too high a proportion of high energy photons to attack their bodies. The lifetime and the limited surface temperature are fully met by a solar type star. Because all stars of any mass eventually die the life associated with them has a finite life (though it may be measured in units of 10^9 y) other stars could offer a home to life forms though they may not be TCL. The fact that the star has a finite life span means that any life associated with it will have at most that time span and probably much less.

Once the requirement of a broadly solar type star is met other requirements can be entertained. A requirement to provide a broadly equitable climate throughout the years is a closely circular planetary orbit around the central star. This is unlikely to be exactly circular so we can relax our requirement to an elliptical orbit with a very small eccentricity, e, say not greater than 0.01 ($e < 0.01$) taking the case of the Earth as an example. To ensure the mechanical stability of the total system other planets in the system would need to have similar eccentricities. Seasons will be provided by the obliquity of the planet which is the angle made by its rotation axis to the plane of its orbit. If this is greater than about 35° the seasons become grossly different and can be reduced from 4 to 2 with extreme differences between night and day. An extreme example is Uranus where the obliquity is 94° giving half the "year" light and half darkness.

It would be necessary that our planet carrying advanced life is not alone in orbiting the star. Other planets? — yes. It is necessary to have some form of protective outer barrier to reduce the incidence of impacts on the planet from outside. This could be offered by the presence of planets of much greater mass with greater orbits, rather like the major planets of the Solar System. It is often supposed that Jupiter offers this protection alone, but it is more likely that the major planets accomplish this as a unit (a little like the stops on a pin-ball machine). As a quartet they offer a gravitational maze, some 25 AU deep, attracting and diverting bodies entering the Solar System from the outside to reduce significantly the impacts on Earth. The filter is not perfect but it can reduce the impact of a major body on an inner planet very substantially. For the Earth as an example observation suggests there is something of the order one major impact per hundred million years, but this figure is very uncertain.[4]

It is important that the planet's rotation axes remain stable. A lone planet is susceptible to instabilities (witness Venus which has turned "upside down" in its orbit). This represents a change of angular momentum. One way to ensure stability here is to have another body (a moon) orbiting the planet as the Moon orbits the Earth. Neither Venus nor Uranus has got a significant satellite. The satellite will also induce tides in the oceans of the planet through the mutual gravitational interaction between the planet and the satellite. It is sometimes conjectured that the motion of the tides themselves were important in providing a gentle stirring motion within the oceans which was important for the development of life on Earth.

6.6. Suitable Exo-planetary Systems

A central question now is which observed exo-planetary systems are likely to be suitable for the development of life as we know it on Earth and particularly can allow advanced life to develop. Modern telescopic technology is not yet able to probe the environments of distant stars. The essential problem is the enormous ratio between the light emitted by the star and that emitted by the planet. There are screening techniques to blot out the radiation from the central star, developed for observing the environments of our Sun, and preliminary applications have been made (see Fig. 1.7). Unfortunately smaller planets cannot be detected although but there are plans for much enhanced telescopes in the future, such as the European Extra Large Telescope and various American orbiters, but this is for the future. Observations of transits is a real possibility as explained in Chap. 1 but the chances of a planet orbit on a distant star lying in the plane of the line of site of the star from the Earth are not very great though not, of course, zero. In spite

---

[4]It is difficult to assess how modern humans would cope with a situation which confronted life 65 Mya with both a major impact event and a very major volcanic lava flash flood. Life is very dependent on the acts of nature. A huge extinction would surely follow, but it is interesting to conjecture what the new regime would be.

of these problems, there are some observations that can be significant and allow an assessment of observations that can hope to lead to realistic results.

It was seen in Chap. 1 that there are several star — planets systems that could possibly fit the requirements set down in Sec. 1.5. We shall interpret the requirements as: (i) a solar type star; (ii) closely circular obits for the planets; (iii) at least one Jupiter mass planet in orbit with a semi-major axis of not less than 2 AU; (iv) a silicate planet of essentially Earth mass with an interior orbit. This last requirement cannot be confirmed observationally at the present time, but such a body will be assumed present if the other criteria are satisfied. These characteristics of the Solar System seem not to be shared generally with the exo-planetary systems found so far. This is clear from Figs. 1.1–1.4. It is also realised that stars with higher metallicity are more likely to have planetary companions than those of lower metallicity. No star system discovered so far fully satisfies the set of requirements which we listed above, but there are four that do to some sort approximation. These four are the first entries in Table 6.1. They were all found by orbital velocity (Doppler) measurements which restrict the planetary mass that can be detected at the present time. Each has got a large planet at some 2 AU from the star (apparently none further away), but Earth sized planets are at present too small to detect by this method. The fifth entry in Table 10.1 is a recent discovery using micro-lensing techniques. It is rather the other way round as far as planets go — it has a planet of about 3.3 Earth masses at a reasonable distance from the star but no major planet beyond.

Table 6.1. Five stellar systems with orbiting planets each having a large semi-major axis and a small eccentricity: and last is a system with a planet of broadly Earth mass, but with no accompanying major planet ($1 \, \text{pc} = 2.06 \times 10^5 \, \text{AU} : 1 \, \text{ly} = 0.307 \, \text{pc}$).

System	Stellar Distance		Stellar mass (solar mass)	Minimum planet mass (M_J)	Semi-major axis (AU)	Orbital eccentricity
	(pc)	ly				
47 Ursa Major	14.8	48.2	1.03	2.41	2.10	0.096
				0.76	3.73	< 0.1
HD 10697	30.0	97.8	1.10	6.12	2.13	0.11
v Andromeda	13.47	43.9	1.3	0.69	0.059	0.012
				1.19	0.829	0.28
				3.75	2.53	0.27
HD 70642	29	94.54	1.0	2.0	3.3	0.1
HD 117207	33	107.6	1.04	2.06	3.78	0.16
MOA – 2007 – BLG-192-Lb	1000 ± 400	$3300 \pm$	0.06	5×10^{-3}	0.62	

Table 6.2. The metallicity and age of the stars systems of interest together with their estimated ages. Gliese 876 is a red dwarf star. An METI message has been sent to Gliesse 777. The ages straddle that of the Sun at 5×10^9 y.

Star	Metallicity	Age (10^9 y)	Star	Metallicity	Age (10^9 y)
47 Urs. Maj.	0.0 ± 0.07	6.03	HD 117207	0.27	6.68
HD 10697	0.10 ± 0.06	6.9	55 Cuc	0.29	4.5
υ Andro.	0.09 ± 0.06	3.3	Gliese 777	1.38	6.70
HD 70642	0.16 ± 0.02	3.8	Gliese 876	-0.12 ± 0.12	—

Two things can be concluded from Tables 6.1 and 6.2. One is that the exo-planetary systems are a long way away — the nearest is about 40 light years (ly) away. The second is that the metallicity is not less than that of our Sun. Linked with this is the comment that the ages of the systems of especial interest are rather greater than that of the Solar System. None of the ages are significantly less.

6.7. Isolation of the Systems

Large distances are a recognised feature of the Universe. The nearest star to the Solar System (Proxima Centauri) is some 4 ly away and this is right next door by astronomical standards. The distances shown in Table 6.1 are altogether larger, the closest being 10 times the distance to the nearest star. This separation raises serious problems of communication. An electromagnetic signal sent to 47 Ursae Majoris today, for example, would take a little more that 48 y to reach there and the same time to be returned. Ninety-six years for the total return journey is more than twice the working life of the person on Earth sending the message: it is greater than the life expectancy. Other planetary systems would require even more communication time. One is Gliese 777 which is 52 ly way. This is a star very much like the Sun in terms of mass and size with only a slightly greater luminosity which would be expected because it is a little more than 20% older than the Sun. A radio message was sent to it by METI (Messaging to Extraterrestrial Intelligence) on 1 July 1999 which will arrive 2051 April. If there is an immediate reply it will be received on Earth in 2103. The message and its answer will have taken a human working lifetime to happen. The case of HD 117207 would be worse. This system is 198.9 ly distant so an immediate return signal would take nearly 400 y to take place — the time between the Renaissance in the West and now. We must realise that such time scales are independent of technological progress: more information might ultimately be sent but, according to the Special Theory of Relativity, the speed of light in a vacuum (and space is a good vacuum) provides a strict upper limit to the speed of communication. Such long time scales for communications present organisational and financial implications in providing continuity through some form of controlling institute. This is well understood on Earth by the Institute for the Search for Extraterrestrial Intelligence (SETI). Adequate finance in the extrasolar "civilisations" is another aspect associated with contact with them: there will need to be a long-term infrastructure with financial controls presumably involving something equivalent to a stock exchange. We can conclude that any centres of TCL (let us call them civilisations) will be very far apart by our standards even using all electronic aids.

It is possible to define some sort of mean distance between possible sites for civilisations. For this purpose, let us suppose that the substance of Table 6.1 is an indication of a general feature in the Galaxy. Let us assume that each civilisation lies within a "cell" of volume v and that the size of each of the cells is the same. If the volume of the galaxy is V then the number of cells in the galaxy is approximately $n = \frac{v}{V}$. If every cell contains a civilisation then n is the maximum number of civilisations that exist. If proportionately fewer cells are populated, the number is less.

Let us take a very rough example. Suppose that the cell is spherical with a radius of 50 ly, so that $v = \frac{4\pi}{3}50^3 \approx 5 \times 10^5 \, \text{ly}^3$. It may, in fact, be larger but take this for the moment. The volume of a typical galaxy (if there be such a thing) could be of the order $V \approx 10^{15} \, \text{ly}^3$. The number of cells in such a galaxy will be of the order $n = \frac{10^{15}}{5 \times 10^5} = 2 \times 10^9$ cells. The number of independent cells will cover the lifetime of the Galaxy so the number of civilisations existing at any time will be fewer. The possible range in numbers would seem to be $1 \leq n \leq 2 \times 10^9$, a very wide range indeed. These sites will not all be occupied at once and the number that are must depend on the average length any individual civilisation lasts and the length the galaxy itself will last with active stellar colonies within it. If T is the active time period of a galaxy and τ is the life time of a technological civilisation the number associated with a galaxy is $N = \frac{T}{\tau}$. These will be spread over its life time so there will be periods when there are few civilisations but most periods there will be none. Those that are there will be widely spaced and essentially isolated. The essential thing is to specify T and τ but we have no clear criteria to do this. For T we might suppose 10^{11} y but it might be more. How long can TLC last? It is usually presumed to be about 10,000 y. Then, $N \approx 10^7$. Not all the possible cells will be occupied even if all the life forms appear at once.

There are considerable variations in the arguments to specify τ. It is worth remembering that on Earth this time interval can be encompassed between the years of about 1920 when radio began to be grown up and today which is a period of less than 100 y. How much longer these conditions will last is an open question — 10,000 y covers the whole time span between the first modern hominid and now. This is certainly substantially longer than individual civilisations have lasted in the past, but the discoveries of one has been past on to those that follow (for instance, we today are still able to study with relevance what the ancient Greek and Roman philosophers had to say).

Alternatively, we can notice that if stars of essentially similar age can be expected to have a similar spread of life and whether evolution develops or not is an equal probability at every step we can ask how many stars we will need to visit in order to have a unit probability of finding TCL. At each evolutionary step the next step might or might not proceed: let us suppose this choice is made with equal probability. If there are n_s essential steps to any particular level of evolution the number, N_e, of places to look before this level will be found is of order $N_e \approx (\frac{1}{2})^{n_s}$. To find microbial life on his bases one must visit two stars. For TCL, there are certainly 30 essential steps giving $N_e \approx 10^9$ systems. This would suggest a single system per galaxy for TCL although microbial life would appear on every other system. Of course, the number of TCLs would be increased or retarded if the advance of evolution was not dictated by the toss of a properly balanced coin, but weighted in favour of further advance or less. Our very simple argument, nevertheless, suggests the number at any time will be very small indeed. The separations, if random, would be very large indeed running into thousands of light years.

6.8.   The Fermi Paradox and Drake Equation

Calculation such as we have just made are, of course, not new. In the 1950s Enrico Fermi, the Italian Nobel Prize winning physicist, asked what conclusions we can draw from the nonappearance of extraterrestrials. If the Galaxy is full of them then why have we not become aware of them? This could be by direct meetings, but more likely from encounters with their automatic space probes or from space radio messages between them. There has been nothing like that reported[5] — Fermi asked where are they all? There are a number of possible answers. An obvious answer is that "they" are just not

---

[5]UFOs represent an interesting problem. Their existence is denied officially worldwide, but there have been so many alleged sittings that there might seem to be some phenomenon there to understand. It will not be linked to extraterrestrials, however, for reasons that will appear later.

there. Another is that they do not want to talk or show themselves. Another is that they have not reached the stage of being capable of space associations. Another is that, although they have a space capability, they choose not to spend their money that way. One could go on getting ever more involved in hypothetical discussions. In 1961, Frank Drake offered a way of limiting the uncertainties and so to quantify the discussion by constructing a mathematical equation which includes the factors expected to determine the number of civilisations that there could be in our Galaxy. This is generally known as the Drake equation.[6]

Drake listed seven factors that must go into the equation. These are

— average rate of star formation in our Galaxy, R;
— fraction of stars which have orbiting planets, f_p;
— average number of planets that could sustain n life, per star , n_l;
— fraction of star systems that actually support life at some time, f_{li};
— number that actually develop intelligent life, n_i;
— fraction that develop a space capability and use it, f_t; and
— average length of time that such a civilisation will actually send messages into space, τ.

The equation is formed by multiplying all these quantities together to give the number of civilisations in our Galaxy for which communication should be possible in principle, denoted by N

$$N = R \times f_p \times n_l \times f_{li} \times n_i \times f_t \times \tau.$$

This is the Drake equation. He made estimates of the various quantities as follows:

$$R = 10/\text{year}; \quad f_p = 0.5; \quad n_l = 2; \quad f_{li} = 1$$
$$n_i = 10^{-2}; \quad f_t = 10^{-2}; \quad \tau = 10{,}000\,\text{y}.$$

This assumes that 10 stars are formed on the average every year in the Galaxy of which one half have orbiting planets. There are two planets in each system capable of supporting life and each has life on it. It is supposed that 1% of these develop intelligent life and of this 1% again develop space technologies. This situation is assumed to last for a period of 10,000 y. These numbers seemed not unreasonable at the time and putting them into the Drake equation gives $N = 10$, a surprisingly small number, but fully in keeping with the numbers that are suggested by the different arguments of Sec. 6.4. The numbers Drake used would perhaps be modified now. For instance, the

---

[6]The equation was presented to a meeting of scientists at Green Bank, West Virginia and so is sometimes known alternatively as the Green Bank equation.

NASA estimate of the number of stars formed per year is 7 rather than 10. The fraction stars with appropriate planets is now likely to be put at most 0.3 instead of 0.5. The number of planets capable of sustaining TCL is now thought to be less than 0.01. The fraction of planets that support life at some point is placed at 0.2 at most. We can take the other values as still valid. Even these more modern numbers are open to objections and later corrections, but give some indication of the modern view. The result of these changes is broadly to reduce the estimated number of TCLs to a number closer to 1.

It can be objected that estimates such as these are based on our belief that any other civilisations must be mirrored generally speaking on the Earth experience and this need not be so. This point of view cannot be rejected out of hand. It must be remembered, however, that the Universe is made from a common stock of chemical elements and the compounds formed from them will tend to be same due to the common action of chemical bonding. On Earth nature tends to use the same solution to separate problems of the same type (biological convergence). It is not unreasonable to accept these common solutions as valid wherever the same chemical structures are found which is throughout the Universe. It would be expected, therefore, that life everywhere will make the same demands on its environment as on Earth so that the Earth experience can be supposed valid everywhere. Perhaps even more than this, the forms of life might not be so different making the Earth a unique example of an acceptable habitat. This would be the ultimate biological convergence and is far from an impossibility. Indeed in practice it is tended to be taken as appropriate in selecting suitable possible sites for TCL. It must be remembered that the evolution of life on Earth has been controlled very largely by changes of the environment and especially the climate. This is an entirely unknown factor in even beginning to estimate the occurrence of life elsewhere.

6.9.   Contacting Other Civilisations: SETI and METI

In spite of the pessimism that must be felt following the conclusions of the last two sections, there is still interest in some quarters in contacting other civilisations in space. In many ways, the most immediately operable activity is to shout out as loudly as possible to attract attention and to listen to see if others are doing the same. There is an organisation that has been doing both shouting and listening since 1984.

The Institute for the Search for Extraterrestrial Life, or the SETI Institute, has got the mission to explore, understand and explain the origin, nature and prevalence of life in the Universe. It undertakes a range of activities and employs 150 staff of all types. This organisation attracts many sponsors ranging from NASA, university and other academic centres, business and private donations. The cause is a popular one. The staff is mainly distributed between two centres into which its mission is divided. One is the Center for SETI with Dr Jill Tarter as its current Head. The other is The Carl Sagan Center for the Study of Life with Dr Frank Drake (of Drake equation fame) as its Head. The SETI Center listens for interstellar "traffic" but so far has found none. Everyone realises that the task is one for the long haul and no early results could reasonably be expected.

There has also been established as an outreach activity a so-called Home SETI where private people can devote the free time of their personal computers to a coordinated search of the heavens for communications. There are some 3 million private individuals taking part in the scheme.

The sister task of being an active SETI is MESI which is Messaging Extraterrestrial Life. Here, the concern is actually directing transmissions of hopefully understandable messages to specific target stellar systems. This has, actually, been going on before SETI was established. The first radio transmission was made at the inauguration of the Arecibo Radio Telescope in November 1977. The message was aimed at the globular cluster Messier 13 and is expected to reach its destination sometime during the year 2574. This has been recognised as a long-term project from the very start. To date 16 star systems have been targeted with a message including Gliesse 777 (sent July 1999 with the arrival date 2051), 47 U Maj (twice, first in July 2001 and then in July 2003 with reception dates 2047 and 2049, respectively) and also 55 Cnc (in the constellation Cancer sent July 2003 and to arrive in 2051).

Not everyone is in agreement with the active rôle or indeed the passive one. It has been argued that shouting to everyone that we are here might attract technologically more advanced alien immigrants who will then take over the Earth (this is presumed a bad thing because the invaders may not be benign). Even contact with a benign civilisation may have its problems if the levels of advances in evolution are not compatible. Another worry with communications is the possibility of viruses infecting our communications systems. For the listening rôle, it has been suggested alternatively that the absence of contacts means that other civilisations are listening as well but do not wish to be known much as Japan of the 17th century did not wish to have communications with anyone else. Listening must fail according to this argument. There is a lot in common between these activities and science fiction accounts of imaginary situations. It has been said that these activities are not really science at all which is concerned with observation, hypothesis and testing. This may be so but it does not make the ambition any less praiseworthy.

6.10.   Space Travel

Forgetting the analyses we made in Secs. 6.4 and 6.5 let us try to meet the Fermi question head on. The problems the extraterrestrials would face in getting here will be the same as those faced by us getting to them. We have to accept that such a journey would not be feasible for us at the present time and probably not for a number of decades. This may, indeed, provide the answer to the Fermi question — perhaps the extraterrestrials are not able to undertake serious space travel yet or perhaps they do not think it a worth while effort and far too expensive. Let us explore these possibilities in the light of what would be necessary for such a trip.

When we set out on a journey we have a destination in mind. The first thing is to identify our objective. We then pack for the journey and for our stay at the destination. If we travel by car we have it serviced and take some form of insurance against it breaking down on the journey and so needing repairs. We need not take extra fuel because there will be garages along the way where we can buy petrol or diesel as we need it. Insurance against personal injuries will amount to arrangements for local hospital care or/and being returned home. We might take something to eat on the way, but here there will be shops so only a minimum fare is necessary. We might take sickness tablets if part of our journey is by sea ferry or any tablets and creams that our living may demand. We know there will be hospitals where we are going which will accept emergency cases. We can always check with home via telephone, mobile or e-mail and can expect immediate response. We are not worried by excessive radiation from cosmic and gamma rays. We know the accidental hazards we might meet and that the insect bites will be manageable. We are not going to require special research to deal with such things. We can take just the amount of currency to meet our fringe needs for where we are going, knowing that our credit cards will be valid and there will anyway be local banks when we get there. Everything is simple and straight forward although the actual packing for the journey tends to be a bit frantic at the last minute. We can simplify things still further by travelling by train or aeroplane and hire a car at the other end. This could be more expensive if a number of people are involved although we know that we can meet the bill whatever way we choose. But in any case we can look forward to a happy break and know that we will be home again quite soon.

How familiar all that is but space travel, when we can achieve it, will be utterly different. We might remember that, apart from the International Space Station and various other excursions into Earth orbit, we have only been able to make four or five "weekend" type trips to the Moon so far. There are plans for humans to visit Mars but not for many years yet and only after there has been a substantial activity of exploration of the Martian surface and environment by automated space probes and landers. How far away will a deep space mission be? — perhaps a hundred years?

There are no shops or breakdown services in space. No banks to allow us to share some extraterrestrial currency. No doctors and hospitals for emergencies. Again the journey will not be a short one and the conditions at the other end can only be guessed at with our current technology. Everything needed for the journey and the return must be taken to begin with and every eventually anticipated as far as one can. All this becomes more complicated the longer the journey because a crucial factor then is the longevity of the crew. We saw in Table 6.2 that the type of distance that we must consider is in excess of 50 ly. Even travelling at the speed of light the return trip would last around 100 y plus whatever time is spent of the planet there. This is beyond the expectation of life on Earth and this expectation would be likely to be less in space for various reasons many of which are not yet known in detail because no one has remained in space for this amount of time. The human body evolved for life on Earth with the strength of the gravitation found there. The physiology of the bone structure, of the fluid flows and the general well being has been based on this. When the environment is one of microgravity (as in Earth orbit) or essentially zero gravity (as in deep space) special steps must be taken and continual exercises performed in order to simulate the absence of gravity to at least some extent. This is an area very much under investigation and with growing knowledge as more astronauts spend more and more time in Earth orbit. A full knowledge of these effects will be essential before any move can be made into deep space. A crucial need is an environment with simulated gravity with a geometry that is different from the form of a simple rotation.

The success of any mission must depend on the effectiveness and efficiency of the propulsion units. Certainly a space ship can cruise in space where there are no effective forces without power consumption when it moves in a straight line and at constant speed (this is the Galileo–Newton law of motion) which happily limits severely the times when fuel must be used. Remember every ounce of fuel needed for the entire journey *and the return* must be carried from the start. This will be the power to reach the chosen economical speed for free travel, the deceleration on reaching the end of the journey and the fuel for the corresponding sections of the return. Being realistic, the average speed is not likely to be greater than $\frac{1}{100}$th the speed of light in free space. With such an average speed (about 6×10^{10} miles per hour) a 50 ly journey would take 5,000 y each way — a 10,000 y return journey at least. It would take 500 y at $\frac{1}{10}$th the speed of light making a 1,000 y plus return journey. This would take the craft some 30 h or so to reach the outskirts of the Solar System. As a comparison, the probes Voyager 1 and Voyager 2 have taken some 30 y to achieve the same distance. These journey times to the stars are very great and would involve several generations of people with all that would mean for medicine, hygiene, schooling, recreation ad so on. Bearing in mind that all the facilities would need to be taken (growing food, water, oxygen, space ship maintenance, fuel, medicine, dentistry are some examples of everyday living), it is clear that a number of very large space vehicles would be needed to transport all these things. The explorers now have to organise a fleet of space ships and the people that leave will not be the ones that return. The financial cost will be high, perhaps too high for the returns that might accrue in the future. It is one thing to go on holiday, but it is quite another to explore the Universe. Comparing it with terrestrial explorations in the past is unrealistic. There the risks could be assessed with some degree of certainty before the beginning and financial backing could be obtained on a certain return. One might be forgiven for thinking that the exploration of deep space is too expensive when so much needs to be done at home on Earth. Might any extraterrestrials take the same point of view? They would certainly have the same problems that we would have if they were anywhere near our level of technological development. The extreme distances between possible civilisations means that each will be essentially isolated communities in the Universe. It will be as if no others were there — every civilisation in isolation without immediate contact.

6.11.   General Validity of Evolutionary Processes

The discovery of the overall evolutionary sequence of life on Earth that we discussed in Chaps. 3, 4 and 5 is accounted for by applying two very important theories. One is the theory of evolution by natural selection proposed especially by Charles Darwin in his now famous book and independently supported by the findings of Alfred Russel Wallace (1823–1918). The theory was championed on the Continent, and especially in Germany, by Ernst Haeckel (1834–1919) who also made several inspired but, as it turned out, erroneous guesses about its extension. He was, however, the first to construct evolutionary trees, something Darwin wished for but never properly achieved. All in all the correctness of the theory of evolution by natural selection cannot be doubted and forms a simple and very beautiful description of the observed evolution of life according to the fossil record. The theory is augmented by the independent analyses of Gregor Mendel (1822–1884) on inheritance characteristics which form the basis of the modern science of genetics. In spite of its many important successes Darwin's theory has, sadly, its deficiencies. In particular it describes what has happened, that is in hindsight, but does not allow for predictions of future developments. Prediction is a vital part of modern theoretical work. The theory gives no indication of the time scales involved in the different aspects of evolution. It does not link animate development with the changes in the physical environment. It makes no comments on the physiology of the different organisms or why evolution has generally been towards the more complex. It makes no reference to the apparent repetition of particular physical features which have often in the past been described as biological convergences. It gives no limitations of the evolutionary process by, for example, accounting for the number and type of variations of species that are observed. These deficiencies would not be present in a modern general theory. It is necessary to extend the theory into a more quantitative realm if this were possible.

A physical basis for evolutionary theories began to take shape in the 1940s through the work of Oswald Avery (1877–1955) and his colleagues who recognised DNA as the essential genetic material. After much work, the final piece of the puzzle was found in the double-helix structure of DNA which was announced in 1953 by James Watson (1928) and Francis Crick (1916–2004) following the work of Rosalind Franklin (1920–1958). This new approach has formed the basis of the modern theory of the evolutionary sequence on Earth and expresses our current knowledge of evolutionary biology. Analyses using DNA have allowed the genomes of a range of organisms to be constructed and have finally allowed the evolutionary tree to be constructed. One result has been the finding of the so-called "missing links" that so bedevilled Darwin. These modern approaches have given rise to the science of evolutionary developmental biology, or evo-devo for short, which attempts to trace the ancestral relationships between organisms and so to understand the essential similarities and differences between them. The approach has led to some surprising results. One has been the recognition of the universality of features of living things and has given an understanding of the old concept of biological convergence. Common features between different species now can arise from the acquisition of a common gene and the acquisition can be a very fast process. Another has been to show the age of some organisms and to confirm that microscopic life roamed in pre-Cambrian seas. Indeed, a few microscopic fossils are now being found very occasionally in pre-Cambrian rocks.

It is now realised that living things have the form they have because of the interplay between genes, proteins and certain molecules. These are constructed from a set of atoms which are common to all life. It has been realised over recent years that some special genes form a restricted group which control specific functions of animate matter and that these genes are common to all life. It seems that there has been a single appearance of a specific function (for instance, eyes through the presence of the PAX 6 gene in the molecular structure) and not a multiple appearance. In this case, eyes were invented only once and the evolution in successive organisms has been due to the different requirements that are to be met to sustain them. Another example is the homeobox genes which are a DNA sequence of genes involved in the regulation of the patterns of development in animals, plants and fungi. Perhaps the most studied so far have been the Hox genes which control the patterning of an embryo during development. It is these that determine whether the embryo is to be, for example, a cat or a camel and ensure that the different parts, head, legs, etc., are assembled in the correct order and places. Some mutants are observed where the process has gone wrong. Here are possible variations. The PAX genes, another set of genes with nine divisions, are responsible for the presence of specific functions such as eyes, ears and so on. These genes are universal to all life and the recent discoveries are pointing to life being more a single construction rather than the combination of many separate developments. The molecular evidence is showing many life forms to be much more ancient than thought up until now and some elementary examples reach back into the pre-Palaeozoic seas.

The control networks of neurons needed to allow a creature to function are becoming recognised as the central feature of life. Early life was primitive and involved only primitive controls. The very earliest life must have been primitive indeed. The development of life can be said to have involved the continual extension and complication of the logic of genetic networks. More complex networks allow the development of new capabilities not realised during the initial development. All these developments have extended enormously the power and perspective of Darwin's theory and have confirmed its essential correctness.

The DNA and RNA are the components that carry the genetic code which determines what a particular organism is and how it will reproduce. A true new inheritance will reproduce the old with complete precision. These are the components that carry the very basic information of what a particular species actually is. It is known, however, that irradiation with high energy photons or cosmic rays will cause mutations.[7] It is also known that bacteriaphages can also transfer genetic information by a process called transduction[8]. There is, of course, always the possibility of errors when transferring information and this applies to reproduce DNA/RNA. Such errors will also cause mutations although some will have no recognisable effects. The way this system operates is still not fully understood, but it is clear that there is a random component. This will vary the time scale for its operation: sometimes it will operate quickly and at others slowly in absolute terms although this will be an insignificant development over the life times of members of a given species. It is, consequently, still not possible to be in anyway certain about the time scale for movement along the evolutionary path.

There appears to be a single underlying logic which has developed to provide life on Earth associated with the basic chemical building blocks of the Universe. The observed homogeneity of the inanimate Universe might lead us to presume that the same feature will apply to animate matter. This gives support to the hypothesis that life will have arisen everywhere in its most elementary form although its development to more advanced forms must depend on the occurrence of favourable circumstances. The case we know on Earth has seen the appearance of microbial life which has lasted strongly for 4.5×10^9 y and is still with us; macroscopic multicellular life is only some 5×10^8 y old; TCL has developed over the last few million years and has developed into life capable of leaving the Earth and entering space over the last 100 y at most. A fluctuation by only a few percent in this time table would leave us yet to achieve the space capability stage on the one hand or push us further forward on the other. It is extremely likely that the genetic logic developed here will have developed elsewhere in the Universe. Our previous arguments suggest that the number of sites occupied by elementary life could be very high making it essentially ubiquitous, but the number of technologically advanced creatures will be very small and very far away from us. Other systems, presuming they exist, will surely differ from each other by the time scales of evolution but the evolution of the logic will most likely lead to the same sequence of events. This rather stresses the mechanistic aspect of life but this is of central importance. The central rôle

---

[7]This possibility was first confirmed experimentally in 1941 for the bread mould *Neurospora crassa* by George Wells Beadle (1903–1989) and Edwin Lawri Tartuan (1909–1975) using X-rays.
[8]This was shown by J. Lederberg (1925–2008) and N. Zinder (1928) using Salmonella. This explains, as one example, how different bacterial species can very quickly show resistance to the same antibiotic.

of neural networks could suggest that life elsewhere need not be very different from that here. It is a single Universe and has been so in the past as now.

6.12.   The Role of Information

The genetic code provides the blue print for reproduction — it is the information, the archive, for the given species and the genome is the contents list. One can compare the structure of two species by comparing the genomes just as one can compare the contents of two books by comparing the tables of contents. If you wish to make a new species you could, in principle, settle the genome.

The genetic code provides the information for the continuance of a species and, with mutations, provides new (random) information for the evolution of the species through natural selection. There is also the hidden new information of existing species that can develop to fill any gap in the spectrum of species that could occur. For instance, the mammalia were able (ready and waiting) to fill a gap left when the dinosaurs became extinct. Information is then used that was dormant, or virtual, before. It may be said that the whole process of the formation of life and its evolution is a process of the unfolding of virtual information. What was virtual before is real afterwards, once it appears. From the point of view of the real world information is effectively being created by life and evolution.

There has been developed a heuristic relation between information and thermodynamic entropy. In a closed system (and the Universe is a closed system), physical changes proceed in such a way that the previous information is steadily lost until a situation is reached where there is no information about the earlier state of the system. This is the condition of thermodynamic equilibrium when the entropy is defined as having a maximum value. The initial state can be described as having zero entropy, but this increases as the heat processes continue. In this sense, entropy can be regarded as a measure of the information of a system that has been lost during its evolution due to the action of irreversible forces such as viscosity, thermal conductivity, electric conductivity and others. The real Universe is dominated by these irreversible effects. It is these that cause mutations in reproduction. The increasing complexity of animate life in its evolution is increasing its information content and so decreasing its entropy. This is in marked contrast to the inanimate world. Whilst the entropy of the animate world is decreasing as more detail is added to its overall structure that of the inanimate world is increasing. The initial hidden details are being lost as more and more are revealed with time. There is a duologue: the hidden details contained originally in the initial structure of the inanimate world are decreasing while the actual structural details of the animate world are increasing. The initial creation must have contained all the information of everything that has been revealed since including the appearance of animate matter under the appropriate circumstances and the random searching revealed as evolution.

Nothing can have been lost or added to the Universe during the 13.7×10^9 y of the present era. This means the central controlling quantities must be conserved and to remain now in toto as they were earlier on and as they will be later. As the inanimate information is revealed with the evolution of the material Universe so the animate information is collected as the animate evolution proceeds. As the information of the material world passes "above the line" so the animate information is also revealed. It is as if this were a single dual process. Certainly they operate synchronously and in parallel and it is tempting to wonder whether they may, indeed, be related events. If this were the case it would have very far reaching consequences.

The idea of the evolution of animate matter as a counterbalance to the development of material matter is exciting in itself but it has difficulties. In particular it is very difficult to place on a quantitative basis at the present time. There are a number of crucial unknowns. A very basic one is that of assigning an information index to both inanimate and animate matter. The full contents of the Universe are not yet known let alone understood. In thermodynamics the negative of the entropy is often regarded as the information content, but the application of this formula to the Universe at large hits theoretical problems. It turns out that assigning an information content to animate matter fares no better. It might seem there that the DNA could be a strong indicator because surely the differences in DNA sequences for mouse, elephant and mankind must be very different. The differences are found to be very small and are unlikely to offer an immediate indication of the huge physical differences between them. There is the possibility that the essential information content is the development of very elementary life itself: it may be that further developments offer only small additional information contents. Our hypothesis of a balance between virtual and real information would then suggest that elementary microbial life is very common in the Universe, but that there is no necessary advantage in the development of more complex life.

There are other difficulties in attempting to view the history of the Universe in any detail. The very early beginnings of the present era are not known through observation then but only by indirect inference now. It is very suggestive to reverse the Hubble expansion observed now to go back in time and reach a stage where the volume of the Universe was very small.

There is no direct evidence at the present time to say that this is not a possibility although equally there is no direct evidence to say that so unlikely a thing is actually true. The conditions that might have existed then are being explored by the new work being done at the European Nuclear Centre in Geneva. This work will take particle energies up to the highest values yet which are believed to be those appropriate to a "big bang" condition for matter. Of course, the rôles of dark matter and dark energy are still unknown. A cosmic background radiation in the microwave region of the electromagnetic spectrum was observed first in 1941 and is now well established to have a black body equilibrium form. Although this radiation is generally interpreted as being that left over from the big bang itself this cannot be more than an hypothesis. In fact, there are other interpretations that would apply equally well. The characteristic feature of a system in equilibrium is that all the information about its earlier, nonequilibrium, state is lost in the move to equilibrium. The radiation itself cannot give a clue as to its origin so to link it to an initial state is no more than conjecture. Any minute departure from complete equilibrium will show that the equilibrium was reached by some thermalisation process but cannot tell which one. The roles of dark matter and dark energy must be crucial here but are not known.

If the initial state of the inanimate Universe is confused the initial state of the animate Universe is no better known.[9] There is really no indication of the initial conditions here making it quite impossible at the present time to quantity the hypothesis of the rôle of information set out above. It can, nevertheless, be useful as a guide to future work to be discarded if it ceases to be useful.

6.13. Role of Automata

Imagine for a moment meeting another planet in a sun-system having travelled through space from Earth to explore it. If there has been no investigation to begin with the whole thing is full of unknowns which may be dangerous. What is the level of radiation given off by the central star? One will have deduced this before one left Earth from the type of star it is — but better to check. What are the gases like that make up the planetary atmosphere? One might have been able to gain some knowledge of this on

---

[9]It has been pointed out that the situation where life can be formed in laboratories worldwide is itself a very dangerous situation. When this becomes possible the most severe restrictions must be in place to control activities of this kind.

the way during the space flight, but are the components corrosive to skin or to space suits or to the space craft itself? You meet some life on the surface. Is it very aggressive or is it placid? How intelligent is it? Is it likely to want to put you, the explorers, in bottles to see what you are made of? Is some communication possible? A fly comes along — is it dangerous or will it fly away again? Alien creatures will need to breathe for the same reasons that we need to breathe — is their breath harmful to us? Are there germs unknown to our physiological systems and so of the greatest danger to us? What is in the soil? — is it safe to walk on without special precautions? We see a lovely looking fruit — can eat it or is it poisonous to us? Is it safe to walk under that tree or near that shrub? — these are not, we hope, giant relatives of the Venus fly trap on Earth. The microbial base is also very important but difficult to explore quickly. Every step is fraught with danger or potential danger without a clear base to begin from. What is one to do?

Manned space exploration has many features in common with a military campaign. One of the most important things is prior information sufficient to allow one to construct a plan of action and to set a realistic objective. Since there are not friendly guides waiting on the planet, it is necessary to get this information for oneself in some other way. The only possibility is using automatic space probes to conduct preliminary investigative campaigns of explorations. It will probably require more that one campaign. Quite apart from sophisticated search equipment there will need to be some sort of visual exploration. There will probably need to be three or four such automated technical missions before the manned mission can be attempted. We are familiar with this because this process preceded the manned landing on the Moon in 1969. For a distant star system this adds enormously to the length of time of the total mission, probably passing into a second or even a third generation of scientists and engineers. This is a construct that we have no knowledge of on Earth and the management of such a total mission is something that will itself have to be learned. The place of the automated probes is very clear, but the distances likely to be involved will provide restrictions of various kinds. Hardware technology advances and quite quickly. Such developments cannot be passed on to a probe after it has been launched. Advances in software can so in those terms the capabilities of the probe can be improved during the mission.

Obviously, a manned mission is the premier achievement and will give the best returns but is the pinnacle that has to be approached properly. This series of events was followed before the Apollo astronauts landed on the Moon and is being followed now as the build up to ultimate landings on Mars. The argument is not for or against automatic probes or human exploration.

The argument is that each is important in an overall context. These various aspects will apply to extraterrestrials, should they exist, as to us.

6.14. Summary

In this chapter, we have drawn all the previous arguments together to try to reach a conclusion about the possible existence of life in the Universe beyond the Solar System. We have made the following points:

1. The composition of the Universe is a single entity which makes no distinction between inanimate and animate matter. All is made from the same chemical elements which make up the periodic table of the elements.

2. The simplest life is microscopic bacterial. It covers all physical conditions on Earth and is virtually impossible to kill. These are characteristics that make it very likely ubiquitous throughout the Universe. It is likely that this form of universal life is more elementary than anything existing now on Earth, having been lost very early in the evolutionary development.

3. The arguments have been presented which show that all life on Earth, including Homo sapiens, has resulted from a single evolutionary process which appears to have been an integral part of the formation of the Earth. The observed expansion of the Universe can be interpreted as showing the inanimate Universe is also evolving according to some evolutionary blue print. It is natural to presume that extraterrestrial life will also follow an evolutionary path and that this will be convergent with that found on Earth.

4. Although an evolutionary template may be present elsewhere it can only be made real if the environmental conditions are favourable. The study of the evidence of evolution on Earth has shown the stochastic nature of the thread of life even though there has clearly been an enormous tenacity of life to survive. This means that TCL need not have arisen everywhere or indeed have developed anywhere other than on Earth.

5. Planetary systems have been discovered around stars other than the Sun but observations so far have only been able to detect bodies of broadly Jupiter/Saturn mass. These systems are generally very different from the Solar System.

6. Multiple planetary systems are rare and so far there has not been the discovery of a system comparable to our major planets. The general rule is a single large planetary body, often a super planet, with an orbit lying close to the parent star. The orbit is not necessarily close to a circular form.

7. It is presumed that smaller planets exist of generally Earth mass. Presumably these will have orbits outside the major body.

8. The number of systems which have sufficiently expansive, closely circular, orbits are only four or five so far. The distances involved are large — in excess of 40 ly from Earth.

9. These distances make communications between them extremely difficult even in principle. The possibilities of actual contact are virtually zero in any short period of time from an Earth point of view and would be possible from an extraterrestrial point of view only if their mean individual life span is many times that of Homo sapiens.

10. The number of sites available in the Galaxy for the development of civilisations is perhaps about 10^9, but the vagaries of the evolutionary process will make it likely that only a small number will be occupied at any given time.

11. Life can only develop once the appropriate chemical elements have been formed. This is likely to have been with the appearance of third generation stars perhaps rather less than 10^{10} ya. This makes the period of life on Earth among the earliest possible.

12. If this is so it is unlikely that any other civilisations will be more technically advanced. This we are unlikely to ever know because simultaneity is not a characteristic of the Universe at large. Even if contact were made the information received would refer to a distant time in the past. As an example: a 400 y or so gap on Earth would show a world very different from the one now.

13. The question what part living material plays in the Universe as a whole can have several answers. A very obvious one involves information. Let us suppose the Big Bang theory is correct so that the Universe had an origin some 1.47×10^{10} ya. The Universe is evolving according to information contained in its initial formation — essentially the laws of physics. The details of the Universe are continually being revealed as

it evolves. This implies a continual loss of "hidden" information as the properties of the material Universe are revealed.

14. Life is an expression of information through its RNA/DNA structure and the evolutionary process provides creatures with ever more information content. In this way information is continually being created.

15. Nothing enters or leaves the Universe which is closed. Information cannot enter or leave. One might conjecture that the information "revealed" in the evolution of the material Universe is balanced by that which is "made" by the evolution of animate matter. If this were so animate matter would play a central rôle in the Universe through the balance between actual and virtual information.

16. There is, unfortunately, no way at the present time of making such a comparison quantitative. The information content of the Universe is most uncertain and there is no direct way of assigning the information content of animate matter on any common scale. The hypothesis may, however, be of some use as a guide in the future.

17. Present indications are that other TCL will be at most rare and apparently separated from us and from one another by distances sufficiently great to make it essentially isolated. There is no logistical or genetic advantage in the close encounter between different sites.

18. The development of life appears to be associated with isolated sites and that is the answer to Fermi's original question of where all the extraterrestrials are. Actually, whether they are there or not is of little concern to the future of any site because the evolution will not be helped by the introduction of extraterrestrial genes.

Further Reading

1. The starting point of the theory of evolution was

   Darwin, C., 1859, *Origin of Species*, John Murray, London,

   while Abbot Mendel's original work involving peas (abortively undertaken to show that evolution need not involve natural selection but could come from simple inheritance alone — which it cannot) was collected in

   Gregor, M., 1865, *Versuche über Pflanzen-Hybride* (*Treatise on Plant Hybrids*), Austria Bundespresse dienst, Wien.

Epilogue

Our journey is finished and we might be permitted an element of speculation. We have found an evolving Universe in both its inanimate and animate aspects. We have made a hypothetical link between the two through information within an anthropic Universe. There is every indication that microbial life is ubiquitous in the Universe but that advanced life is likely to be rare and technically capable life very rare indeed. There is unlikely to be more than a handful of "civilisations" in an evolved galaxy at any one time and these will be widely separated, so widely that they will not be able to communicate with each other in any ordered way.

The evolution of animate matter on Earth appears as a continual advance towards more complex life but retaining also a robust structure of the more elementary life at the same time. The most basic units of living material are the amino acids which form the genes and evolution can be seen as involving the continual rearrangement of the genetic units. The process might be modelled as a continuous and systematic regrouping and development of the many combinations of the basic genetic units. A cumulative selection of all the possibilities of genetic combinations can then be formed giving an ordered compilation of continually growing complexity. The practical realisation of this index as it grows is continually tested and constrained by the friendly or hostile nature of the environment at any time — the fittest survive. It is this that protects the most elementary life which can remain fit through many vicissitudes. The homogeneity of the Universe with a universal chemical composition makes it likely that this general process can be expected to be active elsewhere in the Universe. This developing genetic index of increasing complexity, if it were to exist, could well be expected to be a common feature of the animate Universe elsewhere. Darwinian evolution is given a universal validity and would make our experiences on Earth significant throughout the Galaxy.

Although, on this basis, the genetic material of each cosmic civilisation is most likely to be broadly similar, evolution is not about a cosmic genetic advantage. No such advantage can be expected from physical communication between different cosmic civilisations. It still remains a beautiful world for all life to live in together including others hypothetically living elsewhere in the cosmos and with whom we are very unlikely to have meaningful communication. We might hope, however, to obtain eventually some indirect evidence of their presence. How accurate any of these speculations are we will find out in the future.

Appendix 1

Discovering the Cosmos

Here we give a broad survey of the Universe we live in to put our main arguments in perspective. This account may seem a little cavalier, and may indeed be so, because its main aim is to mirror especially those aspects that relate one way or another to the existence of life. There is another aim. It will be useful for us to have some idea how knowledge of the Universe has been collected in the past both as discoveries and then as parts of a coherent world picture that substantiates a self-contained system of observations and conjectures. This is how science works and the account can act as a guide the status of other more recent findings and of our understanding of the Universe as a whole.

A1.1.   Comments on the Early Solar System

Knowledge of the skies was very rudimentary even at the turn of the 20th century. The northern constellations in the night sky had been assigned during antiquity and still remained essentially unaltered. The constellations of the southern heavens had been added as that hemisphere was first explored and then settled in. The permanent constellations appeared to lie on a rotating sphere with the Solar System at the centre. The five planets visible to the naked eye from the Earth (the Earth itself making six) had been known since antiquity. The planets, together with the Sun and Moon, were presumed to orbit the Earth as centre as was clear for anyone to see. The problem of calculating their paths in the sky, however, presented severe problems. It was certainly clear that the Moon was nearest the Earth and then the Sun followed by the planets moving outside with the orbits in the order we know today. There seemed few problems in describing the movement of the Moon but the planets were very different. There were substantial discrepancies between the observed motions and those calculated assuming the orbits simple circles.[1] Ptolemy (ca.83–ca.168 AD) showed that these discrepancies could be drastically reduced by introducing a set of empirical auxiliary circles called epicycles to supplement the central circular path together with so-called quadrants for geometrical construction. This gave the planets a very complicated wavy set of paths around the Earth, but the use of sets of circles gave it all a clean academic pedigree to the thinkers of the time. There were no thoughts on why the orbiting objects should follow the complicated paths that they were supposed to follow.

Our understanding was moved a stage further by Nicolaus Copernicus (1473–1543) who showed that the picture was simplified very considerably if the Sun were placed at the centre rather than the Earth. The Universe appeared then to be a heliocentric structure. Such a possibility with the Sun as a centre had been conjectured by astronomers long before (even as far back as ancient Greek times), but no one was able to prove it. This made the Earth orbit the Sun and Copernicus set out the ordering and relative distances of the planets that we know today. This was pre-telescope work, but it was amply confirmed by Galileo (1564–1642) who was able to view the Solar System using his newly invented refracting telescope. The planetary orbits were still supposed circular and Copernicus still found it necessary to make some appeal to epicycles to correct small errors in calculating the planetary paths through the sky.

By the end of the 16th century, the orbital details of the Solar System itself were known to quite high accuracy due very largely to the painstaking observational work of Tycho Brahe (1546–1601) supported by the mathematical analyses of his assistant Johannes Kepler (1571–1630). These data were synthesised into three "laws" by Kepler who established that each planet orbits the Sun in a plane elliptic path. Isaac Newton (1643–1727) later gave the laws a firm dynamical foundation with his theory of universal gravitation. It then became clear that the planets move under gravity according to the laws of the conservation of energy and of angular momentum.[2] The path involving epicycles is now unnecessary so Kepler's observational/empirical analysis gives the epicycle structure the death knoll. Instead of the complicated path with circle upon circle, Kepler had shown that a simple plane elliptic path accounted for the planet's motion extremely well

---

[1] The circle was regarded as the pure figure by the ancient Greeks and others and it was assumed that the Universe other than the Earth must be pure.
[2] The discovery of the planets seven and eight, Uranus (1781) and Neptune (1845), required telescopes and so were not known to the ancients.

without artificial contortions.[3] It has always been assumed that the simpler explanation is the correct one because it is supposed that nature chooses simplicity over complication. This principle is sometimes known as Ockham's razor.

Kepler had accounted for the orbital motion of a single planet about a star as centre by formulating his three laws of planetary motion. This also gave a reason for this motion because his laws are an alternative expression of the fact that the planet moves around its star (for the arguments are easily generalised) under the action of universal gravity subject to the conservation of the energy and of the angular momentum of the orbit. This had very considerably improved our understanding of planetary movement. Unfortunately, small variations of the planetary orbits from the calculated plane elliptical orbits were still observed but epicycles were not available to account for these. In his work, Kepler had considered a single planet orbiting the Sun under gravity but Newtonian gravity acts between any two masses whatever they are. The Solar System consists of many masses moving with respect to each other and this will lead to perturbations of the orbits even though the gravitational forcers causing them will be very small in comparison with the effect of the Sun. There are interactions between the planets and they will be variable because they will change as the planets move. The effects of these interactions were shown to fully account for the small deviations from the Kepler ellipses. Understanding these small variations, in this way, also gave added confirmation to the correctness of Newton's description of gravity. It turns out that the difference between the required ellipse and a true circle for each planet in the Solar System is actually very small (Mars is the worst fit but even this is not bad) and that this is the reason why the initial assumption of a plane circular path was not so entirely silly.[4] It is interesting to realise that quite small errors can give rise to very different physical interpretations.[5]

---

[3]The construction of a single theoretical formula or general principle allows a complicated physical situation to be expressed in a concise way that can be expanded to reveal all the variant situations as special cases. Only the single generalised statement need be quoted or remembered.

[4]In medieval times, the circle was regarded as representing heaven whereas earthly things were represented by a square. The problem of reconciling these two things (sometimes called squaring the circle) played an important part in design and particularly of religious buildings. For instance, cathedral circular pillars were often given a circumference equal to the height providing a square area if the surface were unwrapped.

[5]The innermost planet Mercury has a complicated orbit in that it is slowly processing around the Sun separately from a simple Newtonian interaction. The orbit is not easy to observe and so was not used by Kepler. It is interesting to conjecture whether Kepler would have reached his laws if he had worked with orbital data for Mercury. Such chance events can have important consequences.

It was found a little later that the motion of the small components of the Solar System, such as comets and meteors, could also be accounted for, like the planets, in terms of elliptic paths under gravity with the Sun as centre although the paths generally departed considerably from the circular form. Edmond Halley (1656–1742) demonstrated this for comets by predicting the periodic path of a well-observed comet later named after him. This comet had been known from the ancient Chinese records and it had been found to reappear regularly with a slightly variable period of 75–76 y. Halley calculated its orbit about the Sun on the basis of motion under gravity. He found its orbit went far out into the Solar System, beyond the planets known then, and predicted its reappearance in 1758. It did indeed reappear (a little late — the slight variability was found later to be due to the perturbing effects of Jupiter and Saturn) although Halley did not live long enough to see it. This was the first time the periodic motion of a comet had been demonstrated and its return correctly predicted theoretically. Comet Halley last appeared in 1986 and is expected to reappear during 2061. As a matter of fact, it has an important place in history and was often given mystic powers in the past, usually denoting doom. For instance, King Harold of England saw it in 1066 before the battle of Hastings and predicted defeat for himself (presumably doom did not apply to his opponent William of Normandy, the victor — who might not have seen it: joy and woe are woven fine). We now know what the comet looks like and what it is made of. It was photographed close up from space probes during its 1986 appearance (see Fig. A1.1) and it was shown, in fact, to be simply a large irregular piece of rock and ice but a very interesting piece.

Fig. A1.1. Close up photograph of comet Halley taken by Giotto 1986 space probe. The Sun is on the left and the stream of ejected particles is composed of volatiles and ice heated by the Sun (ESA).

The hunt for comets became very popular from the 17th century onwards following the availability to ordinary people of optical telescopes of moderate power. Comets show obvious movement in the sky and often look spectacular with their tails of volatile materials — comet Halley is a very good example. The tails develop only when the orbit brings them close to the heat of the Sun, but if the orbit does not bring them that close there is no discernable tail. Then the comet appears as a blurred region of light which moves perhaps more slowly because its orbit is far away from the observer. The French astronomer Charles Messier (1730–1817) was a very active comet hunter. It is said that he discovered more than 20 comets himself, but is remembered now primarily because of his extensive catalogues of some 103 nebulous objects that he compiled. This collection was published in two parts, one in 1730 and the other in 1781. The objects he listed are still quoted as Messier objects and are referred to by Messier numbers from M1 to M103 (there is evidence that he knew of six more after the publication of the second Catalogue and these are usually included with M numbers up to M109). The members of a Catalogue compiled by a successful comet hunter might be expected to refer to comets but Messier's do not. Rather they are listed as objects that might be thought to be comets at first sight, but which are in fact not and the observer should not waste time on them. This instruction is interesting because it is an admission that there are more objects in motion in the sky than planets, stars, comets and shooting stars. It was many years before the true natures of these various objects were discovered which, incidentally, Messier did not conjecture upon. It was not realised then, but became clear later on, that some of these objects to be avoided were actually collections of stars called galaxies (they would be recognised as such 50 y or so later) and lies vast distances away from us. Shooting stars could also be observed, but these were transitory things that could not be confused with comets. The link between these momentary flashes and comets had not been made. It did seem very likely, however, that all that could be seen in the sky, even with telescopes, was everything that belonged to one unique single entity which is our Universe. The cosmos was of strictly limited extent. In retrospect, this was a rather strange Universe. The Solar System was seen to be the seat of great activity whereas the enclosing envelope of stars, presumably also controlled by gravity, was essentially static. Could this really be the whole story? Are the constellations really immutable? Is everything in the heavens as quiet as it seems to be? Two observations suggested this is not so.

The first evidence against was an argument usually ascribed to Heinrich Olbers (1758–1840) and known as Olbers' Paradox. It is based on a seemingly very simple question — why is the night sky dark? This question seems first to have been asked by Kepler (around 1610) and was considered later in the 18th century by Halley and the Swiss astronomer Philippe de Cheseaux (1718–1751). Olbers' formulation dates from 1828, but it was Johann Bode (1737–1826) who actually published it rather later. The argument leading to the simple question is itself very simple. On a clear night, the observer with average eyes will see between 5,000 and 6,000 stars. A telescope will reveal many times this number. Spread in the sky they will cover much of its area. If each star is radiating at the same rate as the Sun the quantity of energy entering the eye of the observer should be many times that of daylight — and yet the night is dark. Why? This is Olbers' Paradox.

The argument is based on hidden assumptions one or more of which must be wrong. It presupposes that all the stars are stationary and are radiating at the same rate as the Sun. It also presumes there is no gas or dust in the way that absorbs visible radiation and converts it to something else, such as infra-red radiation. These assumptions reveal that the stars are not all the same and that they may evolve. It also admits that there can be gas and dust between the observer on Earth and the stars. The most interesting possibility from our point of view is that the stars may have a significant motion and especially away from the observer. The simple observation that the night sky is dark when it might be expected to be as bright as day has the astonishing implication that the Universe is not the static unchanging place that the ancients thought it is.

There is as second result that leads to the same conclusion. This involved the very careful measurements of the positions of stars to confirm and extend the very reliable measurements of their positions made nearly two millennia before by Claudius Ptolemy. His careful compilation of the positions of the bright stars was published as his *Syntaxis* (written in Greek but translated later into Arabic under the title *Amalgest*). To this end, Tycho Brahe made state-of-the-art measurements of a range of stars and compared his measurements with those of Ptolemy. Tycho found some discrepancies of position, very small it is true but real nevertheless. It looked as if there had been a movement of some stars after all. The amounts of movement varied from star to star and not all stars showed this effect. It did, however, make the idea of independent movement totally acceptable. It seems that not all stars are firmly fixed in the firmament. This was confirmed again in the 19th century by such observers as Edward Bernard (1857–1923) who established the surprisingly rapid proper motion of the Southern Hemisphere star Zeta Reticuli (Bernard's Star). It is a red dwarf star, the second nearest to the Solar System at the present time and the star with the greatest proper motion. Its motion and direction will turn it into the nearest star in about 8,000 y time.

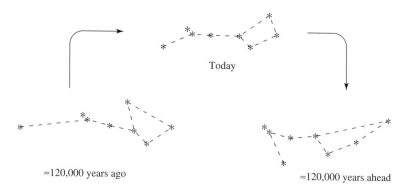

Fig. A1.2. The Great Bear (Ursa Majoris) as it is today together with as it was 125,000 y in the past and how it will be 125,000 y in the future.

Since the work of Bernard and others many measurements have revealed that all stars are actually in individual motion, some random and some organised. These movements have now been measured very accurately, at least for the nearer stars, but the most comprehensive measurements are collected in the survey of the Hipparchus space mission of the European Space Agency.[6] It has turned out that the forms of the constellations have remained so apparently immutable over time not because they are actually stationary and the stars linked but because the stars that make them up are so very far away. This possibility was not considered in early times because it was presumed that the extent of the Universe is very strictly limited and the stars were relatively close by. One example is the constellation The Great Bear or Big Dipper. Its form over a period of 250,000 y about the present is shown in Fig. A1.2. It is seen that while some stars evolve as a group and are actually neighbours, others move independently. This behaviour is typical of all the 88 constellations.

A1.2.   Disquiet in the Heavens: Large Measured Distances

Several interesting identifications of Messier objects were made during the 19th century by William Parsons, 3rd Earl of Rosse (1800–1867). He was an Irish astronomer who possessed the largest telescope in the world at that time which he constructed according to his own design. In 1845, he saw through his telescope the spiral structure of M51 which he believed was a collection of

---

[6]This mission was named after Hipparchus of Rhodes (190–120 BC) who first made such positional measurements. Indeed, there have been many thoughts expressed about the use Ptolemy might have made of the older Hipparchus measurements and doubts about how many new results he might have added. These doubts involve complicated arguments and need not concern us now.

Fig. A1.3. A photograph of the Crab Nebula. It is the remnant of the supernova explosion which occurred in 1079. The small star in the centre exploded to eject the gas cloud that now makes the picture.

many stars — by contrast the English astronomer William Herschel (1738–1822) thought it was composed of gas. This is known as the Whirlpool Galaxy today because of its appearance: it is known to be predominantly stars with some gas clouds. Rosse also saw the form of M1 which is the Crab nebula (see Fig. A1.3). The situation stayed rather like this for a number of years with the Universe regarded as a small compact affair, but with various and varied components. Advance beyond this to increase our understanding had to wait for the availability of better telescopes and associated auxiliary equipment. Although the Rosse telescope had an aperture of 73 in (about 1.8 m) its technical qualities were restricted in comparison with what was to come later. The big, indeed crucial, step forward came with the advent of the 100 in Hale reflecting telescope built on Mount Wilson, California, soon after the end of the 1st World War. This was the largest telescope aperture in the world and the auxiliary equipment supporting the main telescope, for instance the spectrograph, was also most advanced at the time. Astronomy of the heavens beyond the Solar System could now be made as a quantitative study.[7] But first there was a further step to be taken and this was concerned with measuring the distances of observed objects.

It is common knowledge that it is difficult or sometimes impossible to find the distance of an unknown object by simply looking at it. Further information is necessary such as size or precise motion or knowledge of what it actually is. For example, we all realise that a good model aeroplane flying

---

[7]The existence of a single telescope gave enormous power in astronomy to the few designated to control it and use it.

close by is not easily distinguishable from the real thing flying at a distance —
there is a whole entertainment industry based on this model approach. The
distance can, of course, be found by triangulation about a known baseline,
using an effect known as parallax. The angle subtended by the object at each
end of a known baseline is found and simple geometry provides the distance
of the object. This method can be made as accurate as the precision with
which the angles can be measured. In modern form, the distance of a nearby
object can be found more easily by aiming a radar pulse at it and measuring
the time it takes for the pulse to return to the observer. With the speed of
light known (the same for all frequencies of the electromagnetic spectrum)
the distance of the object follows immediately.

These techniques have been applied in astronomy. The distances of the
closest stars began to be measured by parallax around 1838 using the full
extent of the Earth's orbit as the base line. Observations of the chosen star
in the summer and the winter, six months apart, gives a base line of 2 AU
or closely 300 million kilometres. Many accurate measurements of this type
were made around the end of the 19th century and the beginning of the
20th century and gave the first indications of the large distances to even the
nearest stars. There is a close limit, however, to the distance at which this
technique is possible. The smallest angle of parallax that can be measured
accurately is about 0.01 arc sec which allows measurements within 10%
accuracy out to distances of about 100 pc which is 3.086×10^{18} m or about
34 ly. This distance is some 10^{10} times the distance between the Earth and
the Sun. Although this is a large distance there is much more to be measured
beyond it so another technique must be used. We might notice in passing
that we have introduced the idea of a distance within a limit of error. Entirely
precise measurements are not possible — there is always some uncertainty.
What the new method of distance measurement can be was based on a
discovery by Henrietta Leavitt made in 1908. It has proved a central tool in
discovery in modern astronomy.

Henrietta Leavitt (1868–1921) discovered that the star δ Cephei shows
a variable luminosity and that certain other stars showed the same effect.
She had discovered a class of variable stars subsequently called Cepheid variables and went on to find 2,400 other examples. The distances of some of
them could be found by parallax so the details of the variability could be
made clear. They were found to show a constant variability with the particular characteristic that the period of the variability of the light curve over
a period of many days is linked directly to the mean absolute magnitude of
the star. This means that knowing the absolute magnitude immediately gives
the period of variability. The usefulness in astronomy comes the other way

round: if you measure the period of the light curve and measure its apparent magnitude, you can infer the absolute magnitude of the star assuming there is no significant obscuring material in the way. The comparison between the observed and calculated magnitudes gives an estimation of how far away the star is. The important thing is that there need be no reliance on parallax.

Here was a powerful new technique, but it required a powerful telescope to make its application significant. The opportunity to move the studies into the deeper Universe took a new and unexpected turn in the 1920s using with the new telescope on Mount Wilson, outside Los Angeles.[8] It had been dedicated to the study of the deeper Universe and its eyes could penetrate further into space than any instrument before it. There was excitement about what it might uncover and this expectation was not disappointed. It is still in operation today and, although it is now one of the smaller telescopes in the world, it will always retain its mystique as the first of the modern telescopes.[9]

In the early 1920s, Edwin Hubble turned the new telescope onto the spiral galaxy in Andromeda (M 31). He found he could distinguish individual stars quite easily on photographs and among them were Cepheid variable stars. This meant that he could do the astonishing thing of measuring how far away it is as we have just explained. He was able to show that this galaxy is at the enormous distance of closely 2 million light years — the light from it takes 2 my to reach us. He found other Cepheid stars in the galaxy that confirmed the measurement. The Andromeda galaxy must be outside our own system of stars which form our Galaxy. This was a momentous discovery because it was the first time that it had been shown unambiguously that the Universe did not consist simply of a single agglomeration of matter. Our own system of stars appears quite self-sufficient and is our "universe". Hubble found that other spiral structures that had been observed can be studied in the same way and distance even greater than that for M 31 were found. It seemed that matter in the Universe occurs in independent groupings like our own and forms a series of separate "island universes". They are separated by distances of many millions of light years. These early results have been confirmed and extended by all subsequent observations. The surprising conclusion must

---

[8]The location near a centre of population is necessary for various logistical reasons such as daily services for living, transport, hotels and so on, but has the disadvantage that a growing town nearby will interfere with observing in various ways. This has certainly happened for the case of Los Angeles.

[9]When he first saw it, the author was amazed how small it looks. It was certainly amazing to realise how stupendous were the early discoveries using it.

be drawn that the Universe at large is composed of separate self-contained packages. The number of independent galactic packages has since been found to be enormously large, perhaps of the order 10^{11} or 10^{12}.

The method using variable stars is very powerful but its applications are so important that other complementary methods need also to be developed. New classes of variable stars have been able to be isolated (characterised by the details of their light variation curves) and brought into play. One extension came about after it was realised that the Cepheids of Population I stars (type I Cepheid variables) will have characteristics of the light curves rather different from those associated with the less metal rich Population II stars (called type II Cepheids). There are other practical problems to the use of the variable stars for distance measurement. One of the most important is the presence of obscuring material between the observer and the star. It is difficult to correct for this partly because the opacity of the material may not be known and partly because the quantity and geometry of the material may not be clear. This means that cosmic distance are likely always to be approximate at least to some extent. Certainty in the Universe is very elusive. Indeed, much research in the 20th century has showed that this is, in fact, a relative world based on uncertainty. One of the more immediate conjectures arising from this early work was a great surprise at the time although we have since got used to it and accept it. There seemed nothing that made our Galaxy special from the others (except for ourselves). Copernicus had shown there is nothing central about our Earth in the Solar System; it was found that our Solar System had no special place in our Galaxy collection of stars; could it be that our Galaxy was not central within the Universe? And then the almost forbidden question — could it be that we Homo sapiens have no special place within the Universe? This point of view had been suggested by a range of philosophers in the past and some, for instance Giordano Bruno (1548–1600), had been martyred for raising the question — but could it actually be true?

A1.3. A Larger Universe in Motion

We have seen that the Mount Wilson telescope was equipped with all the latest ancillary equipment of the time and especially a spectrograph which allows the light that it receives to be broken down into its different component colours. This means it is possible to analyse the light into the different frequencies that make it up. This capability was to prove crucial in the early years of its use. This led to a second fundamental discovery. Everyone expected the galaxies to be in some form of stationary equilibrium,[10] but Edwin Hubble found that the galaxies are most probably in motion and not stationary.

The actual discovery was that the spectra of the nearer galaxies that could be studied showed a shift of their spectral lines towards the red. There was the one exception of the nebula in Andromeda which actually shows a small blue shift. The cause of the shifts has been strongly debated over the years even though the only sensible and consistent assumption that made any sense is that made by Hubble: the shift of frequencies is a Doppler shift caused by the motion of the galaxies away from us. It seems the Universe is expanding. The one exception is the Andromeda galaxy which is slowly moving towards us. The Doppler shift is a feature of wave motions and is concerned with the relative motion of a source and an observer. The effect of the motion is a compression of the wave for movement towards each other (in optics a shift to the blue) but a lengthening of the wave for motion away from each other. This effect is very well known in connection with sound waves. If an object, such as a train or motor vehicle which is the source of sound, is moving towards you as the observer the frequency of the wave is greater than when it is moving away from you and you hear a change of pitch. The degree of change depends on the speed of the source relative to the observer according to a formula which is given by the well-established theory. Hubble had established that the Universe is expanding and was not stationary as was believed at the time — he had changed our view. He went further. For the small number of galaxies available to his telescope for observation, he found that the expansion is greater the distance between the galaxy and Earth and that this result was independent of the direction of the observation. Since the Earth cannot, then, be supposed to be in a favoured position this means the Universe is expanding uniformly from every point within it. This was revolutionary talking at that time and led to a very new way of viewing the Universe. This result has now been established out to all distances and is a basic observation.[11]

There is a very intriguing consequence. If the Universe is expanding it will have a larger volume tomorrow than today. Because it is expanding it will have a larger volume today than it had yesterday. It will have been

---

[10]This would have contradicted Newton's law of gravitation if it were true unless the Universe is of infinite extent in every direction. A finite agglomeration of matter would fall to the centre of mass.

[11]The inferred movement of the Andromeda galaxy towards us was confusing at the time, but can now be seen as the first example of a proper motion of galaxies around their equilibrium points — rather like a motion reminiscent of temperature fluctuations of position.

smaller the day before and it is reasonable to suppose that, *provided* the expansion has been taking place unchanged all the time, there could have been a time when the volume was indefinitely small, essentially the proverbial mathematical point. This particular time in the past will then be a characteristic of the present phase of the Universe and is often thought of as marking the actual origin of the Universe. Did the Universe really have a "point" beginning at that time in the past? Certainly no one was there to record it but this can be taken as the essential observational evidence for an origin provided that observations do not suggest any peculiarities of expansion in the past. Present indications are that there were none.

A1.4. Basis of Theoretical Descriptions

We have reached a heady result. Can theory help us to understand it through some assumed model of the cosmos? A model must by its very nature be more simple than the real world but, nevertheless, can be sufficient to give guidance in analysing observational data. The modern view of the Universe is based on the General Theory of Relativity proposed by Albert Einstein (1879–1955) in 1915. This is essentially a continuum theory of gravitation based on a geometrical interpretation of space and time taken together to be a combined four-dimensional entity. The time is mathematically distinguished from the three space components, but the combination shows that the space and time components must behave in the same way. He showed the appropriate geometry is that of a sphere of four dimensions, a form that had earlier been explored by Bernhard Riemann (1826–1866). This means that if an observer were able to travel in any direction in space the result would be to arrive back at the starting point, just as would happen on Earth. It can, of course, be checked on Earth though not in space.

The theory requires that expressions of the laws of physics are independent of the coordinate system being used which, of course, is a very proper requirement. This leads immediately to the use of a special mathematics designed for the purpose (the tensor calculus) which guarantees this independence of the coordinates, but we need not delve into technical details here. By contrast Newtonian theory is best expressed in terms of the differential calculus using the flat Euclidian geometry[12] with which we are familiar in everyday life. Relativity theory accepts the fact that the action of a uniform gravitational field on a particle is indistinguishable from that of

---

[12]Euclidean geometry is a special form of the more general Riemannian form when the radius of curvature of the space is infinitely great — space becomes flat.

a uniform acceleration (the so-called Principle of Equivalence). It combines space and time together to provide a four-dimensional framework, but in a way that space and time are nevertheless kept separate throughout.[13] This is a strongly mathematical theory and is substantially non-linear in the sense that different solutions of the mathematical equations cannot be superimposed one with another to form a new solution. This is very different from the Newtonian theory of gravity which preceded it and which, by contrast, is a linear theory so that solutions are superimposable. The differences between the predictions of the two theories turn out to be small when the gravity is weak, but they become very significant for very strong fields. General relativity in its original form is generally restricted to relatively weak fields and will need to be extended to cover the strongest fields associated with the most massive aggregated of matter that we know exists today. Such an extension has yet to be made. It is most likely to involve quantum aspects but how this is to be done is as yet unknown. The present theory reduces to the Newtonian theory in the limit of very weak gravity (when the geometry of space — time becomes flat) and this feature must be retained in any modification.

The Newtonian and Einstein models of gravitation have very different starting points and means of expression and yet, surprisingly, they involve common basic features. The basis of relativity is that the structure of the space and time (space–time) of a region is expressed in terms of an appropriate (4th order) tensor: the state of matter in the region is also expressed by a tensor quantity (of the same order) representing the energy and momentum of its material components. The presence of matter causes a curvature of space–time (expressed by a curvature tensor) while the curvature of space–time is manifested as matter. These two tensor quantities, the space–time tensor and the energy–momentum tensor, are to be equated in every way — they have precisely the same symmetries and are to be treated as entirely equivalent. This means that specifying the space–time continuum also specifies completely the matter while, on the other hand, specifying the matter in its turn completely specifies the associated metrical properties. This resort to geometrical arguments is reminiscent of certain 18th and 19th century analyses in the nonrelativistic classical mechanics. The Newtonian approach is very different. The geometry of space is presumed to be flat Euclidean and

---

[13]Technically, this is achieved by introducing imaginary numbers and treating "imaginary" time in combination with "real" space. It is rather like putting "his" and "her" things in the washing machine. Everything is treated equally during the wash, but can easily be recognised and separated afterwards.

the dynamical behaviour of a mechanical system is expressed in terms of forces. The time is presumed totally invariant and the same for all observers everywhere. The behaviour of the forces is described by Newton's three laws of motion. One force is that associated with the motion itself and is called the force of inertia, while the other is the force causing the motion. The inertia is closely related to the kinetic energy of the system. In many cases of practical interest, the force causing the motion can be expressed in terms of a continuum called the potential of the force: the force is then derived from the change of the potential in space (that is its gradient locally). This is not always the case but is so for gravitation. Systems where it is are called conservative. Using the kinetic and potential energies, the 18th century French mathematician Jean d'Alembert (1717–1783) showed that the motion of the mechanical system can be described fully in terms of the kinetic and potential energies which each then have the form of a continuum. This continuum form of representation is in many ways much more convenient for calculations than the entirely equivalent form of Newton where the forces act at discrete points determined by the distribution of the particles. The point of this diversion is that the approach of general relativity can be regarded in the same way with the space–time curvature linked to the potential energy and the energy–momentum aspects linked to the kinetic energy. The mathematical tensor equations which follow from general relativity are numerous and difficult to solve: those from classical theory are much easier. In fact E. A. Milne (1896–1950) and W. H. McCrea (1904–1999) showed, in the 1930s, that the dynamical solutions of general relativity known at that time can more easily be derived from classical theory. Although there is a formal equivalence between the two theories from their results their physical interpretations are, of course, very different.

A1.5. An Important Cosmic Constant

Energy and momentum are conserved quantities (they are never lost) so the expression representing them (the energy–momentum expression) must satisfy a conservation statement and, because of the dynamic-energy equivalence, this must also be true of the expression for the space–time geometry. An expression of conservation is basically a statement that the quantity concerned does not change with changes of position or time. Any simple constant added to the initial expression will not contribute to any change of it and so will add zero to any change. An arbitrary constant can be added to the expression to give the most general statement of it. In the present case, it will be added to the geometric expression. This constant can take any appropriate form: because the theory was applied at first primarily to cosmological problems it is called the cosmological constant and is denoted by Λ. It has the dimensions of energy and acts similarly to gravity. After being largely ignored at first it is now proving a most important part of modern cosmological theory.[14]

A1.6. Possible Universes

The material Universe is assumed to form a continuum. We know now that if the actual visible structure could be spread out smoothly it would have a mean density of the order of $10^{-27}\,\mathrm{kg\,m^{-3}}$. This is a very small quantity. The first general solutions of general relativity which applied to the Universe were those derived by Abbé Georges Lemaitre (1894–1966), Alexandra Friedmann (1888–1925) and Arthur Eddington (1882–1944). They all showed a dynamical behaviour of motion rather than the static form that was observed at that time in the heavens with the fixed stars and Messier objects. The balance is between the kinetic and potential energies and the different models amount to the result of the excess of one over the other. Many of the models have zero motion at the origin at zero time. Some are not finite but begin at indefinite (infinite) time ago. The time scales vary enormously, but are anyway open to transformations which can extend or contract them. Abbé Lemaitre first showed a solution exists which expands continually from the origin and so foreshadows the more recent "big bang" theory for the origin of the Universe. This correspondence does assume that the theory of general relativity is a sufficient approximation to reality.

Not all the solutions of the equations have been studied in equal detail. Those with an origin for time and space have been considered with some care. The general relativity interpretation requires that both space and time start there so the expansion is from time zero and space zero: both are created continually as the expansion proceeds. This is a difficult concept to appreciate, but means that anything before the origin is outside time and so outside the scope of the theory and should properly be discarded. This problem does not apply to those solutions with an infinite past. This implies that in all cases there is no meaning in a "before" the "origin". When Einstein proposed general relativity the Universe appeared static and his solution had not to include the time. Gravity alone will pull matter together unless the field of matter is of infinite extent so to make it inappropriate to

---

[14]In his early applications of general relativity to model universes Einstein set this constant equal to zero. He said later that this was his greatest blunder.

think of a centre. A centre implies the existence of an outer boundary against which any internal location can be assigned. A finite static universe must involve some counter to gravity — an anti-gravity component. Here is a job for the cosmical constant Λ we met earlier: Einstein used it as a device to oppose gravity everywhere precisely. Just as gravity can be associated with a potential energy so a potential energy can be assigned to the cosmical constant. Just as it was originally designed to counterbalance the effects of gravity so it can be used to *overcome* gravity. It has become of great interest recently in this rôle as a means of explaining the observed acceleration of the outward motion of the galaxies. The presence of the cosmical constant would mean also that a field of cosmic energy is present with opposite effect to that of gravity. We will have more to say about this later.

We must ask now: Is there a solution which mimics observation closely? The Hubble expansion reversed would provide a model in which the initial state is essentially a point from which expansion could have started.[15] This suggests a solution expanding from a point. There are three obvious possibilities depending on the relative strengths of the kinetic and potential energies — the kinetic energy can be greater than the potential energy, or less than it, or the two energies may be equal. In the first case, an expansion once started with proceed for ever. In the second case, an expansion once started will come to rest after a certain time after which the matter will contract — the expansion and contraction will be repeated continually, for ever. It is the only periodic solution. For the last case, an expansion once started will slowly stop and come to rest after an indefinitely long time. It is the only case where the matter eventually is reduced to rest. This description using the familiar kinetic and potential energies has an exact counterparty in terms of the geometrical structure. There are three basic geometries of interest now: that of Euclid (fl. 300 BC), of Riemannian and of Lebachevski (1792–1856). The first is associated with infinite flat space, the second with the closed curved space of a sphere (spherical space) while the third can be associated with the open space of a double hyperbola (hyperbolic space). In terms of geometry the differences arise from different values of a constant k in the theory which may have one of the three values +1 (Riemannian), 0 (Euclidean) or −1 (Lebachevskian). There has been a great amount of effort put into finding which space the Universe actually shows. It has turned out to be flat — but we will have more to say about that later.

---

[15] The work of George Cantor (1845–1918) implies that a classical point can have an infinity of structures. Galileo had previously discovered the infinities, but had not got the mathematical tools to pursue the matter.

So far we have viewed the Universe in very artificial, theoretical terms, but observation shows a very different rich reality. The total Universe is found to cover an unimaginable range of magnitudes. The present epoch (which, from what we have seen, may in fact describe the full history of the Universe that we know so far, but could equally be just a phase in a bigger story) is judged to have lasted for some 1.4×10^{10} y. Its boundaries appear to be expanding at essentially the speed of light now and are believed to have been doing this throughout the present epoch. This would make the characteristic dimension of the Universe of the general order 3.6×10^{26} m (which is the age multiplied by the speed of light). The nature of matter before the present epoch began is entirely unknown as is the nature of the region into which the Universe is continually expanding — both lie outside the Universe as we can observe it and so we cannot find information about it. What cannot be measured cannot exist within our approach.

The continual expansion, should it continue indefinitely, implies a continually decreasing density of matter within it. This implies that in sufficient time the Universe will expand into nothingness unless new matter is introduced into it. This suggests two possibilities. One is that the Universe will end in infinite dilution and total inactivity while the other is that more matter will be introduced at some stage to prevent this. This latter possibility is the basis for the steady state theory of the Universe investigated by Hoyle, Bondi and Gold. All this is very speculative, but the possible conditions at the end must surely influence thought about the beginning.

A1.7.   The Galaxies

The principal visible objects in the cosmic sky are the galaxies. These are mixtures of stars and gas in various proportions. There are five distinct types.

The most obvious and most spectacular are the rotating spiral galaxies of which an example is M101 called the Pinwheel Galaxy and shown in Fig. A1.4. This consists of stars at all stages of evolution and gas (mainly hydrogen). The stars and star forming regions are spread out in (most often) two swirling arms which join to the central condensation. There are often several subsidiary arms not necessarily going to the centre. The arms appear blue because they are the seat of the formation of new stars. The central region is a grand condensation of stars forming a core which is not always accessible to observation. This central spherical region often obscured by an impenetrable dust. This usually involves a very dense concentration of

Fig. A1.4. The spiral galaxy M101, the Pinwheel Galaxy (NASA, ESA).

mass often so dense that it is designated by the porte manteau expression black hole.

There is no standard sized spiral but many have a diameter of about 100,000 ly or perhaps 35 kpc. They will be a few tens of light years thick in the mean. The central condensation is usually a few times thicker. It is usually assumed that a typical spiral contains some 2×10^{11} stars each of about solar mass (that is 2×10^{30} kg). This puts the total mass in the region of 4×10^{41} kg although this could be 4×10^{42} kg. Again, it is usually assumed the Universe contains the mass equivalent of some 10^{11} or 10^{12}, such galaxies giving the Universe an overall mass of 4×10^{52} or 4×10^{53} kg. These figures are so large as to be almost meaningless. Nevertheless, they offer some yardstick that could act as an anchor for further discussion. While we are in the area of speculation we might notice that there is a large lock-up energy associated with this mass. According to special relativity, a mass m represents a "condensation" of energy E given by the expression $E = mc^2$ where c represents the speed of light. For a mass of 4×10^{52} kg and with $c = 2.99 \times 10^8$ m/s this gives $E = 3.6 \times 10^{69}$ J. This is the general magnitude of energy that would need to be available for its formation.[16]

---

[16]Going completely to the extreme this energy would be associated with a temperature $T = E/k$ where $k \ (= 1.38 \times 10^{-23}$ J/K) is the Boltzmann constant giving $T \approx 10^{93}$ K!!! This could be the theoretical upper limit to temperature in the Universe achieved when the Universe was formed.

Fig. A1.5.   The barred spiral galaxy NGC 1300 (HST).

A variant of the simple spiral we have described so far is the barred spiral. They have a similar structure except the spiral arms do not penetrate to the central region, but have their origins on the two ends of a bar across the centre. Many galaxies, perhaps as many as two-thirds, fall in this category having a bar of some kind. An example NGC 1300 is shown in Fig. A1.5. The galaxies often have an active nucleus which emits more strongly in some frequencies than in others. It seems that the bar might act as a channel for the formation of new stars but the details are not yet clear. Certainly, the increased emissions show that considerable activity is going on there.

It has become clear over recent years that our own Galaxy is barred. A representation of the observational data for our own galaxy is shown in Fig. A1.6 indicating the main features arms and the ancillary ones. The two main arms are the Perseus Arm which follows on from the Sagitarrius Arm and the Scutum-Centaurus Arm. Each arm is joined to the central Long Bar. There is a rudimentary Outer Arm, a rudimentary Norma Arm and more recently discovered Near and Far 3 kpc Arms. Also in Fig. A1.6 is the galactic coordinates based on the Sun which is also shown inside the Perseus Arm. The North (x-) direction is the line joining the Sun to the centre of the Galaxy, which is in Sagittarius. The y-axis lies at right angles in the plane of the Galaxy while the z-direction is perpendicular to this plane. Longitude and latitude is marked out as for a sphere. All galactic astronomy uses this coordinate system. It is truly amazing that modern astronomy can present a diagram such as Fig. A1.6 with full confidence that it will be confirmed by future work.

Fig. A1.6. A representation of our Galaxy with observational data drawn from several sources. The main Scutum-Centaurus and Perseus Arms are clear with the secondary Outer Arm, Norma Arm and Sagittarius Arm. The coordinates are galactic coordinates centred on the Sun and the centre of the Galaxy and the plane of the Galaxy (NASA, R. Hurt).

A different type of galaxy is the elliptical galaxy. These appear as feature-less blobs of light and can vary in profile from the spherical to the strongly elliptical. Their composition is primarily stars there being very little gas. They are dominated by old stellar populations and consequently appear reddish ion colour. Each can contain anywhere between 10^8 and 10^{11} stars or even more each of broadly solar mass. In fact the largest galaxy known, M87, is an elliptical galaxy (Fig. A1.7). Its diameter is about 120,000 ly across with a mass of about 10^{12} solar masses (that is about 2×10^{42} kg). It is about 60 million light years away. The smallest known galaxy is also an elliptical. It seems that these galaxies have been less common in the early Universe. They have no net angular momentum but, although the galaxy as a whole is not in rotation, the individual stars are in a random motion within the galaxy.

Galaxies are known to collide with each other. One example is shown in Fig. A1.8. Such an encounter shows the vicious strength of gravitational force between the two bodies. The encounter takes place very slowly by

Fig. A1.7.   An elliptical galaxy.

Fig. A1.8.   Two spiral galaxies NGC 2207 and IC 2163 in the process of collision. It is thought that eventually the left-hand galaxy will incorporate the right-hand galaxy. The encounter is 40 million light years away. The filamentary lines are due artefacts of the photograph (D.M. Elmegreen (Vassar Col) *et al.* and Hubble Heritage Team (NASA)).

our standards taking several million years to complete. This is, neverthe- less, rapid in cosmic terms. Although the distribution of stars within each galaxy is too sparse to allow collisions between the stars themselves the structure of each galaxy can be totally disrupted. This is seen in Fig. A1.9. Depending on the details of the collision the two galaxies can move apart after the encounter, as in a glancing blow, or one may absorb the other to form a large agglomeration of stars. The result of the encounter may well be

Fig. A1.9. The colliding pair of galaxies NGC 4676 showing the way that such encounters can destroy the original structures of the galaxies (ACS Science and Engineering Team, NASA).

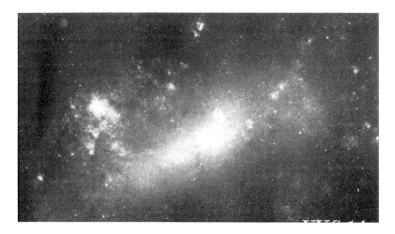

Fig. A1.10. An irregular galaxy — the Larger Magellenic Cloud (LMC). This is a small companion of our Milky Way (HST).

two galaxies with little or no structure. About one quarter of the observed galaxies are of this category and are called irregular galaxies. An example is shown in Fig. A1.10. This is the Large Magellanic Cloud which is a small companion galaxy to our own Milky Way Galaxy. It has a mass of about 10^{10} solar masses and is at a distance of about 50 kpc which is alternatively about 160,000 ly. It is, in fact, the third heaviest of the companion

Fig. A1.11.   The lenticular galaxy NGC 5866. It lies 44 million light years away close to the constellation Draco (the dragon) (HST, NASA).

galaxies to our Galaxy, only the Andromeda galaxy (M31) and the Formax galaxy (M33) being more massive. It is seen in the figure that the structure resembles a barred spiral with the arms largely dispersed. This might suggest that it was once a barred spiral in its own right. It is visible to the naked eye in the Southern Hemisphere bordering the constellations Dorada and Mensa.

The spiral galaxy is an important basic galaxy form and it gives rise to a variant when most if not essentially all the contained gas and dust has been used up or lost. The result is the loss of the normal central "bulge" giving a lenticular form. This is well shown in Fig. A1.11 which is an image of the galaxy NGC 5866. This galaxy is most remarkably seen edge on and the lack of the central condensation is very clear. Other examples are found face on, or nearly so, and can show a spiral form.[17]

A1.7.1.   *Radio galaxies*

While essentially all galaxies radiate across the full electromagnetic spectrum some galaxies emit more strongly than others in certain wave lengths and especially in X-rays, γ-rays and especially in the radio band 10–100 MHz. These emissions represent the release of enormous quantities of energy and the sources are not always understood. Those emitting strongly in the radio frequencies, and called the radio galaxies, are known more about. Their

---

[17]They have been called nonspiral spirals.

visual profile is of a central source with jets of gas being ejected with a strong confinement in each direction along an axis. The radiation is strongly polarised and has the form of synchrotron radiation where electrons at essentially relativistic speeds are radiation due to acceleration in the magnetic field. These very high energy events (some involve energies comparable to the conversion of a whole galaxy into energy) are not yet properly understood, but appear to be associated with conditions in the earlier periods of the history of the Universe.

A1.7.2. *Clusters of galaxies*

The visible galaxies do not live alone, but each is a part of one gravitationally bound cluster or another. The cluster may contain as few as 10 galaxies, but the larger ones will contain many thousands. The Local Group, of which our own Galaxy is a member, is one of the smaller groups with more than 40 members, mostly small. The largest member is the galaxy in Andromeda (M31) with our galaxy as the second largest. The third largest is the galaxy in Fornax (M33). The Large Magellenic Cloud is fourth in the list. Larger clusters include the Virgo cluster, the Hercules cluster and the Coma cluster. A photograph of part of the Coma cluster is shown in Fig. A1.12.

The individual galaxies in a Group have their individual motions. There is typically a spread of velocities of some 150 km/s. This can be thought of as akin to the random motion of molecules in a gas but it is much less marked. The cluster undergoes the normal Hubble expansion as a group, but the

Fig. A1.12. Part of the Coma cluster of galaxies. All types of galaxy are present set in random orientations (NASA).

larger groups affect the expansion locally. For example, the Great Attractor is a massive concentration of matter some 500 million light years away is dominated by the Norma cluster. This affects the local Hubble expansion of the Universe.

Clusters of galaxies can also be members of larger grouping called Super Clusters. On the largest scale the Universe is found to gather into "filaments" and "walls" of galaxies giving the appearance of "foam" containing what appear to be vast voids.

A1.8. Visible Gravity Is Not Enough: Dark Matter

There is now much more than purely circumstantial evidence that the visible matter is not the only material component in the Universe or indeed is even the most abundant. Evidence for matter and energy we cannot see directly has arisen in several unconnected circumstances.

Although we only see a snapshot of the Universe at large (in general it takes millions of years for anything to happen) it is supposed that the appearance of classes of object means that these have a stable structure. Thus, the spiral galaxies are presumed to be continuous features of the sky and it should then be possible to understand their mechanical equilibrium. They are in rotation with a distribution of mass throughout that can be estimated with some precision. When this was first done some years ago it was found that for most galaxies the kinetic energies of matter and the associated gravitational potential energies did not match to establish the equilibrium that is believed to be observed and that there is too little gravity in the outer regions. Some galaxies are in balance but most are not. This suggests that mass exists in the outer regions that is not actually visible to us but which makes a gravitational contribution that can be recognised. This was surprising to say the least. A sketch of the observed velocity of visible matter with distance from the centre of the galaxy is shown in Fig. A1.13. The measured motion of the visible disc from the centre outwards is marked (2) while the speed necessary to provide a mechanical equilibrium of the visible matter is marked (1). These are seen to differ radically in the outer regions. They can be reconciled only if matter is present in the outer regions that is not visible.

The same problem of an apparent deficit of matter has been found in other connections. It was found that the mechanical equilibrium of clusters of galaxies could not be understood on the basis of the observed matter alone. As one instance, for clusters of galaxies such as the Coma cluster of Fig. A1.8 to be in stable equilibrium the kinetic and gravitational potential

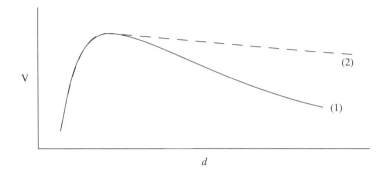

Fig. A1.13. The relation between the rotational velocity v at distance d from the centre of a typical spiral galaxy. (1) as predicted from the observed image assuming an equilibrium between the kinetic and potential energies; (2) the measured velocities. There appears to be a mass deficit in the outer regions that can be accounted for by the presence of dark mater.

energies must balance, but it was found that again there is not enough visible gravitational energy to achieve this. It seems that unseen matter must be added to each galaxy grouping to describe equilibrium. Other effects also showed that something was wrong.

One consequence of any continuum theory of gravitation like general relativity is that light beams are affected by matter through a gravitational interaction. The presence of matter distorts the space surrounding a region of mass and so also the path taken by the light (technically a nullgeodesic line). This effect was first shown for the Sun during the 1919 eclipse expedition organised by Sir James Jeans. The stars in a patch of sky which was to be occupied by the Sun six months later were photographed most carefully. When the Sun actually occupied the space it should cover the stars behind it so a field of stars should be hidden during the eclipse. It was found, however, that one or two stars right on the edge of the solar image were in fact visible when they should not be. The Sun's gravity had bent the light rays around its rim to make the stars visible. In this way the Sun acted like a lens but with a difference. This gravitational lens has not got a focal length or a focal point. The inner most rays are bent more than the outer ones unlike a normal glass optical lens where the outer rays are bent the most. The study of the geometry of the bending can provide information about the warping of space and therefore of the mass of the object causing the bending. This leads to an important astronomical application.

If a mass lies behind another object so as to be hidden from an observer from Earth the gravitational lensing can show the presence of the hidden mass. Remember this lensing does not produce a concentrated image like an

Fig. A1.14. An example of gravitational lensing in Abel 1689. The complex arc structure, which is part of the lensing image, is clearly visible (HST; NASA; ESA: H. Lee and H. Ford).

optical lens does, but a series of circles and arcs depending in the relative geometry of the object. In other cases, there can be multiple images of the same object sometimes displayed in the form of a cross called the Einstein cross. An example of an arc structure is shown in Fig. A1.14 produced by a source of light hidden behind Abel 1689. The arc structure is readily seen. The analysis of such images allows an estimate of the masses involved to be made and these have very often given surprising results. In most cases, the mass necessary to provide the image is much greater than the visible light will provide. The hidden mass is often 10 or 100 times greater than the visible mass. Such multiples are not uncommon.

These various mismatches of energy can be explained if there is a sea of matter which is gravitationally active, but does not interact with the full electromagnetic spectrum. It is, apparently, completely transparent to light, radio waves, infra-red and ultra violet radiations as examples. This unseen matter has been called dark matter. Its nature is at present entirely unknown. It is clearly not of the form of ordinary atomic matter (called baryonic matter) and represents a major unknown in the Universe at large. Determining its nature is a major current research activity. It is embarrassing to have to admit that the quantity of dark matter could be ten times or more that of the matter we can see.

A1.9. Dark Energy

If the problems with dark matter were not enough worse was to come. The introduction was simple enough. Supernova explosions in galaxies show a characteristic light curve and a characteristic maximum intensity and have been used for some time as distance markers in the most distant parts of the Universe. Accurate measurements of the light curves of very distant supernova explosions (specifically of type 1A supernova) in the later 1990s showed red shifts that were larger than predicted by the luminosity. The two sets of measurement could be reconciled if the expansion of the Universe is accelerating. The data leading to this discovery were repeated and, as surprising as it was, found to be correct. The standard cosmological model based on general relativity predicted otherwise. The new measurements suggested that there is some force within space which counteracts the effects of normal gravitational attraction.

Actually the simple form of the Newtonian law of gravity had given difficulties from the start for very small and very large distances. Even in the 17th century some modification of the law had been proposed of the form $F = -\frac{GMm}{R^2}\exp\{-\alpha R\}$ where F is the attractive force between the two masses M and m and α is some empirical constant. If α is suitably small the law will be unaffected at small values of R to very good approximation, but will decrease at very large distances like $\frac{1}{R}$ rather than as $\frac{1}{R^2}$. The new observations (obviously not available in the 17th century) require a more subtle explanation.

One essential question when trying to unravel the history of the Universe is to know the type of special curvature involved. In particular, what is the geometry of space — time? More specifically, what is the ratio of kinetic to potential gravitational energy? With more and more data becoming available it was felt in the 1990s that this age old question could be raised once again but now for a definite answer. It was expected that the geometry would be Riemennian with a curved space specified by the three internal angles of a triangle adding up to less than 180°. To every ones' surprise this was not so: three angles added up to 180° making space–time flat and Euclidean. If this is true then there is a great deal of mass yet to be accounted for. We know such mass already in dark matter. Adding dark matter to the inventory certainly helped but it is not enough. Indeed it was only about 30% of enough!

There is another sea of energy — that driving the accelerated expansion of the Universe. Relativity theory tells us that the energy has got mass and so can contribute to our sum. This energy driving the expansion and the acceleration is called dark energy. There is as yet no agreement about its nature although it exerts a negative pressure. A very popular identification is with the cosmological constant of the field equations of general relativity. This will provide a homogenous and constant energy density which will not diminish on further expansion since it is an intrinsic property of space–time itself. Other explanations would not have this property and this may be a basis for distinguishing between theories in the future. It appears that the quantity of dark energy is such as to provide this critical mass to make the space–time geometry flat Euclidean. It is still early days, but already there seems a general consensus on the portions of the various energies involved. A recent estimate proposed by NASA is given in Table A1.1.

It seems that the visible Universe that we started out with is but a tiny fraction of what is actually there. Most of the Universe appears to be composed of material that we know virtually nothing about.

We must stress that this is the current guess and is likely to change as more data become available. We are entering very new territory now. It is known since the work of P.A.M. Dirac in the 1930s that special relativity leads to the recognition of a vacuum energy underlying all baryonic matter and is associated with various sub-atomic particles, many transitory in time. There seems to be a major "basement" underlying the world which is yet unexplored. There is much still to be discovered about the gross Universe on every size scale.[18]

Table A1.1. The approximate percentages of various sources of energy within the Universe (after NASA).

Source of energy	Percentage of whole (%)
Heavy elements	0.3
Neutrinos	0.3
Free H and He	4
Stars	5
Dark matter	20
Dark energy	70

[18]The philosophers in Ancient Greece were concerned with what lies "beneath" the observable Universe, but had no rational answer to the question. We have the same problem with the same ignorance, but we call it now the quantum vacuum. Dirac showed in the 1930s that this is a real region underlying matter with real energy, but we have very scant knowledge of its properties and have no idea if there is yet another layer of existence underneath that. At the present time there seem to be a large duplication of different vacua, one for each sub-atomic elementary particle. How do we apply Ockham's razor?

A1.10.   Summary

In this Appendix, we have reviewed briefly the broad structure of the Universe at large stressing the historical perspective. The aim is to gives some perspective to the scope of the Universe as the backcloth for our studies. We have been interested in the way the information was obtained to gain some knowledge of the way immensely complicated material is reduced to a form we can understand. We work from the Earth outwards.

1. The development of our knowledge of the Solar System is treated first leading to three laws of planetary motion (a form of the conservation laws of energy and angular momentum) and the law of universal gravitation. These laws must apply anywhere in the Universe.

2. The problem of the measurement of distance is considered including parallax (triangulation), Cepheid and other variable stars and supernovae. These methods lead to the present recognition of a broken up Universe composed of some 10^{11} or 10^{12} separate galaxies, or "island universes" each on the average containing some 10^{11} stars of broadly solar mass.

3. Using spectral characteristics interpreted as an example of the Doppler effect, Hubble established the general expansion of the nearer galaxies away from the Earth following the Hubble law that speed of recession is directly proportional to the distance. This discovery has now been extended to apply to the furthest reaches of the Universe. It appears that the Universe is everywhere expanding in an organised way.

4. This Universe, so strange at the time, was a part of the then new concept of only relative knowledge. The consequence for matter on the grand scale was the theory of generalised relativity. The theory of gravitation relies on a continuum representation using a spherical Riemannian geometry with space and time combined in their analyses into a single space–time. Mass is now a manifestation of the bending of space–time while the bending itself is a manifestation of mass. Thus, gravity was interpreted as a bending of space–time rather than a force involving direct action at a distance.

5. From the beginning theory was designed to reduce to the already very well-established Newtonian form in the limit of very weak gravity. There is one difference: the gravitational influence surrounding a body now

travels at the speed of light in a vacuum rather than instantaneously as
in the old theory.

6. One consequence of the new theory following directly from the conservation statement take now is the appearance of a new constant called the
   cosmological constant and traditionally denoted by the Greek symbol Λ.
   It has the dimensions of energy and acts to counter the attractive effects
   of gravity.

7. Possible model universes based on general relativity are considered some
   with an origin at the origin of time and others which have lasted from
   an indefinite time in the past. The effect of Λ is seen clearly.

8. The present theory has the Universe being formed some 13.7×10^9 ya
   from a virtual point.

9. The different types of the galaxies are considered and the joining of
   galaxies into clusters with individual galactic collisions is considered.
   A representation of our own Galaxy based entirely on observed data is
   shown making it clear it has a barred spiral structure.

10. The study of the rotational stability of galaxies and of gravitational
    lensing leads to the recognition of an invisible dark matter component
    to the Universe.

11. We have no knowledge of dark matter except that it responds to gravity
    but does not take part in electromagnetic interactions.

12. It is found that the expansion of the Universe is now accelerating. This
    unexpected result implies the presence of an unknown dark energy. Its
    nature is entirely unknown, but it is thought that eventually it will be
    found to be a manifestation of the energy associated with the cosmolog-
    ical constant Λ.

Note

The detailed structure of the inanimate Universe that carries the capability of supporting living things has been assumed as part of the observed
Universe — Einstein called it the "divine inputs". It is certainly true that,
for example, extremely small changes in atomic properties such as the pro-
ton/electron masses or the strength of intermolecular forces would make life
as we know it impossible. There is also a fascinating link between cosmic
and atomic numbers. This had led to suggestions of parallel universes where

there may be matter or energy differently arranged from the one we know and which would not support life. These hypothetical matters have their philosophical interest but to explore them would take us too far beyond our present purpose which is based on observation. There is a literature about these very basic questions and references can be found in:

Dirac, P. A. M., 1937, *The Cosmological Constants*, Nature, London, **139**, 323.

Barrow, J. D., 2002, *The Constants of Nature: From Alpha to Omega*, Jonathon Cape, London,

which link the atomic and cosmic structures providing interesting links.

An attempt (not successful) to account for the structure of the Universe within itself was made by

Eddington, Sir Arthur, 1948, *Fundamental Theory*, Cambridge University Press (published posthumously). This has given rise to attempts to construct a general unified theory and a theory of everything, which have so far not proved successful.

Appendix 2

Further Comments

In Appendix 1, we considered the large scale structure of the Universe. Now we turn to a number of the components to describe them in a little detail and to stress the chemical composition of these lesser bodies neglected before. It will be seen that the composition of both inanimate and animate matter involves the same chemical elements.

A2.1. An Overall Cosmic Abundance of the Elements

It has been possible to find the composition of various bodies in the Universe and so to gain some knowledge of the relative distribution of the chemical elements. The spectroscopic studies of the surfaces of stars, and especially the Sun, and of dust clouds in galaxies have been especially important in this connection. We list in Table A2.1 what is believed to be a good approximate composition of the Universe as a whole. It is seen that the simplest atoms (hydrogen and helium) predominate overwhelmingly. We can see that the first and second chemically most active elements (helium is chemically almost inactive) are hydrogen and oxygen. These react violently together to form water so this can be expected to be the most abundant molecule (combination of atoms) in the Universe and in the appropriate temperature range the most abundant liquid (it is sometimes said that either petrol or beer will be the second most abundant). It must be stressed that this refers to the visible matter only and cannot include the major component which is dark matter.

A2.2. Forming Condensed Bodies

The visible material is colleted into galaxies most of which contain a great deal of hydrogen gas and some heavier elements. Two different energies are involved. The gas is in motion, and so has kinetic energy: it also includes

Table A2.1. The universal distribution of the chemical elements.

Element	Relative abundance by number (Si = 1)	Relative abundance by mass
Hydrogen	3.18×10^4	0.980
Helium	2.21×10^3	
Oxygen	2.21×10	
Carbon	1.18×10	0.0133
Nitrogen	3.64	
Neon	3.44	0.0017
Magnesium	1.06	
Silicon	1.00	
Aluminium	0.85	
Iron	0.83	
Sulphur	0.50	0.00365
Calcium	0.072	
Sodium	0.06	
Nickel	0.048	
The remainder		

the gravitational interactions between different parts of the gas, the so-called self-gravity. The actions of these two energies oppose each other, the kinetic energy tending to expand the gas while the gravitational energy tends to make it contract. The temperature is crucial because the kinetic energy increases with increasing temperature although the gravitational energy remains unaffected. The temperature of a galactic cloud is typically 10 K which corresponds to a typical energy of a hydrogen atom in the cloud of the order of 10^{-22} J. The attractive energy between two hydrogen atoms to form a molecule (the interatomic force) is of the order 10^{-20} J so molecules will not be broken up. The density of the cloud is usually such that the gravitational energy associated with it will be several orders of magnitude less than this. Left to itself the gas will generally remain a stable entity indefinitely and in particular will not condense.

This equilibrium will be disturbed, however, if anything happens to compress the gas, increasing its density and so also increasing its gravitational energy without seriously changing its temperature (that is its energy). One event that will change the density markedly is if the gas happens to be near a supernova star explosion. Now supersonic waves that are produced in the explosion will compress the gas in the immediate neighbourhood and it is then possible for the gravitational energy to become strong enough to cause the individual atoms to condense to become material of higher mean density. The process was first considered by Sir James Jeans (1877–1946). He introduced the criterion that a mass of gas will condense if the time, τ_f, for its free fall under gravity is less than the time, τ_s, for sound waves to travel through it that is $\tau_f < \tau_s$. In terms of the parameters specifying the condition of the gas, and explicitly the density ρ, the mass M, size L and the temperature T, we have $\tau_f = \frac{1}{\sqrt{G\rho}}$ and $\tau_s = \frac{L}{C_s}$ where G $(= 1.67 \times 10^{-11}\,\mathrm{m^3 kg^{-1} s^{-2}})$ is the Newtonian constant of gravity and c_s is the local speed of sound. The free fall is possible if there is no opposition. The opposition will come from the pressure of gas. The speed of sound is directly proportional to the pressure and inversely proportional to the density. In this way, the speed of sound is a measure of the pressure to resist gravity. The Jeans criterion tells us that if the pressure force, expressed by the sound speed, offers too little resistance to gravity, expressed by the free fall due to gravity, then the gas must collapse. Developing this argument leads to expressions for the Jeans mass, M_J, and Jeans length R_J which represent the smallest values beyond which condensation will occur. Explicitly, it turns out that

$$M_J \approx \frac{AT^3}{\sqrt{\rho}} \quad R_J \approx \sqrt{\frac{kT}{G\rho M}},$$

with $k = 1.38 \times 10^{-23}\,\mathrm{J/K}$ being the Boltzmann constant. If $R > R_J$ or if $M > M_J$ then gravity wins and condensation occurs. This general process is accepted as being the method by which many larger bodies form including galaxies and stars. In the case of stars it appears that bodies of mass in excess of about 10^{28} kg, or about $\frac{1}{100}$th the solar mass, can form by condensation. There is an upper limit for the mass: this is likely to be no greater than the order of a very few 100 solar masses or about 10^{32} kg. The reason for the upper limit is because rapid condensation will increase the temperature of the main body of the gas. This will increase the kinetic energy and so oppose condensation. Virtually all stars will fall between these two limits of 10^{28} kg and 10^{32} kg making four orders of magnitude. It is believed that planetary bodies are formed by a different mechanism but we will consider that elsewhere.

A2.3.   The Detailed Constitution of the Sun

It is seen from Table A2.1 that the principle components of the visible Universe are hydrogen and helium gases. These will collapse to form a star given the right opportunities. It is observed that the majority of condensations are stars not unlike the Sun in mass and size (mass $= 1.99 \times 10^{30}$ kg and radius 6.6×10^5 km) to within a relatively few percent. They have luminosities of the order 5×10^{26} J/s giving a surface temperature of about 6,000 K. The central temperature is estimated to be of the order 1.5×10^7 K. Our

Table A2.2. The principle chemical elements associated with the Sun and their percentage abundances.

Atom	% by number	Atom	% by number
Hydrogen	92.1	Neon	0.0076
Helium	7.8	Iron	0.0037
Oxygen	0.061	Silicon	0.0031
Nitrogen	0.030	Magnesium	0.0024
Carbon	0.0084	Sulphur	0.0015
Remaining elements			

nearest star is the Sun and as an example we list in Table A2.2 the principle chemical elements in terms of the percentage of the number of atoms. It is seen that for many purposes it can be approximated as a hydrogen body. These data have been measured at the surface (because one cannot directly measure the interior composition) and the list is believed nevertheless to be a faithful representation of the whole.[1]

A2.4. Stellar Constitutions in General

There is a very wide range of stars in each galaxy, but perhaps 95% of stars in our Galaxy are very similar to our Sun in mass and size. This is also true of the other galaxies whose composition has been catalogued.

A2.4.1. *Hierarchy of stars*

This wide range of stars is generally set into a category of five different types. This represents an evolutionary sequence of stars. The first category involves the smallest stars called dwarf stars though they might not seem so — the Sun is a typical member. It is not that they are actually small (there are smaller stellar objects as we will see below) but they get larger as they evolve. These stars burn hydrogen and this is how they spend most of their lives. They move to become sub-dwarfs with increasing luminosity as they get towards the end of their lives. The third category includes the normal giant stars. This stage will include low mass stars (again like the Sun) having swollen to become giants stars of high radius at the end of their lives. It is also a stage reached by high mass stars as they move to become larger; Acturus and Aldaberon are examples here. The fourth category contains

---

[1] The element helium was first discovered in the Sun — hence helium from Helios from the Greek for Sun.

bright giant stars with Alphard being an example. The fifth category contains the super-giant stars. It is usually divided into two sub-categories 1a (the very brightest) and 1b (the rest). These stars are very rare, perhaps one star in a million is in this category. 1b stars are Betelgeuse and Anters while an example of a 1a member is Rigel.

A2.4.2. *Expressing stellar evolution*

A method of expressing the evolution of stars in panoramic form was devised independently by the Danish astronomer Ejnar Hertzsprung (1873–1967) and the American astronomer Henry Norris Russell (1887–1957) and known as the Hertzsprung–Russell diagram. There are, in fact, three closely related diagrams. The basic idea is to compare the stellar luminosity with the surface temperature of the star. This is shown in Fig. A2.1. The variations arise from the particular choices of the indexes to achieve this. Thus, the luminosity can be the measured (relative) luminosity of the star or the (calculated) absolute luminosity of the star, though this presumes one knows the star's distance from the observer. The surface temperature (always represented "backwards" from right to left instead of the usual left to right) can be specified in terms of a bolometric temperature, a spectral temperature or a stellar class.[2] Whichever combination of the variables that are chosen the

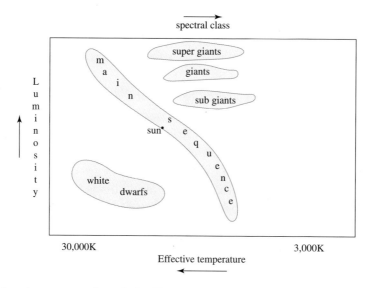

Fig. A2.1. A representation of the Hertzsprung–Russel Diagram. The location of the Sun is also given.

---

[2]These are O (very hottest), B, A, F, G, K, M (the coolest). This sequence is often remembered by the line: Oh, be a fine girl — kiss me.

observed stars are found to lie on well-ordered paths and regions indicating the evolutionary sequence. It should be noted that the professional literature uses the term evolution for the life history of a star and this must not be confused with the theory of evolution for living organisms.

The star will begin its life at the bottom of the main sequence (on the right in Fig. A2.1). This is a confined region snaking across the diagram but it has some width. This arises from the small differences of composition from one star to another. Although they are all composed overwhelmingly of hydrogen with some helium, the amount of heavier elements will vary by a very small amount depending on the precise composition of the dust clouds from which they were formed. Each generation of stars adds a quantity of the heavier elements to the environment when it reaches the end of its life. The Sun appears to be a third generation star, but other solar type stars may be younger. The heavier element component of the star is (inaccurately) called the metallicity. In Table A2.2, the elements that are not hydrogen and helium form the metallicity. This small variation of composition affects the magnitude of the radiation in a very small way and it is for this reason that the main sequence line in Fig. A2.1 has a certain width. The metallicity in a particular star is measured relative to the Sun and it is expressed logarithmically. The planetary system of interest to us in our search for extraterrestrials is composed of silicate materials and will accompany a star composed partly of the heavier elements. Stars of low metallicity (relative to the Sun) may not have sufficient metallicity to be associated with such planets. We shall, in what follows, be interested in solar type stars with a nonzero metallicity.[3] Apart from the metallicity, the compositions of the stars generally can be expected to be closely similar to the solar composition. It is possible, therefore, to discuss the main sequence stars as a single group, dealing with a representative member.

A2.4.3. *Comments on solar-type stars*

Because of the high internal temperature, all the atoms in the interior are fully ionised, the free electrons behaving like a gas. The stars are distinguished by having a mass to ensure a core region of sufficient material density and sufficiently temperature to allow nuclear fusion to occur with hydrogen being converted to helium. The interior is in radiative equilibrium which

---

[3]The very first stars in the present epoch are presumed to have been formed without any heavier elements (that is with zero metallicity) and these are called Population III stars. Those with very low metallicity are supposed old and are called Population 1 stars while stars like the Sun are population II stars.

means that the radius is held at an equilibrium value against gravity by the pressure of the radiation passing from the centre to the outside. The outer regions at least have the chemical components mixed by convection because the escape of heat to space is greater than can be carried by simple conduction alone.

The life time of a star is determined by its mass, M, the life time, t, being proportional to $1/M^4$ for a star like the Sun. The earliest stars, Population III, are believed to have been very massive — perhaps as much as 100 solar masses which is about 2×10^{32} kg. The corresponding life time would have been about 1 million years. The life time for a star of solar mass is of the order 10^{10} y. With an age of 4.75×10^9 y it is seen that our Sun is about half way through its expected life. The present range of masses for stars of solar type lie between about 10^{32} kg as the upper limit with about 10^{29} kg or about 1/10th a solar mass as the lower limit. The lower mass will have a life perhaps as high as 10^{11} y. The stars reach the end of their lives when their energy source runs out (see Appendix 4 for details). In summary, we have seen that stars such as the Sun formed by the collapse of a cold gas cloud (typically with a temperature about 10 K) of sufficient mass (in excess of about 1.6×10^{29} K) to compress the central region to a temperature and pressure where the thermonuclear fusion of hydrogen can occur. These will typically be a temperature about 10^7 K and a pressure of order 10^{12} bar or 10^7 atmospheres. The radius is maintained by the pressure of the radiation inside acting against gravity.

A2.5. Brown Dwarf Stars

An entirely different object is formed if the quantity of gas is insufficient to form a very hot central core. The cold gas with a mass less than 0.08 solar masses (about 1.6×10^{29} kg) but more than 13 Jupiter masses (about 2.6×10^{28} kg) still collapses and the constituent atoms are ionised, but the central region never attains a sufficient temperature to allow hydrogen fusion. The equilibrium radius is achieved when the quantum degeneracy of the electrons, in which they reside in their individual cells, resist the reduction of the cell volume beyond a characteristic amount. This is the same mechanism that determines the equilibrium radius of a white dwarf star. Typical central region values from a theoretical model give the central temperature a few times 10^6 K (about a factor ten lower than for a main sequence star) and a corresponding pressure of about 10^{11} bar which is about 10^6 atmospheres. These conditions are not sufficient to allow general hydrogen fusion, but can support deuterium (heavy hydrogen) fusion and also lithium fusion. Here the collision between ^{7}Li and a proton produces ^{4}He. The result is a small quantity of energy through fusion but the main contribution arises from the contraction of the initial gas to form the star. The radii of the stars of this type are found to be closely the same within 10–15% although the surface temperatures may vary more widely, between about 750 K and 2,000 K. The stars cool quickly and can be expected to have a "radiating" life of only a few million years.

The observational history of brown dwarfs is relatively short. It was only in 1995 that the first member was verified, in the Pleiades cluster of stars. An X-ray flare was observed to occur on LP 944-20, a body of 60 Jupiter masses. Radio emissions have also been observed from this body. This suggests the presence of magnetic fields of substantial strength which vary with time. Some rather exotic examples have also been found even including a member whose spectrum gives a strong methane signature. Observations are consistent with the surface of the brown dwarfs being encased in a substantial turbulent atmosphere.

A2.6.   Ice as a Planetary Material

The especial abundance of water makes it inevitable that it should feature strongly in the inventory of planetary materials. The satellites of the major planets are composed broadly of water as are the Kuiper bodies, the Oort cloud bodies which include between them the comets. Water is, in fact, the most extraordinary substance as we will now explain. Water (H_2O) is a tasteless and odourless substance which is inherently colourless in small quantities. Its molar mass is 18 g/mol and it has a density of 1×10^3 kg/m^3 in the liquid state but 0.9 kg/m^3 in the solid state. The fact that its solid form is less dense than the liquid form makes it unique among ordinary substances and has been of central importance for life. In fact, pure water at ordinary atmospheric pressure has a maximum density at 277.13 K (3.98°C). This means that ice formed at a water surface will float allowing only the surface region to be frozen solid while the underneath can remain liquid. The melting point is 273.15 K and the boiling point 373.15 K, giving the liquid form a temperature range of 100 K. This is unusually large. The triple point (the unique point where the pure solid, liquid and vapour forms coexist in a stable equilibrium) is 273.16 K at a pressure of 6.0373×10^{-3} atmospheres (611.73 pascals). The specific heat capacity for the liquid is 4,200 J/kg K, a value second only to that of ammonia. This is associated with a high heat of vaporisation (the energy required to convert liquid to vapour) of 40.65 kJ/mol. The detective fiction genre would suggest that water is a good conductor of electricity because many unfortunate people have been killed by having an electrical appliance dropped in their bath. Unfortunately, pure water is, in fact, an almost complete insulator although normal water contains a range of salts (including bath salts) and CO_2 which provide the commonly ascribed conductivity. This would be the condition to expect most often in planetary problems. There are three other properties worth noticing here. One is cohesion where water is attracted to itself to form drops when it falls on a surface, as in dew. This is associated with a very high surface tension. A second is capillary action where water will move up a narrow tube until this force is balanced by gravity. The third property is a very marked ability to act as a solvent. Substances that dissolve readily in water are called hydrophilic, or water loving, and include most salts (substances such as oils and fats that do not dissolve are called hydrophobic or water fearing).

These various properties come about because of the special atomic structure of the water molecule. The two hydrogen atoms are not set symmetrically about the oxygen atom but the hydrogen atoms are set at an angle of $104.45°$. The result is for the water molecule to be strongly polar with a net positive electrical charge on the hydrogen side and a corresponding negative charge on the oxygen side. Indeed, it is often convenient to regard the water molecule as a positive hydrogen ion, H^+, bonded to a negative hydroxide ion, OH^-. The same chemical properties apply to heavy water where deuterium (composed of a hydrogen atom and a single neutron) replaces the normal water. The peculiarities of the chemical structure can now be appreciated. The hydroxide end of one molecule is attracted to the hydrogen ion end of another, the whole forming a loose lattice structure.

Ice has a complicated structure occupying 14 distinct states as the pressure is steadily increased although the pressures may need to be very high to see the full range. For this reason it will be seen later that the most exotic forms are not likely to be of interest in planetary problems. The various forms of ice are described in the Tables A2.3 and A2.4.

A2.7.   Icy Bodies

The ices of the last section refer to pure water. In practice water is never pure but is mixed with minerals of various kinds. In planetary problems the impurities are very largely silicates. There are some impurities from other ices. Principle among these is ammonia (NH_3) and carbon dioxide (CO_2). From the gross point of view, however, these other ices, present in only very small quantities, can be disregarded in comparison with water ice.

Table A2.3. Forms of ice at lower pressure.

Lower pressure ices		
Type of ice	Temperature range	Comments
Amorphous	Around 0°	Lacks crystal structure — quick cooling of water
Ice Ih	Normal and lower	Hexagonal crystalline structure
Ice Ic	Metastable cubic crystalline structure with oxygen atoms in diamond structure	Formed between 130 and 150 K but stable to 200 K. Then reverts to ice Ih
Ice II	Rhombohedral crystalline structure	Forms from ice Ih through compression in the temperature range 190–210 K. Heating transforms to ice III

Table A2.4. The ten phases of higher pressure ice.

Higher pressure ices		
Ice	Temperature/pressure	Comments
Ice III	Tetragonal crystalline ice. Least dense of high pressure variants but more dense than ordinary water	Formed when water is cooled to 250 K under 300 MPa pressure
Ice IV	Metastable rhombohedral structure	Requires A nucleating agent
Ice V	Monoclinic crystalline structure	Cool water to 253 K at 500 Mpa
Ice VI	Tetragonal crystalline structure	Cool water to 270 Kat 1.1 GPa
Ice VII	Cubic structure	Hydrogen bonds form two lattices which interpenetrate
Ice VIII	Similar to Ice VII but hydrogen atoms have fixed positions	Forms from Ice VII when cooled below 278 K
Ice IX	Tetragonal structure and metastable	
Ice X		Forms at about 40–46 GPa
Ice XI	Orthorhombic form of hexagonal ice	
Ice XII	Tetragonal dense crystalline phase — metastable	Heat high density amorphous ice from 77 K to about 183 K at 810 MPa

A2.8.   Comments on the Composition of Animate Matter

An approach to a description of structure of living matter is best made in terms of the constituent atoms. It is an extraordinary thing that the constitution of animate matter is very simple in basis terms and follows closely the distribution of the chemical elements familiar in inanimate matter. We can see this from the following comments.

The basis of all cellular life on Earth now, and some viruses, is deoxyribonucleic acid (DNA) supplemented by ribonucleic acid (RNA). These are closely related polymers although DNA forms a double spiral whereas RNA generally does not.

Generally speaking, the strands of DNA can be very long whereas the strands of RNA are generally short. The DNA is stored inside the nucleus of the cell in eukaryotes such as plants and animals but in the cytoplasm of the cell in bacteria and archaea.

DNA is comprised of four pair bonds linked in a special way involving the four bases cystosine linked with guanine and adenine linked with thymine. These form a helix enclosed by sugar and phosphate molecules. Each of these bases has got a simple atomic structure. RNA has a closely similar pairing of molecules except that thymine of DNA is replaced by uracil. The structures are listed in Table A2.5. It is seen that they are all composed of a small range of atoms and especially of H, C, O, N and P. In this respect they are more elementary than stars which include a wide range of elements. The planets are certainly substantially more complex. It is seen that most of the basic molecules appropriate to animate matter are constructed from the very abundant elements H, C, O and N. The same is true of the other central components of animate matter. In the same table are the chemical structures

Table A2.5.   The chemical details of the components of DNA and RNA: of a typical amino acid; of a typical protein; and of ATP.

Molecule	Chemical formula
Cystocine	$C_4H_5N_3O$
Guanine	$C_5H_5N_5O$
Adenine	$C_5H_5N_5$
Thymine	$C_5H_6N_2O_2$
Uracil	$C_4H_4N_2O_2$
Typical amino acid	$H_4C_2O_2N$ R R = amino acid
Typical protein	$C_{400}H_{620}N_{100}O_{120}$ + + S, P
ATP Adenosine triphosphate	$H_{10}O_{12}C_4N_5P_3$ $3PO_3$; H_6CO_3; $H_4C_5N_5$

of ATP (the universal energy source for animate matter), a general protein and amino acids.

It is clear from Table A2.5 that only the simplest chemical elements are involved with animate matter with the exceptions of some sulphur and some phosphorous. The latter is used as an energy source because of the exceptional release of energy when phosphorous is ionised. The animate molecules are longer than the inanimate ones and in this way can lead to greater complexity in the macroscopic structures it creates.

A2.9. Conclusions

It can be seen from the arguments presented here that the Universe is a chemically single homogeneous structure in its basic ingredients. In particular, animate matter is composed of exactly the same chemical elements as inanimate matter and both forms are described by a common chemical distribution of the elements. Chemically speaking animate matter is only slightly more complicated than that constituting the stars and appears much simpler than the molecules constituting planetary materials. This implies that animate matter will have formed naturally within the Universe once sufficient quantities of the rarer chemical elements had been assembled not only to form the animate matter, but also especially the planetary body on which it can evolve to technologically capable levels.

A2.10. Summary

This Appendix is concerned with the bodies and the gas/dust that constitute a galaxy.

1. It has been found possible to specify in general terms the elements that make up the Universe and to give their relative abundances.

2. A brief account is given of the way gas clouds condense to form stars. The stars disintegrate when they reach the end of their fuel supply and form heavier elements in the process. These form new gas clouds richer in heavier elements and so form new stars with a larger abundance of the heavier chemical elements.

3. Our nearest star is the Sun and an account is given of its constitution. The constitution of stars more generally is also set down.

4. Brown dwarf stars are bodies with masses below about 10^{29} kg that therefore do not have enough mass to become a normal main sequence star but still radiate energy. These bodies are considered. Their surface temperatures are low and the luminosity small.

5. Rather smaller than brown dwarf stars are the super-plants. These are essentially hydrogen bodies with masses between about 3×10^{28} kg and the lower limit for brown dwarfs of about 10^{29} kg. They do not radiate energy appreciably and are, as the name suggests, very large planets.

6. Going down the scheme we met the gas giant planets, the major planets Jupiter and Saturn and perhaps Uranus and Neptune. These are composed of hydrogen and helium principally, but Uranus and Neptune also contain considerable amounts of water.

7. Next we consider the terrestrial type planets composed of the heavier elements in the form of silicate and ferrous mixtures. A list of the main chemical components is given for Earth and a restricted list for Venus.

8. Water ice is common in the satellites of the major planets and the surface rock has very largely this composition. The properties of ice are considered in this chapter.

9. Finally we consider the general composition of animate matter.

10. This chapter shows the common composition of the Universe in both inanimate and animate forms. It appears that all that is observed can be accounted for in terms of a single general composition and no special materials are needed to produce animate matter. The Universe in this regard behaves like a single self-contained entity.

Appendix 3

The Strange World of the Atom

We have seen that the fundamental bricks of the Universe are atoms, but these are far too small for us to see and touch. It is of interest to see how evidence for their reality was eventually found and how their properties came to be laid bare. The study turned out to be associated with some extraordinary conclusions about the degree to which it is possible to obtain unambiguous knowledge of the Universe.

A3.1. Early Thoughts about Atoms

The existence of atoms had been talked about for many thousands of years before they were actually shown to exist. The first question went back at least to the ancient Greeks who asked whether it is possible to subdivide a small volume of matter indefinitely or whether there is a minimum volume, however small, which cannot be divided further because this is the ultimate essence of the material. The Greeks divided into two camps neither of which found a definitive answer to the question but this began to be found in the 17th, 18th and 19th centuries. The final proof of the existence of atoms came much later, in 1908.

The first hints that there are units of matter came with John Dalton (1766–1844) who showed that the relations between chemical molecules could be represented by the combination of a set of basic components which he regarded as atoms. These ideas were supported later by Joseph Guy-Lussac (1778–1850) and Jacques Charles (1746–1823) with their laws of the behaviour of very dilute gases. Amedeo Avogadro (1776–1856) deduced the number of atomic units in a volume of material equal to the atomic weight in grams: in modern terms we use kilograms and the Avogadro number then is 6.022×10^{26}/kg. This is an extraordinary result not only because it counts the atoms, which were at that time very hypothetical for most people, but also it does more — since the volume of a gram molecule is known from experiment, knowing Avogadro's number allows a mean volume for an atom to be calculated — what an extraordinary result. The atomic nature of matter was later supported by the work of Dimitri Mendeleev (1834–1907) who showed in 1869 how the properties of the chemical elements show a periodic similarity expressed by his periodic table of the elements. Not all the chemical elements were known at that time and he had to leave spaces but these have been filled subsequently. It is now known that there are 92 permanent types of element, but unstable elements beyond this number have been synthesised in the laboratory. The total number has now reached 117 known chemical elements.[1] The periodicity of properties (accounting for such chemical similarities as those between lithium, sodium, potassium, rubidium and caesium as an example) is a strong indication of the atomic nature of matter. The reason for the precise chemical properties of an atom have also been identified and it is now realised that there is some ambiguity in each type of atom through isotopes but we will see what this means later.

A3.2. Atomic Structure of Gases

Guy-Lussac showed from experiments that the pressure, p, of a dilute gas increased or decreased proportionally with the temperature, T, which is written $p \propto T$. Another relationship was discovered by Jacque Charles that the volume, V, of a dilute gas is directly proportional to the temperature, which is $V \propto T$, if the pressure is kept constant. This means the volume goes up as the temperature goes up and vice versa. These results lead to the conclusion that $pV = $ constant if the temperature is held constant: if the temperature varies as well then we have instead $\frac{pV}{T} = $ constant, a relation usually known as Boyle's law (Boyle (1627–1691)). It was quickly realised that these strictly macroscopic laws must have an atomic basis if there really is a microscopic underpinning of the observed world.

The first thoughts of this atomic basis for gases were those of Daniel Bernoulli (1700–1782). Independently arguments were presented by John James Waterston (1811–1883), John Henapath (1790–1868) and James Clerk Maxwell (1831–1879). From all these ideas was developed the first kinetic theory of gases. The concept of the conservation of energy was devised by James Prescott Joule (1818–1889) while the elementary theory was pulled together and refined by Rudolf Clausius (1822–1888). The crux of the theory is that the macroscopic pressure arises from the motion of the constituent

[1]Element 117 (a member of the fluoride family) has not been synthesised but element 118 (a member of the inert gas elements) has been.

atoms hitting the side of a container while the temperature is a manifestation of the kinetic energy of the atoms themselves. The restriction to a dilute gas ensures that the atomic motions are very close to the ideal of straight line trajectories without collisions with each other or interactions of any kind. Regarding the gas as a collection of spheres in motion Maxwell was able to derive an expression for the distribution of the energy of the system of particles. The Maxwell distribution predicts the number of particles possessing energy in a given small range. It shows that all energies are present from the lowest to the highest although particles with the higher energies are very rare. This result has had many applications and an important one from our point of view involves the retention of an atmosphere by a planet.

Real gases are not quite like this because there is some interaction between the constituent atoms. The degree of interaction depends on how close the atoms are together so the interaction increases with the density of the gas. This new situation can be taken into account empirically by assuming the Boyle law statement to be modified appropriately. The effects of continually increasing the pressure in the gas, and so increasing its density, suggests the use of a series of terms, each term becoming of increasing importance as the pressure increases further. The first member of the series will provide Boyles law. The series of terms gives an expansion for the pressure in terms of the density or the inverse of the volume. If n is the number of atoms per unit of volume then Boyle's law is written

$$\frac{p}{kT} = n,$$

for a very dilute gas and for a gas of higher density there is the series of terms

$$\frac{p}{kT} = n + B_2(T)n^2 + B_3(T)n^3 + \cdots .$$

This expression is called the virial expansion. The quantities B_j are called the virial coefficients for the gas where j takes the numbers $j \geq 2$. The second virial coefficient B_2 accounts for interactions between pairs of atoms, the third virial coefficient B_3 takes account of interactions between three atoms and so on. In this way, the various interactions between the increasing numbers of atoms are accounted for. The development of the virial expansion was particularly the work of Maria Goeppert Mayer (1906–1972). The values of the virial coefficients can be calculated if the law of force between the atoms is known. A pioneer in representing the interaction by a formula was John Lennard-Jones (1894–1954) who took account of a strong repulsion when the atoms are very close but an attraction when they are further apart. But we are ahead of ourselves.

A3.3.   Statistics and Thermodynamics: Entropy

The very large value of the Avogadro number had a remarkable consequence. Ludwig Boltzmann (1844–1906) and J. Willard Gibbs (1839–1903) were the first to accept it to be of sufficient importance to account for it explicitly. The result was the changing of our view of the world for ever. The reason is simple to see now but was not generally conceded at the time. The astronomically large number of particles in real matter makes it impossible to follow the trajectories of all the particles separately and so to make any account for their various separate interactions. If the interactions are so important what is one to do? The answer given by Boltzmann and Gibbs is to treat the problem in terms of statistics. There was a general revulsion of the idea of a statistical world. According to the statistical approach, a number of microscopic states will be compatible with a single macroscopic state so the most probable macroscopic state for a system of particles will be that for which there is the largest number of microscopic states. It is this most probable state that will determine the macroscopically observed property. This was in essence the approach proposed and pioneered particularly by Boltzmann and also independently Gibbs. In these terms they developed the theory of what is now known as statistical mechanics. Using this general approach they were able to explain the rules of thermodynamics in statistical terms. This was highly controversial at the time and was not readily accepted generally then but has become the cornerstone of our description of the world about us. One of the most important concepts in thermodynamics is entropy, first devised by Clausius, and Boltzmann's major achievement was to give this a statistical interpretation.[2] To appreciate this achievement it is necessary to make a diversion.

Thermodynamics is concerned at its roots with the conversion of the various forms of energy into work. The great working engines in the 18th and 19th centuries were driven by steam. Nicolas Léonard Sadi Carnot (1796–1838) was the first person to consider quantitatively the conditions which will provide the maximum efficiency of a steam engine — yes it all started with steam. The argument runs as follows. Heat energy flows from a hot to a cold reservoir. A working engine must then have a hot reservoir to provide heat to drive it against a cold reservoir to receive the rejected heat. Rather surprisingly Carnot found that, although the efficiency of the engine is greater the higher the temperature of the hot reservoir, not all the initial

---

[2]The statistical formula for the entropy devised by Boltzmann (as a sum of states) is inscribed on his grave stone.

energy in the hot reservoir can be converted into usable work. Some heat must always be rejected to the cold reservoir — the engine must have both a hot and cold reservoir. Put another way, every working engine must have a hot exhaust pipe. Although the efficiency depends on the difference of temperature between the hot and cold reservoirs, there are practical limits to the value of this difference. Obviously the temperature of the high end will have limits determined by the nature of the material and of the fuel.[3] The lower temperature will be determined by the environment. Carnot found, however, that this lower temperature can never be actually zero although it may be very close to it. This is the origin of the discovery of the absolute zero of temperature. Carnot's arguments were presented by considering the workings of an ideal engine without friction and with an ideal working fluid. It was also shown by Lord Kelvin (1824–1907) that Carnot's arguments lead both to the recognition of an absolute temperature and to an absolute temperature scale both independent of the working fluid so long as it ideal. This is the Kelvin temperature scale which can be reproduced experimentally to a very high approximation. A very important quantity was defined by Clausius — this is the entropy. It is a measure of the unavailability within a system to do work. Work can be done only by transferring an ordered molecular state to a random one so one can regard entropy as a measure of the randomness of the molecules in a material sample. Physical processes in real systems spread energy among the molecules and so increase the entropy. The system is able to do no work when there is complete disorder among the molecules and then the entropy has got a maximum value.

The question then arises: how does one express the entropy? This was first done by Clausius. The definition involves the introduction of a very small quantity of heat δQ into a system at a constant temperature T without friction (reversibly that is): the introduction of heat increases the molecular disorder into the system and so increases the entropy S by the elementary amount $\delta S = \frac{\delta Q}{T}$. It is seen that the entropy increase is larger the lower the temperature and smaller the higher the temperature. The physical reason is clear: the higher the temperature the greater the random motion of the molecules; the greater the molecular motion, the less effect a small addition

---

[3]This analysis did not have any major effect on the efficiency of steam engines at the time, but it did lead to an important new power source — what is now called the Diesel engine. It works on the principle of explosive compression rather than spark ignition of fuel and was patented by Rudolf Diesel (1858–1913) in 1897. The essential feature is the high temperature is much higher than that for the steam engines of the time. Actually the first engine of this type but involving heat combustion rather than pressure combustion had been patented in 1891 by Herbert Acroyd Stewart (1864–1927) of Bletchley, England. The diesel engine was rather bulky in its original form and its present wide popularity has come about only after it was found how to reduce its size.

of energy will have. There are other definitions of entropy but all follow from the most basic expression due to Boltzmann. He introduced the full statistical approach using an analysis involving a combination of position and momentum of each molecule and called the phase. The combination of position and momentum of a particular molecule is called its phase. The distribution of the phases of a system can be plotted on a graph on a piece of flat paper with the momenta on one axis and the positions on the other. With all the phase points displayed in this way we are said to have moved to a description in phase space. A system of molecules at an equilibrium temperature will show the full range of phases, as Maxwell rightly showed for gas molecules, but many will centre about a mean value giving a very nonhomogenous distribution of the points representing the molecules of the system. It is this mean value that is taken to provide the macroscopic behaviour. Clearly, one given macroscopic state will encompass many microscopic sub-states. Boltzmann now made a very bold assumption. He introduced a quantity called the thermodynamic probability, denoted by W, and linked the macroscopic entropy S to W through the relation $S = k \log_e W$, where the logarithm is to the natural base e. Here $k = 1.3806505 \times 10^{-23}$ J/K is a constant of nature now named after Boltzmann. This is the expression for entropy which caused so much controversy but was sufficiently dear to Boltzmann that he had it inscribed on his grave stone in Vienna. With the entropy defined the remaining thermodynamic functions can be deduced from the well known and established relations between the various macroscopic thermodynamic quantities and the entropy. This provides a clear link between the macroscopic properties and those of atoms — unfortunately, the existence of atoms had not been achieved at that time.

A3.4. Entropy and Information

It has been discovered that there is a strong link between the entropy of a system and information about it: the greater one, the smaller the other — and the other way round. Whereas the entropy of a system in equilibrium has a precise meaning information is a more ambiguous quantity. It depends on the context and has been used especially in electronic engineering to express the content of messages. Here we see it in a rather different light.

The entropy of a system begins with a very small, essentially zero, value but as it changes under various constraints the value of the entropy increases. In the microscopic statistical world the initially placed phase points move to most probably values. The initial details are continually lost in this process until in the end, when the most probable positions are reached giving the entropy its maximum value all indications of the initial value have been lost. When a system is in equilibrium it has lost all evidence of its initial state. An example of this is any cooking oven. To cook a particular dish the temperature of the oven must have reached a pre-arranged temperature and must maintain this temperature for the time of the cooking. It is of no matter whether the oven in fired by gas or electricity or burning wood. It does not matter how the oven was lit — by a match, an electric spark or by a flame. It does not matter whether the temperature has been achieved by heating from cold or cooling from an even hotter one. The equilibrium will be the same for each.

We cannot have any information of the past when a system is in thermodynamic equilibrium. We can properly say that such information has been lost entirely. In this way as entropy is increased information is reduced. This relationship has prompted the nomenclature of regarding information of this type as being negative entropy. This is an important concept. One must be most careful not to attempt to associate information which is alleged to have been obtained from an equilibrium situation. We shall return to this topic in Chap. 6.

The equilibrium condition is reached through the action of processes which redistribute energy and which in the real world are dissipative processes. For instance, heat is transferred from one region of a system to another — in principle it could be achieved reversibly but in practice it is not. Thermal conduction means that ordered macroscopic heat flow is reduced by a loss to random microscopic atomic motion. The same is true for momentum in a fluid (associated with the process of viscosity). The energy lost in these processes is not recoverable and the processes themselves are said to be irreversible. Although these various irreversible processes make theoretical descriptions more complicated to describe, they play a vital part in everyday life. One example is normal friction. We rely on friction to be able to walk — without friction, the streets would be as slippery as an ice rink.

We can notice that the action of irreversible forces can provide a condition of mechanical equilibrium which prevents entropy increasing to its maximum possible value. This is achieved by the expenditure of energy of a particular kind. Energy is now being used "wastefully" from a pure point of view. Such sustained irreversibility is of great importance in the real world.

A3.5. Atoms and the Brownian Motion: Irreversibility

We have seen that the idea that the matter we see is underpinned by a sub-world of microscopic particles is much older than the thinking of such pioneers as Dalton. When this idea first emerged is not clear, but certainly the Roman poet and philosopher Lucretius (ca. 99 BC–ca 55 BC) knew of it when he wrote *On the Nature of Things* in 60 BC. There he described the random nature of small dust particles in the air illuminated by a shaft of sunlight. He thought the dust particles were agitated by a sea of smaller particles that one cannot see. A similar conclusion was drawn by the Austrian physician Jan Igenhousz in 1785 as he described the irregular motion of coal dust particles on the surface of alcohol (how he was watching that is difficult to guess). The classic observation of such irregular motion was described by the Scottish botanist Robert Brown (1773–1858) in 1827 and he has given his name to the effect. He described the constant irregular motion of pollen dust on the surface of water and thought at first the dust must be alive but found this was not so. Although there was general agreement that there was something in the fluid causing the irregular surface motions there was no idea what nor was any mathematical analysis attempted. This came rather later.

In 1880, the characteristics of the Brownian motion were found by Thorvald N. Thiele (1838–1910) when he wrote a paper involving the least squares. The classical basis of the Brownian motion is often described in terms of a drunken man trying to walk home. He takes one step in a random direction and another equal step in again a totally random direction and so on: he does not get very far. In fact the mean distance X he will travel in the time τ is proportional to $\langle X \rangle \propto \sqrt{\tau}$ rather than $X \propto \tau$ as in ordinary mechanics (describing the movement of a sober man). This square root behaviour is also a feature of least squares. The mathematics was taken up again by Albert Einstein (1879–1955) who published a definitive paper on the subject in 1905. The same problem was treated independently by Marion von Smoluchowski (1872–1917) in a paper published in 1906. Subsequently the whole theory was put in a definitive form in 1949 by Chandrasekhar (1910–1995).

Consider a semi-macroscopic particle of radius a suspended in a liquid of shear viscosity η at constant temperature T. Einstein derived the formula for the mean square deviation of the position of a semi-macroscopic particle during a time interval τ in the form

$$\frac{\langle X^2 \rangle}{\tau} = \frac{kT}{3\pi\eta a},$$

where k is the constant of Boltzmann ($= 1.308 \times 10^{-23}$ J/K). It is seen that the agitation, described by the value of $\langle X^2 \rangle$, is greater the higher the temperature, the smaller the radius and the lower the viscosity of the liquid. This is exactly the expectation for a particle agitated by smaller ones.

Using the mean square allows account to be made of positive and negative movements. Einstein further gave the expression for the Boltzmann constant

$$k = \frac{R}{N_A},$$

where R is the gas constant and N_A the Avogadro number. Jean Baptiste Perrin (1870–1942) subjected these formulae to experimental tests by observing under the microscope the erratic motions of suspended colloidal particles. He confirmed the formulae completely in the process deriving values for k and N_A that are very close to the values derived by other ways. The atomic structure of matter had been fully established and the work of the early pioneers like Dalton, Avogadro and Mendeleev fully vindicated.

These studies of the Brownian motion led to the development of the theory of irreversible process. An early contributor was Paul Langevin (1872–1946) who developed a semi-microscopic theory for which the acceleration of a particle is the sum of two terms, one statistical and the other describing a drag for a continuum fluid using Stokes' law. This proved the prototype for a range of theories describing irreversible behaviour which is a feature always shown either weakly or strongly by the real world. Understanding the nature of irreversibility has proved a long and difficult task which is not yet finally completed.

Actually a device was developed by C. T. R. Wilson (1869–1959) which shows the atomic particle to the eye or more accurately the track of the particles. This is the Cloud Chamber. Dust free super cooled supersaturated air (or alcohol) is placed in a closed container whose volume can be expanded very quickly. The passage of a charged particle through the container will ionise vapour particles along its track. A rapid expansion of the volume will cause condensation on the ionised particles and make them visible. The tracks result from the loss of energy by the particles entering the chamber. If the chamber is permeated by a perpendicular magnetic field the tracks of different charge will be deflected in opposite directions. If the strength of the field is known it is possible to determine the sign of any electric charge on the particle. The particle mass can also be inferred if the electric charge is known. This gave the first practical demonstration of the presence of ionised atoms and parts of them. The volume available for study was, unfortunately, very small and became insufficient as the studies moved to particles of higher and higher energies. Variations were developed and the very important one was the hydrogen bubble chamber by Donald Glaser (1926). This again has proved to be too confined for the very high energies of interest today and new techniques have been developed for this purpose such as the wire chamber and the spark chamber but we need not pursue these matters further.

A3.6. Understanding Atoms

Having recognised the existence of a range of very small (microscopic) entities called atoms the problem was to understand what there are like. They were thought for a long time to be rather like small spheres. This was the picture used to develop the theory of gases but it soon became evident that they are more complicated than that. A new theory was required and this was the quantum theory because it appeared that conserved quantities like energy and momentum could not have a continuum of values (as in the already existing classical theory), but only certain discrete values called quanta. The recognition of this new situation was first made by Max Planck (1858–1947) in 1901 in connection with black body radiation.[4] He introduced the new constant of nature denoted by $h = 6.67 \times 10^{-34}\,\mathrm{J\,K^{-1}}$ and named Planck's constant of action in honour of him. In particular the energy E associated with the frequency ν is given by $E = h\nu$. In classical theory E could have any value but not in the new theory.

It is well known how experiments, and especially those of J. J. Thomson (1856–1940) and Lord Rutherford (1871–1937), led to the recognition of the modern atom as a central positively charged nucleus surrounded by negatively charged electrons. At first it was thought that the electrons behaved as simple particles but later it was realised that this picture is too simple. Neils Bohr (1885–1962) was able to account for the spectral emissions of atoms through a set of discrete electron orbits selected by applications of the conservation of angular momentum using the Planck constant of action. This was the "solar system" model. The different orbits had different energies associated with them and the movement of electrons between orbits either radiated or absorbed energy in small packets or "quanta". This provided for the first time an almost exact calculated description of the spectrum of atoms. The model had its drawbacks. More specifically, it led to a flat atom but of course atoms are three-dimensional. The full dimensionality could be achieved by appeal to the wave properties. Louis de Broglie (1892–1987) combined quantum ideas with special relativity and found that a particle moving with momentum p has an associated wave length λ according to the expression $p\lambda = h$. This is not an ordinary wave but an imaginary one — it was later recognised as describing a probability. The electrons in an atom could now be replaced by these de Broglie waves to give a full three-dimensional structure. More than that, the different energy levels can be seen as the

---

[4]It is interesting that the study that led to this discovery was an industrial commission for finding the conditions of use in an electric light bulb giving maximum illumination with minimum energy.

fitting of a series of harmonic wavelengths to provide the electron orbits. The matter was taken further by Erwin Schrödinger (1887–1961) who constructed a mathematical equation for determining the wave magnitude ψ for a given event. He interpreted ψ as the amplitude of a probability wave and so provided a way of calculating the *probability* of a given situation, not its exact condition. The full probability is obtained from the quantity $|\psi^2|$ measuring the magnitude of the square of ψ. Schrödinger used his equation to calculate successfully the properties of several important physical effects which ensured that the introduction of the strange probability wave has physical validity.[5] Several strange consequences became evident. One is that there is no certainty that a particle in a closed box will be in one particular place: it could be anywhere within the volume, but there is a larger probability that it will be in certain places rather than in others. One could say that it is in many places at the same time but is most likely to be in one particular place if you look for it. Probability has overtaken certainty in this very strange world. This new situation led to considerable debate.

But is all this true? Various experiments (and especially those conducted by J. J. Thomson's son G. P. Thomson (1892–1975)) made it clear that electrons are not simple particles but can also behave as if they are a wave.[6] This interpretation would certainly be consistent with a spherical atom. However, the electrons do not appear as a particle and a wave at the same time — it is either/or. They are somehow both in a way we do not yet understand. This extraordinary ambiguity between waves and particles in matter is not understood but must be expressed mathematically in some way: it is, in fact, expressed by an uncertainty principle set down by Heisenberg (1901–1976) and called the Heisenberg uncertainty principle. This arose from an abstract mathematical procedure for describing this quantum world, matrix mechanics.[7]

The specification of a particle at any instant is through its position and its momentum. This is the way that the classical theory of particle motions, developed in the 18th and 19th centuries specified the motion. This also proved valuable in quantum mechanics. A wave has not got a precise position but it has got a frequency and an energy. The Uncertainty Principle refers to the particle and says that, as a matter of principle and not as the result of the

---

[5]Technically ψ is mathematically a complex quantity but we need not discuss this. It means ψ does not describe a "real" wave motion.

[6]It is said that J. J. Thomson received a Nobel Prize for showing the electron is a particle while his son received a Nobel Prize for showing that it is not — it is a wave.

[7]Schrödinger later showed the complete equivalence between his wave mechanics and Heisenberg's matrix mechanics.

techniques of observation, the uncertainty in the location of a particle, Δx, is related to the uncertainty in determining its momentum, Δp, by $\Delta x \Delta p \approx \frac{h}{4\pi i}$ where $i = \sqrt{-1}$ is called the imaginary unit (see the Compendium). The special theory of relativity relates position to time, t, and relates momentum to energy, E. This leads to the companion expression $\Delta E \Delta t \approx \frac{h}{4\pi i}$ giving the uncertainty in the measured energy of a particle which is observed to exist for time Δt. If a particle is at rest its position can be determined to very high accuracy since if $\Delta t \to \infty$, $\Delta E \to 0$ so there is no uncertainty in the location and its energy is precisely observable.

A3.7.   The Limits to Certainty

This view of probability was disturbing and gave many problems in interpretation. It was appealing to think that the statistical and probabilistic approaches to the Universe were temporary stop gaps that would be replaced as we understood more. Certainly Einstein took this view about quantum mechanics — might he yet turn out to be correct?

Thinkers in the past had based all their arguments upon the assertion that it is possible to recognise absolutes and develop arguments based on this. For example, Isaac Newton had no doubts when he introduced absolute distance and absolute time as coordinates for his description of the world. He even had a body α — which was the hypothetical centre of all things — where God sits. It was plainly evident that the world is "flat" so the geometry summarised by Euclid was the obvious, indeed only, description to use. His laws of mechanics and gravitation describe an absolute world in which the movement of bodies has an inevitability that used to be the characteristic of a railway timetable. There was the question of how quickly information could pass and, since this was then almost entirely by the passage of light signals, the speed of light was the crucial factor. Galileo attempted to measure this in a series of well-known experiments using lanterns on nearby hills of known distance apart but without success. He concluded not that the speed was indefinitely great but much more accurately that it is faster then he could measure. In these terms the local world could be described in terms of simultaneity of occurrences. It would be possible, if one had a chair at a suitable distance away (with perhaps a telescope) to see all the different aspects of the Universe working away and operating *together*. It is not surprising now, looking back, that this vision provided difficulties.

One problem of gravitation occurred at an indefinitely large distance away. If the spread of particles were not of infinite extent there would be a centre and so a centre of the mass which would attract every body to it.

This means the set of particles would be unstable in time against collapse. If this is not observed to be the case in the Universe something must be wrong with the description: gravity may not be correct at large distances; Euclid's geometry may not be the correct form — perhaps indefinitely large distances do not exist because the geometry is circular so things come back on themselves. Could even simultaneity be wrong?

The concepts of a point and of infinity have led to the most remarkable discoveries. These have interested thinkers from the earliest times even before the ancient Greeks but it seems to have been Galileo in modern times who realised that there can be more than one infinity and that a single point will have an infinity of segments. He was not tempted to pursue these recognitions mainly because the mathematics of the time was not sufficient for the job. Such an investigation was done, however, by Georg Cantor (1845–1918), the inventor of set theory. He found a range of infinites by considering different sets of numbers, each with an infinity of members. Some groupings can be shown to be more numerous than others providing an hierarchy. He sought a relation between these different infinities and called the existence of this relation his continuum hypothesis. The hypothesis has never been proved true nor has it been proved false. This work is now recognised to be of the greatest importance but was vilified by many leading mathematicians at the time including such respected figures as Henri Poincaré (1845–1912), Leopold Kronecker (1823–1891) and the great David Hilbert (1862–1943). The reason was that this work undermined the uniqueness of mathematics and showed it to be an empirical study just like the physical sciences. It is often said to be the language of science, placing it on the same basis as normal languages, and this has proved to be true. It is not possible to treat mathematics as a self-contained structure that can answer any question within the structure using prescribed techniques based on a finite set of clear axioms. Hilbert thought that if you set down a finite set of axioms then the rules of mathematics would allow any theorem to be proved absolutely. This is known now not to be so.

The matter takes on a more general form through the work of Kurt Gödel (1906–1978). He, too, was clear at the beginning that mathematics and mathematical logic has a firm foundation and that the confusion introduced by Cantor was apparent and not real. After a great effort he produced, in 1931, a short and elegant proof that this is, in fact, not so. He found that in making deductions from a finite set of axioms there is always one conclusion that cannot be proved right or wrong. There is, it seems, incompleteness in mathematical logic which is inherent within it. This is called Gödel's Incompleteness Theorem. The result has a wider validity than just in mathematics and applies to logic as well: this discipline again has not got a firm foundation. One very simple example that is often quoted as an example concerns a small town with a single barber. The statement is: each day the barber shaves all those men who do not shave themselves. This sounds very simple until one asks who shaves the barber? Does he shave himself or does the barber shave him? There is no answer within the given statement. A second statement of confusion comes from the person who says: I am a liar? If he is a liar then he I telling the truth — but if he telling the truth you cannot believe him because he is a liar. Perhaps a more difficult statement is: I always tell the truth. This need not be a lie to be untrue — truth often has many components and these can be offered in random order. There is not a lie anywhere here but the whole construct is meant to deceive without actually telling lies. The Gödel theorem is of crucial importance to analysis but is still not widely known or appreciated even though it was enunciated rather more than 70 ya. An important practical example of it would seem to be the Heisenberg Uncertainty Principle. The opposing concepts of particles and waves highlight the confusion in this case.[8]

One somehow feels more confident about things if one can compute something. Of course, there were no computers in Gödel's time but the idea of computing came to the mind of Alan Turing (1912–1954) who devised a hypothetical computer called a Turing machine to explore incompleteness. It was hypothetical then, of course, but none the less useful.[9] Turing found to his surprise that there will always be programmes that cannot provide the result to a problem — some problems are not calculable. If a computer receives such a problem it will go on calculating for ever, that is it will not ever halt but will go on calculating for ever without a result. This will certainly be true if the calculation meets an irrational number and no accuracy limit is placed on the result but Turing's Halting Problem is more general than that. No difficulty here, you might think — just put on one side those problems that will not halt and concentrate on the other problems that will. Unfortunately Turing also showed that there is no way of testing whether a given calculation will halt or not before starting the calculation. No general algorithm exists. In practice, of course, one computes to a given accuracy and stops after an assigned approximation is reached so these problems do not emerge. This means our knowledge is at best approximate and never exact.

---

[8]This was pointed out to the author by Mr. D. G. J. Cole.
[9]The first programmable computer was constructed during World War II by Tommy Flowers (1905–1998) of the British General Post Office Research Station at Dollis Hill in London. It was used at the British Government Code and Cipher School (GCHQ) at Bletchley Park particularly to break the German Enigma, Fish and other military and official codes.

The Gödel Incompleteness Theorems and the Turing Halting Problem have not yet been properly assimilated into our thoughts.

A3.8.   A World View

The relative status of mathematics and logic rather than having absolute credentials has very important limitations for our deeper understanding of the Universe. We try to find basic "truths" and generalise information to form "laws". One such generalisation is based on the recognition of energy in its many forms and its conservation. Another example is mass and its conservation. The fact that the special theory of relativity asserts that they are manifestations of the same entity does not help. That the mass, m, is an alternative expression for the energy E through $E = mc^2$ says nothing of the nature of either. The general theory of relativity relates the gravitational field surrounding a point mass with the bending of a continuum formed by space and time. This sounds great except one is not told what space and time actually are. The bending idea is, however, given credence by its effect on a light beam. A bent environment will cause a light beam to be deflected in a continuum theory of gravitation. The bending will be only local, of course, since it involves the environment of the particle. The form of the bending is dependent on the geometry that is involved[10] and the theory is based on the spherical Riemannian geometry.

It was known in the early1930s that sub-atomic matter is underlined by a further layer known generally as the vacuum. The first indications came from the work of *P. A. M. Dirac* (1902–1984) by generalising the Schrödinger equation to be compatible with special relativity. Dirac found that every Fermi particle has got an anti-particle which is the same as the actual particle in every detail but one. Thus, applying his argument to the negative electron he predicted the occurrence of anti-electrons, the same as the electron except that it carries a positive electric charge rather than the negative one of the electron. This predicted positive electron, or positron as it was named, was found experimentally later in atmospheric cosmic rays showers by *C. D. Anderson* (1905–1991) in 1932. These particles which underlie all matter are called elementary particles and it was found that they all have got an associated anti-particle. For some it will be the charge; for others the spin

---

[10]There are two distinct stationary paths associated with a given geometry, a geodesic curve (which is the usual "shortest distance between two points" in that geometry) and a null-geodesic curve. For Euclidean geometry these two paths coincide but this is not the case for other geometries. Free material particles follow the geodesic path while light beams follow the null-geodesic path. For flat Euclidean geometry both particles and light beams follow the same path.

direction but there is here a very rich array of symmetries specifying a beautiful "underworld" which we will not, however, need to explore here. Suffice it to say that underlying everything is a vacuum with interesting properties. A volume devoid of everything material that we can find is still not empty. It will still allow gravitational energy to pass and, of course, electromagnetic (e.g. light). This vacuum state is not yet understood. For instance, it has not been possible to visualise a single vacuum carrying many activities but instead each activity appears to require its own vacuum. This seems unlikely to be a final solution.

There are some unusual features associated with the existence of anti-particles. Since they are a mirror image of themselves they would merge with the release of rest energy plus kinetic energy were they to meet. This will generally appear as two equal photons, the interaction satisfying the conservation requirements for mass, energy, angular momentum and electric charge. Alternatively, a particle and an anti-particle can emerge from the vacuum for a time compatible with the energy time relation of the Heisenberg uncertainty relation. This means the heaviest particles can emerge but only for a very short time. Thus, two uranium atoms may appear but only for about 10^{-9} s after which they will merge back into the vacuum. A hydrogen pair would last out of the vacuum a 100 times longer. An interesting situation would occur if one of the atoms were to be removed by some process during that very short time. The remaining one could not enter back into the vacuum alone and so would be stranded in the real world. The vacuum would lose energy as a result.

A3.9. Summary

This Appendix has been devoted to the problems of finding and exploring an unknown atomic world which cannot be seen directly. Atoms are the basic building blocks of the Universe and it is therefore important to know something about them for that reason alone. The result is the recognition of fundamental limits to certainty and this includes mathematics and logic as well. This is a fundamental appendix.

1. Early thoughts about atoms led to comments on the atomic structure of very dilute gases. The extension was made to dense gases.

2. The special consequences which follow the very large value of the Avogadro number led to the acceptance that the everyday (macroscopic)

properties of matter can be linked to the collective behaviour of atoms only statistically.

3. This led to a recognition of macroscopic thermodynamics and especially of the statistical theory that must be constructed to account for it.

4. The entropy was introduced and its links with information were recognised.

5. Irreversible behaviour of matter was considered through the so-called Brownian motion of a semi-macroscopic particle suspended in a liquid.

6. It was realised that the atoms and their components might appear lie particles in many circumstances but that they equally behave as a wave in others.

7. This wave–particle duality very strange and leads us to accept limits on the accuracy of measurements of position, momentum, energy and time when discussing particles. This indeterminacy is a fundamental property of matter and is not a simple consequence of the unavoidable inaccuracies of measurement. This applies to macroscopic matter as well as atomic but there the uncertainties are hidden by the inaccuracies of measurement.

8. Following this it was found that there are limits to certainty in the Universe and that this has been recognised for over 100 y. We considered the Incompleteness Theorem of Gödel and the first concepts of computing through Turing's Machine leading to his Non-halting theorem.

9. The Appendix finishes with some comments on a world view resulting from the knowledge gained at the atomic level.

Appendix 4

Sources of Energy

All aspects of the Universe involve activity, nothing is static, and activity requires energy to drive it. The Universe has got its own sources of energy, different for the different components of different scale and purpose. These are of great potential power and great flexibility and they are, either actually or potentially, universal. Separately, the move of hominids to evolve to an advanced form with the ability to develop and use advanced technology has also required energy sources of appropriate density but now devised by the hominids themselves. They have had, of course, to use materials provided by the inanimate universe as a basis which has also included fossilised animate remains. The advance of mankind into space requires bespoke sources of energy of a form not needed before. The development of energy sources is, in this way, an integral part of advanced evolution. Something similar will need to have been done by extraterrestrials, if such exist, so a review of energy sources is very pertinent to our studies. Such a review is the subject of the present Appendix. This again will show the self-contained nature of the Universe which appears to have used every available asset to the best advantage.

A4.1. Conditions Inside a Star

The most widespread cosmic energy source is that which lights the cosmic sky and is in the stars. There are some 10^{11} of them in each galaxy and, with 10^{11} or 10^{12} galaxies, and so some 10^{21} or 10^{22} in the observable universe. The number varies from time to time as stars are born and die but the numbers we quote can be taken as mean values. Stars go through the cycle of birth, maturity and death while there is material (essentially molecular hydrogen) to make them. It will take a very long time for all this gas to be used up, but when this happens no more stars will be born and there will be an end when they die. This will, of course, be billions of years in the future.

When the stars go, so will the light. All was dark before stars were formed; all will be dark again when they have all gone.

At its very simplest a star is a large mass of hydrogen and can be supposed to be just that in many calculations to a quite good first approximation. If there were a big bang origin for the Universe the very first stars must really have been just that — hydrogen mixed with some helium which was formed presuming the Universe was created from an initial hot fire ball. It is usual now to call such bodies Population III stars and they are characterised by containing no heavy chemical elements (called euphemistically metals). They remain hypothetical at the present time because none has yet been discovered although some stars are known where the concentration of nonhydrogen elements is very low indeed giving them a closely hydrogen-only composition. As an example, for the case of the Sun the hydrogen content at the present stage of its evolution is about 70% by mass ($\approx 1.4 \times 10^{30}$ kg) with helium being about 26% and the remaining heavier elements accounting for about 3%. Such low levels of metals, spoken of as low metallicities, are broadly common for stars like the Sun (called now solar-type stars). Some stars, however, have even lower metallicities but none actually zero. The great preponderance of hydrogen in such stars makes it useful in a preliminary analysis to explore the interior as if it were, in fact, composed simply of hydrogen.

The local equilibrium conditions inside the star are determined by the pressure there. The gravitational force of the material acts to reduce the volume and so to increase the density. The increased density causes the pressure to increase and this opposes the gravitational collapse. This balance is a characteristic of hydrostatic equilibrium. The deeper one goes the greater the weight from the material above due to the gravity of the material and the greater the pressure to contain it. The pressure in the central regions will become very high by ordinary standards. Now, the pressure, p, is the force per unit area across a unit plane surface. For a body of mass M and radius R the force at the centre due to gravity is of the order $p_c \approx \frac{GM^2}{R^2}$ where G is the universal constant of Newtonian gravity. The area has got the dimensions R^2 giving pressure, a force per unit area, the dimensions $p_c \approx \frac{GM^2}{R^4}$. As an example, for a solar-type star, then $M \approx 2 \times 10^{30}$ kg and $R \approx 7 \times 10^8$ m so with $G = 6.67 \times 10^{-11}$ m kg s this gives the central pressure, as to orders of magnitude, of $p_c \approx 1 \times 10^{15}$ bar $\approx 1 \times 10^9 N \approx 1$ billion atmospheres. A more careful analysis gives $p_c = 2.5$ billion atmospheres. This is a high pressure by any standards and is associated with a high density and temperature. The number of atoms per unit volume under these conditions is of the order $n \approx 10^{32}$ giving a material density $\rho \approx 1.67 \times 10^{-27} \times 10^{32} \approx 1.67 \times 10^5$ kg m^{-3}. The central temperature, T_c, can be estimated. The central gravitational energy is of order $E_g \approx \frac{GM^2}{R} = k\,T_c$ where $k = 1.38 \times 10^{-23}$ J/T is the Boltzmann constant and T_c is the associated central temperature. Then $T_c \approx \frac{GM^2}{Rk}$. This leads to a value of the order $T_c \approx 10^7$ K, a very high temperature. Further, it seems that the temperature is higher the greater the mass but the precise dependence is not clear because R is also related to M for equilibrium. Under these extreme conditions the hydrogen will behave as an ideal gas. A proper analysis suggests that the central temperature is proportional to $M^{\frac{1}{4}}$ which is not, in fact, a very strong dependence on the mass. We conclude that all stars (except the very heaviest) have very similar central temperatures. The variations have, however, been found to be important.

A continuously acting source of energy at the centre of the star will release a range of energy photons and these will diffuse outwards to the surface, which is cooler, and then be radiated into the surrounding space. Each of the astronomically large number of atoms per unit volume at the centre contributes a free electron to the local volume which is available to scatter photons and so to change their energy. With such a dense concentration of scattering sites, the photons will not be able travel far (a fraction of a millimetre in fact) before being scattered with the loss of some energy. This form of scattering of photons by electrons is called Compton scattering.[1] Such continual scattering will have reduced the energy greatly by the time the photon reaches the surface. Each photon in this way follows a path dictated by chance so the path actually followed by different photons will be different. This means they will reach the surface with different energies. Some will have lost most of their energy while others will have lost relatively little. As in all statistical cases most photons will have lost a closely similar proportion of the initial energy so there will be a clear mean photon energy reaching the surface. This will characterise a mean surface energy and so temperature. Photons of essentially all energies will be emitted from the surface into the surrounding space with a mean energy showing the path that most photons will have taken. This spread of energies is the origin of the so-called black body spectrum of the radiation. All wavelengths will be present from those for γ-rays, X-rays, ultraviolet and infrared light, visible light and the range of wireless wavelengths. The shape of the plot of the intensity of a given frequency against intensity (and called the frequency or energy spectrum) can be appreciated in very broad terms. The contributions

---

[1]This was discovered in 1923 by Arthur Compton (1892–1962) and was a classical experiment — it showed that the photons must be particles and not waves. A wave scattering would not change the energy. The inverse effect is also possible and has important consequences in astrophysics.

of those photons that have paths that are very short or very long will be relatively small: most photons will have had similar paths between the two giving a maximum contribution at mid-frequencies. The spectrum curve will, therefore, have a hump in the middle tailing of to zero at both the very high frequencies and the very low frequencies. The actual shape is called a black body curve and its details will depend on the various interactions between the photons and the atoms.[2] Because the motion of the atoms themselves will be different for different temperatures this will affect the interactions, the spectrum curve will being different from one temperature to another. Generally speaking, the lower the temperature of the body the less difference will appear between the photon paths and the less pronounced the maximum of intensity will become. Experimentally determined black body curves are shown in Fig. A4.1.

This will not be a very rapid process. The driving force is the gradient of temperature throughout the star and this is very small. With a central temperature of about 10^7 K and a surface temperature of about 6×10^3 K the mean temperature gradient across the radius of about 7×10^8 m is $\frac{\Delta T}{L} = \frac{10^7}{7 \times 10^8} \approx 1.4 \times 10^{-2}$ K/m $\approx 1 \times 10^{-4}$ K/cm which would require care to be measured accurately in the laboratory. Calculations show that the mean

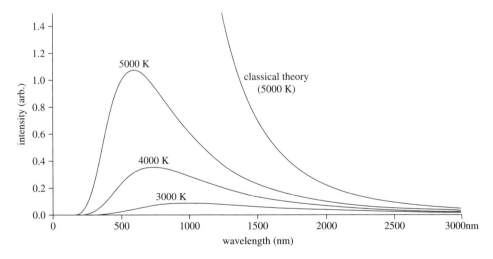

Fig. A4.1. The relation between intensity of radiation (arbitrary units) and the temperature for three temperatures of the emitting surface. The explanation of the curves was a triumph for the then new quantum theory. The prediction of the old classical theory is also shown. The classical theory approaches the new quantum theory in the limit of high temperatures.

---

[2]Max Planck successfully reproduced this curve in 1901 by introducing the then revolutionary concept of the quantum of energy.

time for the photons to reach the surface is some 700,000 y or very nearly one million. Some photons will take a longer time to reach the surface region while others will take a much shorter time. Once near the surface the material density is then very low, collisions extremely infrequent and the photon trajectories are essentially unimpeded straight lines. This happens over a very small region and it is this that gives rise to the observed surface of the star.

A4.2. Actual Stellar Energy Sources

What has been said so far is independent of the energy source provided it is dependent on a high temperature. The source of the energy of a star had been controversial for many decades although it was accepted that it must operate at an essentially constant level for a very long time (in our case the Sun must be at least as old as the Earth so must have lived for the best part of 10^{10} y). Actual burning and gravitational collapse were quickly ruled out as possible sources of stellar energy partly because of a lack of intensity of the resulting radiation that would be produced and partly because the process would last too short a time. The true source of the energy was eventually discovered by Hans Bethe (1906–2005) at the end of the 1930s and is the fusion of the lighter atomic nuclei to form steadily heavier ones. For the Sun and solar-type stars this is the conversion of hydrogen into helium.[3] The key was the enormous central temperature of a solar-type star which we saw in the last section to be of the order 10^7 K. This discovery led to the resolution of two earlier intransigent problems — the origin of stellar energy and the origin of the chemical elements.

These thermonuclear processes are dependent on the temperature so it is necessary to consider a number of different cases separately. Their speed of reaction generally increases with the temperature which means that heavy stars consume their fuels faster than lighter ones. This will be seen later to provide an important constraint on the form of the stars that can provide light and heat for the development of advanced life.

One thing should be made clear now. The nuclear reactions we will see in what follows are rather like a chemical reaction in that they are reversible, the initial ingredients which give rise to the final ingredients can themselves be obtained by reversing the process. The essential point is that, like some chemical reactions, the direction of the process depends critically on the

---

[3]Such stars lie on a single broad line in the illumination — temperature diagram (the so-called Hertszprung-Russell diagram), the single line being called the main sequence.

temperature. In general terms, the reactions go the way we set out for temperatures up to 10^9–10^{10} K. Above about this energy barrier they reverse. This has been realised to have a very important astrophysical consequence as we shall see later.

A4.2.1. *Solar-type stars*

For stars of broadly solar mass and radius detailed calculations show that the central temperature reaches up to 1.5×10^7 K. The lowest mass for such a star is about 2×10^{29} kg or about $\frac{1}{10}^{th}$ the mass of the Sun. Here hydrogen, the lightest element, is converted into helium, the second lightest, with the release of energy. The conversion process of hydrogen to helium in these early type stars is schematically as follows:

$$H_1 + H_1 \rightarrow D + \text{positron} + \text{neutrino},$$
$$D + \text{proton} \rightarrow He_3 + \text{gamma ray},$$
$$He_3 + He_3 \rightarrow He_4 + H_1 + H_1.$$

Here D is deuterium (proton + neutron which is H_2) and He_3 is an isotope of helium (with two protons and a neutron). The central ingredient of burning is hydrogen nuclei. Four hydrogen atoms combine (fuse) to form a single helium nucleus. There is a loss of mass in the process. Each free hydrogen nucleus has a mass of about 1.67×10^{-27} kg so that four would have a mass of about 6.68×10^{-27} kg. The mass of a helium nucleus is measured as 6.5915×10^{-27} kg. This 0.726% higher so the hydrogen atoms have "lost" 0.726% of their mass in the fusion — astronomers in the past misleadingly said the process is 0.7% efficient. The mass can be converted to energy units according to the mass–energy relation of special relativity $E = mc^2$. The mass difference between four hydrogen nuclei and one helium is $m = 4.87 \times 10^{-29}$ kg: $c^2 = 8.94 \times 10^{16}$ m^2. Then, $E = 4.35 \times 10^{-12}$ J is released per fusion and this energy eventually escapes into space from the surface of the star.

Once a generation of stars, like our Sun, has a component of heavier elements a slightly different and more efficient thermonuclear process can come into play. This involves carbon and nitrogen as catalysts. The components are now hydrogen, nitrogen and carbon which react to form helium, leaving the nitrogen and the carbon "untouched" at the end ready for use in a further cycle. The details of this more general mechanism are as follows:

$$H_1 + C_{12} \rightarrow N_{13} + \text{gamma ray},$$
$$N_{13} \rightarrow C_{13} + \text{positron} + \text{neutron (radioactive decay)},$$
$$C_{13} + H_1 \rightarrow N_{14} + \text{gamma ray},$$

$$N_{14} + H_1 \rightarrow O_{15} + \text{gamma ray},$$
$$O_{15} \rightarrow N_{15} + \text{positron} + \text{neutrino (radioactive decay)},$$
$$N_{15} + H_1 \rightarrow C_{12} + He_4.$$

The present Sun is believed to use this second scheme called the carbon–nitrogen cycle.

A4.2.2.   Details of the fusion process

We can make an estimate of the number of fusions involved per second in the conversion of hydrogen to helium whichever process is involved. The combined number of fusion processes provides the total energy that is radiated and which is the luminosity of the star. For a solar-type star this is of the order 4×10^{26} W. The number N of fusions required to provide such energy is the total energy radiated divided by the energy per fusion. This is then $N = 4 \times 10^{26}/4.35 \times 10^{-12} \approx 10^{38}$. The life of the star is likely to be about 10^{10} y $= 3.14 \times 10^{17}$ s. During this time, there will be of the order of $10^{38} \times 3.14 \times 10^{17} \approx 3 \times 10^{55}$ fusions involving 1.2×10^{56} hydrogen nuclei, remembering that each fusion involves four hydrogen nuclei. The total number of hydrogen atoms initially in the star is of the order $2 \times 10^{30}/1.67 \times 10^{-27} = 10^{57}$ atoms which is ten times the required number of fusions. It seems that only some 10% of the available fuel is used up before the process stops.

An order of magnitude value for the central temperature can be deduced from the self-gravitational energy. We will take the Sun as an example. Using the same notation as before, the self-gravitational energy, E_c, at the centre is

$$E_c = \frac{GM^2}{R}\frac{m_p}{M} = \frac{6.67 \times 10^{-11} \times 2 \times 10^{30} \times 1.67 \times 10^{-27}}{6.6 \times 10^8} = 3.24 \times 10^{-16}\,\text{J}.$$

This energy is equated to that for equipartition of energy, that is $E_c = \frac{3}{2} kT = 1.36 \times 10^{-23}\,\text{T J}$ giving

$$T = 1.5 \times 10^7\,\text{K}.$$

This, in fact, is essentially the correct answer quoted previously so our approximate approach has been surprisingly accurate. It is, however, the order of magnitude that is the essential thing. At this temperature the hydrogen atoms have, in the mean, sufficient energy to penetrate the repulsive barrier presented by each of their positive electric charges.

For a main sequence (solar-type) star there is an empirical relation between the radius, R, and the mass, M, of the form

$$R = AM^{3/4} \quad \text{with } A = 1.32 \times 10^{-15} \, \text{m} \, \text{kg}^{-3/4}.$$

The exponent here of $3/4 = 0.750$ is sometimes taken to be 0.80 in calculations but we will use 0.750. The important feature is that the increase of the radius with increasing mass is rather less than linear. The central temperature now is given by the expression

$$T_c = \frac{2}{3} \frac{GM^{1/4} m_p}{Ak} = 3.55 \times 10^{-23} M^{1/4}.$$

It is interesting to notice that the small exponent on M implies that the effect of mass on the temperature is not great. Expressed alternatively, the central temperature is only weakly dependent on the mass so all main sequence stars have comparable central temperatures. This implies that they all have the same energy production mechanism and broadly the same life times.

The effective temperature of the surface of a solar-type star is a little under 6,000 K while we have seen its central temperature is of the order 1.5×10^7 K. The total conversion process of hydrogen into helium takes place within the central part of the star, typically within the central 25% of the total volume. The encasing 75% acts as a blanket to maintain the central conditions through gravitational pressure. The mechanical equilibrium for the star is maintained by the balance on the material between the inward pressure due to gravitational compression and the outward pressure due to the radiation passing out along the temperature gradient of a temperature falling outwards. There is a small modification in practice because the outer region in fact tends to impede the flow of radiation: the heat flow then relies on the convection of the material. This gives rise to the cellular structure of the surface which is so characteristic of photographs of the Sun's surface. The sub-surface convection is the seat of surface phenomena and especially of magnetic fields, sun spots, flares and other associated phenomena. These processes radiate energy into the surrounding space especially in the form of short wave length radiation like ultraviolet light and X-rays so harmful to living creatures.

The supply of hydrogen at the centre is not inexhaustible. When the central fuel has been converted to helium in a solar-type star the central energy source is extinguished. The order of magnitude of the life of the star can be found from its luminosity, L, which is the amount of energy radiated to space from the surface per second. This energy must have been produced in the centre because no energy is lost during the passage from centre to surface. For a solar-type star we can set $L \approx 4 \times 10^{26}$ W. Each fusion releases

4.35×10^{-12} J which involves the loss of four hydrogen atoms. Consequently there are $n \approx \frac{4 \times 10^{26}}{4 \times 10^{-12}} = 10^{38}$ fusions per second which means that 4×10^{38} hydrogen atoms are lost per second. The initial stock of hydrogen atoms is $\frac{2 \times 10^{30}}{1.67 \times 10^{-27}} = 1.2 \times 10^{57}$. We have seen above that the active nuclear fuel in the core is about 10% of the total giving an initial fuel of about 10^{56} hydrogen atoms. The fusion process, therefore, can continue for about $\tau \approx \frac{10^{56}}{4 \times 10^{38}} = 2.5 \times 10^{17} s = 2.5 \times 10^{10}$ y. The Solar System is some 4.550×10^9 y old which makes the Sun now about half way through its life — it is middle aged.

A4.3.   The Death of the Star: The White Dwarf

When the central heat source ceases to operate the inward action of gravity is no longer balanced by the outward pressure gradient due to radiation. The star suffers a massive collapse, its radius decreasing to only a fraction of its former self. The further evolution of the star both as the energy source is declining and after it has gone is complicated with more than one possible outcome. In general terms perhaps some 10% of the original mass is ejected as gas during the collapse. This collapse is called a planetary nebula, see Fig. A4.2. The expelled gases often make the most beautiful and complicated patterns and can be made very colourful in false coloured light. One outcome is the formation of a white dwarf, a star of very high density. It is an object with only a little less than solar mass, but with a radius as small as a few tens of kilometres. The material mean density is of essentially nuclear magnitude and the interior is very special being in what is called a degenerate state. The atoms are broken down into nuclei and free electrons and are so compressed together as to revert to their quantum form. This can be expressed by saying that each electron moves into its own restricted individual volume, usually called a cell, and no two electrons will occupy the same cell. This characteristic is recognised by calling the electron a Fermi particle.[4] As the cell is compressed by gravity it is opposed by a force arising from the kinetic energy of motion of the electron — in graphic terms the electron moves ever faster to provide a counter pressure in its cell. This describes a quantum process and the degeneracy comes from the fact that this is entirely independent of the temperature: it is a nonthermal effect. Any thermal energy the white dwarf may have is radiated away until it takes on the temperature of the space surrounding it which is about $3\,\mathrm{K}$ (the temperature of the cosmic background radiation).

---

[4]This is because it follows the so called Fermi statistics which will allow only a single occupancy in a cell.

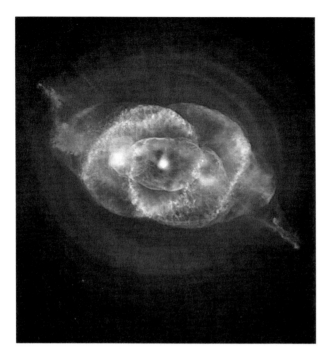

Fig. A4.2. The very beautiful Cat's Eye planetary nebula is listed as NGC 6543. It is a false colour composite picture: it was photographed in visible light by the Hubble Space Telescope while the X-ray emission was photographed separately by the Chandra satellite. The circles outside are an optical effect. The complex wind patterns are clearly evident. This is an end point of a solar-type star and our Sun will experience this convulsion in some 5×10^9 y from now when the Solar System will also be totally disrupted (NASA).

There is a limit to the size of the initial star to allow this process to occur. This is set by the magnitude of the gravitational force which itself is set by the mass of the star. The electron in its cell cannot move faster than the speed of light in a vacuum which determines the greatest resistance it can offer against gravity. After that the star must collapse unchecked. This situation was first recognised and treated by S. Chandrasekhar (1910–1995) who showed that the limiting mass is about 1.44 solar masses, that is about 2.85×10^{30} kg. This is called the Chandrasekhar limit. The condition will apply to stars like the Sun but the end point of heavier stars gives new phenomena as we will see later. It is to heavier stars that we now turn.

A4.4. Energy Sources for Heavier Stars

The nucleons forming a nucleus interact partly through the electric charges of the components, which provides a negative force for the protons, and through a positive (attractive) strong nuclear force which acts between all the protons and neutrons. The strong nuclear force is of short range, shorter than that for the electric repulsion between two protons. The balance between these two forces, one of attraction between all the components of the nucleus and the other repulsive between the protons alone, supplies the stability to the nucleus.

A4.4.1. *Binding energy*

The nucleus remains a stable unit because the strong nuclear forces of attraction between the protons and neutrons outweigh the electrostatic positive charges of repulsion between the protons. This is brought about because the number of neutrons is sufficient to allow an appropriate number of nucleons to lie close together. Generally, there are more neutrons than protons in the nucleus but by only a small margin. For the lighter elements the ratio is of the order 1.1 neutrons per proton, but this ratio increases to as much as 1.8 to 1 for the heaviest nuclei. The strong nuclear force is sufficient to ensure general stability for the lighter elements but not for the heavier ones. Most atoms have at least one isotope where the unstable nucleus achieves stability by the ejection of a sub-atomic particle and often more than one. These are the radioactive nuclei.

The forces inside the nucleus involve a considerable energy called the binding energy. This is usually expressed in million electron volts (Mev) and differs from one nucleus to another according to an ordered pattern. The usual representation is a graph of binding energy per nucleon in the nucleus against atomic number (number of protons in the nucleus). This binding energy curve is shown in Fig. A4.3. It is seen that the binding energy increases at first with increasing number of protons corresponding to an increasing energy but for heavier nuclei the energy decreases. This means that energy must be supplied to move along the curve of increasing number of nucleons until the maximum is reached which the curve shows occurs with the iron nucleus. At iron the curve has reached a rather flat maximum and decreases with increasing atomic number. This means that the elements beyond iron will release energy as they move to the next element on the right. The movement from hydrogen to iron involves adding nucleons to a nucleus and so is a fusion process. After iron the atoms tend to break up and so show fission — the release of energy is not now through fusion. The internal conditions involve grouping of the very stable helium nucleus (the alpha particle when ejected), but other decay processes are possible as neutrons decay to become protons and electrons (electrons cannot exist freely in the

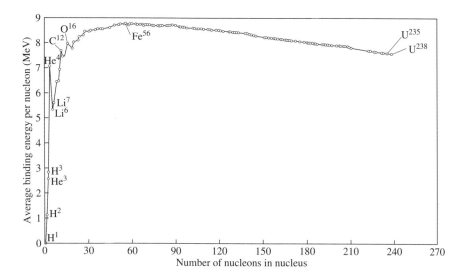

Fig. A4.3. The binding energy per nucleon plotted against the atomic number.

nucleus) or alternatively protons change to become neutrons and positive electrons which can be ejected.

The binding energy curves show that the chemical elements can be formed by combining nuclei up to the case of iron but beyond that the building process must stop. Up to and including iron the fusion process releases energy making the binding of the heavier nucleus more stable. Beyond iron energy must be supplied to bind the heavier nuclei together. The formation of the elements up to and including iron can proceed naturally in stars providing the temperature is right. The formation of elements beyond iron requires special conditions and will be considered later.

A4.4.2. *The extended reactions*

The formation of the more massive elements in stars involves the presence of higher temperatures which result from higher masses. The greater mass leads to a greater self-gravitational energy and so to a higher temperature in the central core region. As a region of fusion runs out of fuel the volume can still be hot enough to allow ash of a previous fusion to ignite and so itself become a source of fusion. The internal temperature increases with depth and the central region has a higher temperature than the contiguous shell. Shells of burning then result (it is often quoted like the interior of an onion) with hydrogen as the outer shell progressing inwards. This is shown diagrammatically in Fig. A4.4. The precise nuclear response to mass depends on the temperature reached. We have seen how hydrogen burns to form a helium ash for temperatures of a few tens of millions of degrees. If the temperature is some 10 times higher, say 10^8 K, the burning of helium ash to form carbon and oxygen becomes possible according to the scheme

$$3 \, {}^4\text{He} \rightarrow {}^{12}\text{C} + \gamma,$$
$$ {}^{12}\text{C} + {}^4\text{He} \rightarrow {}^{16}\text{O} + \gamma.$$

The carbon component of ash itself will burn if the temperature is a little higher, say about 5×10^8 K. The burning scheme then is not unique. Among the possible processes are:

$$ {}^{12}\text{C} + {}^{12}\text{C} \rightarrow {}^{24}\text{Mg} + \gamma,$$
$$ {}^{12}\text{C} + {}^{12}\text{C} \rightarrow {}^{23}\text{Na} + {}^1\text{n},$$
$$ {}^{12}\text{C} + {}^{12}\text{C} \rightarrow {}^{20}\text{Ne} + \alpha,$$

where ${}^1\text{n}$ denotes a proton and α denotes an alpha particle, that is a helium nucleus with four nucleons. In this way carbon burning can lead to several new elements.

Higher temperatures still lead to further burnings. At about 1×10^9 K, Oxygen burns again to give a range of other nuclei:

$$ {}^{16}\text{O} + {}^{16}\text{O} \rightarrow {}^{32}\text{S} + \gamma,$$
$$ {}^{16}\text{O} + {}^{16}\text{O} \rightarrow {}^{31}\text{P} + {}^1\text{n},$$
$$ {}^{16}\text{O} + {}^{16}\text{O} \rightarrow {}^{31}\text{S} + {}^0\text{n},$$
$$ {}^{16}\text{O} + {}^{16}\text{O} \rightarrow {}^{28}\text{Si} + \alpha.$$

Here, apart from the gamma photon, proton and alpha particle, ${}^0\text{n}$ denotes a neutron with no electric charge. Finally, silicon will burn at the higher temperature of 2×10^9 K. This leads to a chain of new elements and examples are:

$$ {}^{28}\text{Si} + \gamma \rightarrow 7 \, {}^4\text{He},$$
$$ {}^{28}\text{Si} + 7 \, {}^4\text{He} \rightarrow {}^{56}\text{Ni},$$
$$ {}^{28}\text{Si} + {}^{28}\text{Si} \rightarrow {}^{56}\text{Ni}.$$

Nickel will burn to form, successively, Cobalt and Iron:

$$ {}^{56}\text{Ni} \rightarrow {}^{56}\text{Co} + e^+ + \nu$$
$$ {}^{56}\text{Co} \rightarrow {}^{56}\text{Fe} + e^+ + \nu.$$

Here e^+ denotes a positive electron and ν denotes a neutrino, a particle of zero mass that carries energy. These examples show that the interiors of the stars are very active in forming the nuclei of the chemical elements up to iron and act as alchemists' laboratories.

A4.5. Death of a Heavier Star

The move to a white dwarf star was described in Sec. A1.3 involving a star of mass no greater than about 1.44 solar masses, but the thermonuclear processes described in Sec. IV.4 involve heavier stars. These have different and more violent ends and it is not relevant to our arguments to look at these in detail. This can be gleaned from the Hertzsprung — Russell diagram of section. There is one end point that, however, is relevant and that applies to the initial stars between about 1.44 and 4 solar masses.

Once the Chandrasekhar limit had been passes the electrons and hydrogen nuclei were compressed together by gravity forcing them to unite and form neutrons, electrically neutral particles with the mass of a proton. These are also Fermi particles with a single occupancy per cell. Again gravity can be resisted providing a body with broadly solar mass but with a radius of a very few kilometres (for a solar mass closely 3 km). These bodies are among the most dense in the Universe so dense, in fact, that the escape velocity is very close to the speed of light. One type has got a very strong magnetic field and is rotating very fast with periods in the millisecond range. These are pulsars with a distinctive radio emission. The energy is thought to arise from one of several causes including the energy of slowly decreasing rotation and energy from the accretion of matter. The radiation is released in each direction along the rotation axis giving rise to the analogy with a double light house light beam.

The explosive nature of many of the end points of stars has the important consequence of providing ideal conditions for the formation of the heavier chemical elements. Energy is now available to allow the fusion processes to take place beyond the iron maximum of the binding energy curve. These are reversible processes but at the high energies involved there is a net resultant residue of the heavier elements (see Sec. 4.7).

A4.6. The Production of the Chemical Elements: Nuclear Synthesis

It was seen in Sec. A4.2 that the conditions inside stars are ideal for the formation of a range of chemical elements up to and including iron. It is seen from Fig. A4.4 why this simple process terminates at iron. How, then, can heavier elements be formed? Is some form of "ladder" possible? What bearing has this for the Universe as a whole and especially for us interested in the occurrence of life? It is, after all, these heavier elements that help form planets and animate materials.

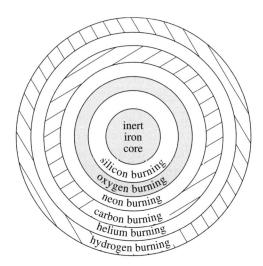

Fig. A4.4. A representation of shell burning in a hierarchy of massive stars with increasing mass. Hydrogen burning is the simplest and defines main sequence stars. Then the inner helium core burns and the ash of that can burn to form carbon which itself can burn in heavier stars to form neon. Neon burning will leave an oxygen ash which can burn to form silicon which can then burn to form iron where the process stops. These elements are thrown out into space at the end of the fuel when the star disintegrates. They are then available to be incorporated into new stars or into planets or into life.

There is one immediate problem with a fission ladder for producing the chemical elements which raises its head virtually straight away. Helium can be produced from hydrogen, but there seems no simple way of moving from helium to produce the next members of the Periodic Table, lithium and beryllium. The binding energy curve Fig. A4.2 shows that the helium nucleus has a higher binding energy than either lithium or beryllium, next along the chain. The problem of getting to carbon is essential because once that is achieved the formation the remaining elements up to and including iron can proceed without difficulty. The major quantity of lithium now is believed to result from the collision between heavy elements (already formed in space) and cosmic rays. Actually it is known that lithium is formed in stars but is very readily destroyed by the extreme conditions found there.[5] The details of completing the ladder linking hydrogen to the heavier elements were solved

---

[5]One of the positive features of the Big Bang theory for hot origin of the Universe is the possibility that some lithium was formed very soon after the beginning. This would allow the fusion process to proceed after the first stars have finished their lives.

by Salpeter (1924-) and Hoyle (1915–2001). The process has several stages. First helium nuclei must produce beryllium according to:

$$\,^4_2\text{He} + \,^4_2\text{He} \rightarrow \,^8_4\text{Be} + \gamma.$$

Then the beryllium nucleus is supposed to fuse with another alpha particle to form carbon nucleus according to the scheme

$$\,^8_4\text{Be} + \,^4_2\text{He} \rightarrow \,^{12}_6\text{C} + \gamma.$$

This can provide beryllium in principle but the question is whether it can have sufficient stability for it reasonably to be expected to occur in stars. Salpeter showed that his could happen for an interior temperature of 10^8 K where the material density would be about $10^8 \, \text{kg/m}^3$. Beryllium decays into two helium nuclei (the reverse process to above) in about 3×10^{-16} s but at this temperature and at the associated density it is possible for a third helium nucleus to fuse with a beryllium atom before the decay. It is also possible that some beryllium will be left in a stable form.

All this is fine except, as Hoyle pointed out, the double process would be unacceptably slow under reasonable stellar conditions. The way out of the dilemma was shown by Hoyle to suppose that the carbon forms with greater than normal probability which would be possible if the reaction was a "resonant" reaction in which the carbon forms in an appropriate excited state. The ground state energy of beryllium happens to be closely similar to the energy of two alpha particles. Again, the combination $^8\text{Be} + ^4\text{He}$ has almost exactly the energy required for a possible excited state of ^{12}C. The excitation energy could allow a "tunnelling" out so that the carbon nucleus could cascade downwards to the stable, ground level, of carbon. It seems that this could happen at a temperature of 10^8 K which is the magnitude of the central temperature of a 100 solar mass star. Such an excited state of carbon was not known at the time but was later confirmed experimentally by Fowler, the exact excitation energy being 750 MeV. This is an example of Hoyle's remarkable intuition in these matters. The practical result is the combination of three helium nuclei to form carbon according to this so-called triple alpha reaction. We can summarise it in its practical form as follows:

$$\,^4_2\text{He} + \,^4_2\text{He} \rightarrow \,^8_4\text{Be} + \gamma,$$
$$\,^8_4\text{Be} + \,^4_2\text{He} \rightarrow \,^{12}_6\text{C}^* + \gamma.$$

where the asterisk denotes an excited state for carbon. This triple process can be completed by

$$\,^{12}_6\text{C}^* \rightarrow \,^{12}_6\text{C} + \gamma + 7.367 \, \text{MeV},$$

this amount of energy being released. Once carbon is produced in a star the nuclear reactions already sited can go ahead. Beryllium has been formed and lithium will then arise from the alternative process as outlined above. Nuclear resonances play an important role in nuclear reactions: for instance the nuclear spin rules are responsible for the formation of unusually large quantities of carbon and oxygen, both vital for living materials. These essential chemical elements appear to have a special place in the creation process.

The path to the formation of elements from beryllium to iron in stars has been laid out but this still leaves the production of the heavier elements beyond iron. This can also involve the interiors of the stars but now under rather different conditions. The elements now can be built up one at a time by the addition of a proton but free protons are hard to come by. It will be remembered that neutrons are radioactive when free (with a half-life of about 20 min) becoming a proton and an electron. This conserves the zero charge of the original. Again with zero electric charge a neutron can more easily penetrate a nucleus. This transformation also takes place inside the nucleus, but now an electron is not sufficiently energetic to remain in the nucleus and must be ejected (this is called beta decay). These processes can be repeated by adding neutrons to existing nuclei and so continue to provide nuclei up to the stability limit for a collection of neutrons and protons which is 92 protons. The controlling factor for the working of such a process is the intensity of the neutron source. There are two possibilities because there may be a very prolific source or a sparse one. These two limits prescribe two separate processes called the r-process (for rapid) and the s-process (for slow).

If an atom is surrounded by many neutrons it may be able to absorb a number of neutrons before any of them decay to form protons. A neutron decay to form a proton increases the number of protons in the nucleus by one which means a new chemical element is formed one further along the atomic number scale. An example is offered by iron itself. If there are many neutrons in the environment of an iron nucleus the r-process can operate which mean that an iron nucleus could rapidly absorb one, two, three or more neutrons before any neutron will decay. Because the number of protons in each nucleus is the same each nucleus is one of iron but there are different isotopes. Adding neutrons to ^{56}Fe one at a time gives successively the isotopes ^{57}Fe, ^{58}Fe and ^{59}Fe. The first two isotopes are stable, but the third one is subject to radioactive decay through beta emission (the loss of an electronic charge). If ^{59}Fe does not absorb another neutron (as it would not in the s-process) it will decay to:

$$^{59}\text{Fe} \rightarrow {}^{59}\text{Co} + e^-,$$

since ^{59}Co possess one more +ve charge than ^{59}Fe. ^{59}Fe might absorb more neutrons to form ^{60}Fe and ^{61}Fe which is radioactive. The following decay reaction occurs in about 6 min:

$$^{61}\text{Fe} \rightarrow {}^{61}\text{Co} + e^-,$$

giving a different isotope of cobalt. This example shows both how the r- and s-processes operate and also how each can give rise to different isotopes of the same element. The process can be followed up to the stability limit of ^{92}U and ^{94}Pl. A study of the range of isotopes present in a given physical system can suggest the conditions of the formation of the material comprising that system. For the Solar System it appears that the s-process is most likely to have provided the majority of the material for its construction.

It turns out that not all nuclei can be constructed easily this way and this is especially true of those rich in protons. There is reason to believe that these nuclei are formed by a proton capture process (P-process). A source rich in protons could arise from an accumulation of neutrons which have sufficient longevity to decay into protons. These studies are currently being applied to the detailed study of the evolution of the chemical structure of our Galaxy where it may eventually be possible to cast light on the formation process.

A4.7. Brown Dwarf: Deuterium Burning

Naturally occurring hydrogen is a mixture of two components, "light" hydrogen with a nucleus of a single proton and heavy hydrogen, or deuterium, with a nucleus comprising a proton and a neutron bound by the strong nuclear force. This combination involves a binding energy. The mass of a free proton $= 1.67262 \times 10^{-27}$ kg while the mass of a free neutron is 1.674927×10^{-27} kg making a sum of 3.34754×10^{-27} kg. The measured mass of a free deuterium is $M_D = 3.343583 \times 10^{-27}$ kg which is lighter by the amount 3.96565×10^{-30} kg. The mass of two free deuterium nuclei which together would form two protons and two neutrons is $2M_D = 6.687166 \times 10^{-27}$ kg. This, of course, is a helium nucleus (alternatively called an alpha particle) with a free mass $M(\alpha) = 6.5915 \times 10^{-27}$ kg. The mass difference is $2M_D - M(\alpha) = 8.57 \times 10^{-29}$ kg and this is the binding energy which is made available upon merging. The energy associated with this mass is found from $E = mc^2$ giving $E = 8.57 \times 10^{-29} \times 8.940 \times 10^{16} = 7.662 \times 10^{-12}$ J per fusion and this energy is now available. This is very nearly twice that obtained from simple hydrogen fusion. On the face of it, it might seem that deuterium burning would be more effective than proton burning as a source of stellar energy. Unfortunately, the concentration of deuterium is small in solar-type stars because of the ratio $D/H = 2 \times 10^{-5}$. This reduces the effective energy per fusion to 1.532×10^{-16} J which is substantially inferior to the proton–proton reaction. It is, of course, available in a normal star along with hydrogen burning but its quantity is negligible.

There is a situation, however, where the deuteron–deuteron cycle is very significant. The close combination of a proton and a neutron reduces the effective positive electric charge at a distance from the nucleus making it relatively easy for two deuterium atoms to come sufficiently close to fuse. In fact the energy appropriate to a temperature of about one million degrees will be sufficient for this to happen. This is found in the core region of a mass of hydrogen and deuterium of about 2.65×10^{28} kg or about 13 times the mass of Jupiter. This situation continues as the mass increases until it reaches about 2×10^{29} kg when the ordinary proton — proton cycle becomes dominant. This region of about an order of magnitude in the mass defines a low luminosity object usually referred to a brown dwarf. The basic reaction is:

$$D + D \rightarrow {}^4He + \nu.$$

Fewer fusions are involved compared to the normal star burning the more common isotope ${}^1H$ with the consequence that the energy radiated is smaller by this fraction. The luminosity, therefore, will be reduced by the same ratio, i.e. 2×10^{-5}. This will lead to a surface temperature of only about 2,000 K. This is a relatively low temperature making the dwarf a very low luminosity object. The internal conditions will lead to a convective motion making the interior composition essentially uniform. The appearance of convection throughout the body of the brown dwarf distinguishes it from a normal solar-type star where convection occurs in the outer regions only. The internal turbulence will lead to a rather disturbed surface. Brown dwarfs will end up as essentially helium objects which will cool down slowly. The very low temperature of these stars will make them unattractive to anything other than the most elementary life.

A4.8. Gamma Ray Energies

A series of very small cosmic energy sources of the greatest power have been discovered and observed over the last decade or so. They have proved elusive to investigate. The initial discovery came from the military and followed the observations of the American satellites of the Vela group (vela is Spanish for vigil or watch) designed to test compliance with the 1963 Partial Test Ban Treaty banning the testing of nuclear weapons in space. The first satellites of the series had orbital distances between 63,000 and 70,000 miles kept a limited watch on space beyond the Earth's van Allen radiation belts . Later members of the series also viewed the electromagnetic spectrum for activity in the atmosphere as a whole so that atmospheric tests could be detected as well. The signature that the satellites were watching for is a double flash, one very fast and bright followed by a less bright flash of longer duration. This is a signature unique to nuclear explosions. Very high energy photons in the γ-ray energy range would be expected to accompany the flash. Photons with this range of energies were indeed detected but the signature was not that of a nuclear explosion.[6] This result that the γ-rays were not of nuclear origin provided great puzzlement although it was concluded early on that this was not an indication of military tests but was much more likely to arise from some unknown natural causes. Unfortunately, the observed flash was of too short a duration for the satellite to record details, the configuration not being suitable to observe the characteristics of the signals being detected. These appeared at random so it was not possible then to gain more information of this unexpected phenomenon. In particular, it was not known whether each flash resulted from a large explosion nearby or a massive explosion much further away (the old problem!). The random nature of the observations suggested a source outside our Galaxy, but a lower energy source was assumed rather than a higher which meant that it was supposed the flashes were extra-galactic but nearby.

So the situation remained for a decade or so but was changed dramatically in 1996 with the launch of the Italian/Dutch satellite Beppo — SAX. This is an X-ray satellite observing radiation principally in the energy range 0.1–200 kev. With this satellite it was possible to observe the full emission and not just the initial flash. It confirmed the isotropic distribution and went further to find on the average a rate of 10 flashes per year. It also found that each flash is associated with an afterglow. More detailed information building on this has followed from the Swift satellite launched in November 2004. Of especial importance has been a link with larger Earth-based telescopes that can be alerted quickly to observe the occurrences in detail. It seems that there is a range of time scales for the flash from a few milliseconds up to a

---

[6]One event of the correct type seems to have been recorded on 22 September 1979 but was later doubted to be anything other than of a natural cause.

few minutes. Measurements of red shifts confirmed that these events have very high red shift and therefore are at very large distances, generally of the order of a few billion light years. This gives them an age of half or more of the age of the present phase of the Universe. Such enormous distances mean the explosions are of enormous proportions. They must be the result of a huge collapse of a massive object. The short time scales mean that the physical dimensions are small. Typically one is looking at the collapse of an object some 100 km across in a time much less than a few seconds.

If the energy is passing through a spherical surface such as a collapsing spherical body, the energy involved must be similar in magnitude to the total rest energy of the Sun, which is about 10^{47} J. The energy we observe could, alternatively, have been channelled in each direction along an axis (rather like the beam from a lighthouse) in which case the energy source could be lower, perhaps 10^{46} J, although this is still a formidable compact energy source. In this case the object must be in strong rotation and may also have a very strong magnetic field. The predicted population of such sources will also depend on the mechanism for emission. If the emission is through the whole surface, the observed population will be the actual distribution because we will be seeing the whole object. If, alternatively, the emission is bipolar then the actual distribution will be substantially greater than that observed because we will see only a relatively few "searchlights".

The locations of energy sources as powerful as these are obviously of the greatest interest to investigations of the occurrence and establishment of living creatures. Such a source near a star with an orbiting planetary system would play havoc with the atmospheres and surfaces of the planets. This in its turn would affect advanced life very adversely although microscopic life might well survive. The same would be true were the explosion within a galaxy. Such very strong energy sources appear primordial: in observing the γ-ray burst it is expected that this release of energy was restricted to the early history of the Universe but one can never be sure.

A4.9.   Energy for Space Travel

Objects on a planet are held there by the force of gravity. Breaking away from the planet to enter space requires that gravity be overcome. This is achieved by motion: the kinetic energy of the object that is to escape must be greater than the gravitational (potential) energy holding it there. Suppose m is the mass of the body that is to escape and v is the speed that is necessary to achieve this then the kinetic energy $E_k = \frac{1}{2}mv^2$. Opposing this is the gravitational energy $E_g = \frac{GMm}{R}$ for the planet of mass M and distance R

from the centre. For the small body to escape it follows that $E_k \geq E_g$ which means $\frac{1}{2}mv^2 \geq \frac{GMm}{R}$ from which we obtain for the speed $v \geq \sqrt{\frac{2GM}{R}}$. This expression for the speed necessary for escape is universally called the escape velocity but this is a loose use of words because the energies involved are all scalar quantities (having magnitude but no direction) providing a speed v. The speed becomes greater the ratio $\frac{M}{R}$. In order to achieve an escape, then, it is necessary to supply the object with sufficient speed. We might notice that the mass of the object to be expelled does not appear in these formulae. The energy to reach ejection speed does, however, the greater the mass m the greater the energy needed to reach a given kinetic energy for escape. But escape from what?

A4.9.1. *The required speeds*

Escape from a planet or planetary body requires an appropriate escape velocity for that planet. The planet is, of course, in orbit about a central star. Depending on the exact speed of escape the body will go in orbit either about its planet or about the central star. To escape from the plenary system the body must then have the appropriate escape velocity for its place in the stellar orbit. As an example taking the Earth, a rocket must first be given the energy necessary to achieve an orbit and then in addition the energy to escape the Sun's gravity from its orbit. Only two then will it be able to leave the Solar System. Only Earth launched automatic space probes have been able to achieve this, Voyager 1 launched in 5 September 1977 and Voyager 2 launched slightly earlier on 20 August 1977. To date they are both in excess of 110 AU away which is closely 2×10^{10} km. There is still communication between the probe and the Earth team at JPL in California and it takes 14.62 h to receive a message from the probe. The probe has entered the heliopause where the Sun ceases to have any influence so the probe is actually entering interstellar space. It has taken 11 y to achieve this. These figures apply to the escape of a body from the surface of a spherical planet not in rotation. If the planet is rotating the rotation speed of the planet will add or subtract from these values. For the Earth this amounts to a speed of 6.8 km/s at the equator where the effect is a maximum. This assistance is valuable and explains why the two major launch sites on Earth (in Florida for NASA and in French Guiana for ESA) have been chosen where they are. The value of the escape velocity is lower the greater the altitude (remember it decreases as $\frac{1}{r}$) from the surface so these represent maximum speeds of escape because they can be reduced to some extent by launching with the rotation at some altitude (Table A4.1).

Table A4.1. The escape velocities associated with
each member of the Solar System.

Object	Escape velocity from surface (km/s)	Escape velocity from Sun at this orbit (km/s)
Sun	617.5	
Mercury	4.4	67.7
Venus	10.4	49.5
Earth	11.2	42.1
Mars	5.0	34.1
Jupiter	59.5	18.5
Saturn	35.5	13.6
Uranus	21.3	9.6
Neptune	23.5	7.7
Milky Way from Solar system	≈ 1,000	—

The speeds to escape from the particular objects are collected in column 2; the speeds necessary to escape from the
Sun at these orbit distances (and so from the Solar System)
is given in column 3. The velocity necessary to escape from
the Galaxy is also given.

A4.9.2. *Achieving an escape*

Altitude is important for another reason. The speed of sound at the surface
of a planet will be less than the escape speed and perhaps very much less.
This means a body travelling at that escape speed will be moving hyperson-
ically with a very high Mach number. Such a motion will generate enormous
friction forces which in their turn could heat the body up — perhaps even
to the point of glowing and being destroyed. The case of the Earth will show
this clearly. The speed of sound at sea level and at normal temperatures
is approximately $340.3\,\text{m/s} = 0.340\,\text{km/s}$. This will vary somewhat with
height through temperature and pressure but not by a significant amount.
The Mach number for the escape is the ratio of the speed of the rocket to
the speed of sound: in this case $\text{Ma} \approx \frac{11.2}{0.34} \approx 33$. This is an extraordinarily
high number and would be outside practical usefulness. The answer is to
achieve the (slightly lower) escape velocity higher in the atmosphere where
there is essentially no air resistance to make a high Mach number irrelevant.
The approach will be first to achieve an Earth orbit and then move to an
escape. There will be no such problem with the Moon which has not got an
atmosphere.

A4.9.3. *Fuels*

The pioneering work on rocketry was performed in the early years of the 20th century very largely by three men, one Russian, one American and the other Romanian/German. These were Konstantin Tsiolkovsky (1857–1935), Robert Goddard (1882–1945) and Hermann Oberth (1894–1989). The recent American activities have relied very much on the work of Werner von Braun (1912–1977) who was a pupil of Oberth's. These people set the scene of technical requirements and possible fuels that was to be developed later into the advanced activity that it is today. If extraterrestrials are to achieve space flight they must retread the same ground. It is not our intention to explore this work here in any detail but one or two general points should be mentioned.

Rocket propulsion relies on the working of Newton's 3rd law of motion: *to every action there is an equal and opposite reaction.* The particular application now involves a box closed on all sides except one. A gas is maintained in it at high pressure. There is, then, a pressure on all the walls except the missing one where a jet of gas is ejected. This means the pressure on the opposite wall is uncompensated so the box will move in the opposite direction to the jet. The propulsion of the box (rocket) is not dependent on air, as is an aircraft propeller engine because the rocket carries its own fuel. Not being associated with air indeed helps the movement because there is no air resistance. Once the rocket is in free space, so there is no force acting on it, it will follow the 1st law of Galileo/Newton: *a body not acted on by any force whatever will move in a straight line at constant speed or will remain at rest if it were so initially.* Here we have really summarised all the problems — how to contain the fuel, which fuel to use, how to carry the fuel in appropriate quantities and which trajectory path should be taken.

Tsiolkovskiy is known primarily for his theoretical work. He realised that any escape from Earth would require a very high ratio in the masses of fuel to payload. For a single stage rocket 90% of the total mass could be propellant fuel at the launch site. This remains true and a major problem. He also realised that the process of launch would need to involve more than one propulsion stage so that empty fuel containers could be shed to lose weight. Generally three stages have proved to be the best mix, including the payload stage, although sub-space activities need involve only one stage. Modern practice is to fill the different stages with different propellants. These may be solid, liquid or gas, or hybrid. Plasma has been experimented with more recently.

The final stage of a launch vehicle, which may weight anything up to 1,000 tonnes, is often powered by solid fuel as are stages designed to boost payloads into higher orbits. The solid materials can be homogeneous or composite. The former might be composed of a single compound (usually nitrocellulose 0 and have a burn time of rather more than 200 s. These are often used to separate a stage when it has done its job and now ceases to burn. Composite fuels will be a heterogeneous powder. This will be a mineral salt as an oxidiser, often aluminium perchlorate. This may be between 60% and 90% of the total propellant mass. The remainder is a polymeric binder, usually either polyurethane or polybutadiens which also acts as a fuel. The total has the consistence of a hard rubber. The Titan, Delta and Space Shuttle launch vehicles use solid fuel strap on modules to assist take off. These supplement liquid fuels. Thus, the Apollo– Saturn rocket used liquid oxygen and kerosene as its first stage and liquid oxygen–liquid hydrogen for the second stage. The Space Shuttle Vehicle uses liquid hydrogen–liquid oxygen as a contribution to lift off and for orbit insertion.

Mass is the main drawback with conventional fuels. This is associated with take off but for flight controls or, for interplanetary activities, landing and taking off at the destination. Anything that can reduce the weight of propellant to be launched will be most welcome. One such development is ion propulsion. Here ions are produced by electron bombardment and then are ejected at high speed to form a nozzle to move the rocket vehicle forward.

Long-term missions will require a long-term source of energy and especially of low intensity. This could be provided by the heat from the radioactive decay of a uranium/thorium salt. This is already used for automotive missions within the Solar System in the outer regions. Here the solar radiation has too low intensity to be used as a source of solar power. Solar collection is certainly valuable up to the distance of the Earth from the Sun and is, for instance, used as a power source in the International Space Station.

A4.10. Domestic Energy

We are particularly concerned with intelligent advanced life in the Universe and one important aspect of technologically capable life is the ability to produce sufficient energy and work as and when needed. This requirement will apply to advanced life anywhere. Controllable energy sources then become extremely important both for personal use and for "industrial" uses. The range of energy sources available on Earth will be typical of those expected on other planets and we consider these now. We will see that seven energy sources are available. Each can, in fact, be designated a cosmic source with a magnitude over which we have no ultimate control. Again, each is associated with the action of gravity either directly or indirectly. They differ in the intensity of the energy they can provide. Taking the Earth as the example, there is an evolutionary sequence leading to full technological maturity. In general terms, each step forward in the development of technological capability has involved more complex energy being available. We will consider these various sources in turn.

A4.10.1. *Direct stellar energy*

The source of this energy was considered in Section A4.2.1. The Sun radiates energy across the electromagnetic spectrum at a rate of about 3.90 W continuously — the Solar Luminosity. At the distance of the Earth (1 astronomical unit away) this has the mean value $1.4\,\mathrm{kWm^{-2}}$ on the surface with the Sun as centre. Not all this flux of radiation is absorbed by the Earth. To begin with about 30% of the radiation incoming to the Earth is immediately reflected back into space — this is the albedo of the Earth. This means only some $980\,\mathrm{Wm^{-2}}$ (or about $1\,\mathrm{kWm^{-2}}$ which is often taken in calculations) actually enters the Earth environment. Unequal quantities are absorbed by the atmosphere, and the surface (land and oceans). The surface absorbs about 80% of the radiation that is some $800\,\mathrm{Wm^{-2}}$. The remaining $200\,\mathrm{Wm^{-2}}$ enters the atmosphere. Because the planet is rotating the area covered by the incident radiation is $4\pi R^2$ where R is the radius of the planet: the "frontal" area for the radiation is πR^2. The incident radiation is 4 times that falling on each average area. This means the mean radiation is near $200\,\mathrm{Wm^{-2}}$. This is a mean value but the actual value will vary from pole to equator being largest at the equator. One acre contains $4046.85642\,\mathrm{m^2}$ so the mean energy received per acre is $8.094 \times 10^5\,\mathrm{W}$ or approximately $804\,\mathrm{kW}$. This value will not be available on a daily basis for two reasons. There is day and night which restricts collection to half a day. To obtain power in the dark it is necessary to have available a reliable chemical battery system to store the energy. If a greater power level is required several acres can be involved. All this sounds fine but it obviously requires a considerable technology if it is to be useful. It requires the components to be refined and the chemical produced. Making a battery is no easy task and making one of very high capacity is something that we have still to achieve here on Earth. Extending the collection area will involve conducting wire and a range of electrical connectors and ancillary equipment. Sitting in the Sun to get warm is one thing but building a technological community on it is another. This is not to say that it could not have a contribution to make but to an already developed technology.

A4.10.2. *Wind power*

The wind shows great strength from time to time and can certainly be used to provide energy locally. It involves rather less than 1/4 the direct solar energy so must be a minor player in an energy spectrum. Again its variability must make some battery/storage system essential for its general use. The difficulty with harnessing this source is that of arranging a collector well above the ground. The air is a viscous fluid and so will show a Prandtl boundary layer effect near the surface. The effect of the surface is alleviated to some extent by open stretches of land which makes the seas and the prairies attractive places for energy collection. Again an already existing high technology is necessary to make this an attractive energy source.

Wind interacting with the water surface produces sea waves which again contain energy that might be collected. Here again the collection process must involve advanced technology and is hardly likely to be useful to more primitive communities.

A4.10.3. *Palaeo-matter: Coal*

The accumulation of solar energy over long intervals of time provides a very effective high density energy source. It is effectively the fossilisation of photosynthetic materials where energy has been stored via ATP hundreds of millions of years ago. Substantial quantities of this energy can be released now by heating the fossilised material in the presence of oxygen. The effect is to release heat through the production of CO_2. This is the coal source which obviously has a definite, though perhaps long, life time for operation. Obviously the availability of this energy depends on the production of coal from substantial forest life in the past. This may not be available to any other civilisations on planets other than our own.

A4.10.4. *Palaeo-matter: Oil/natural gas*

Decayed plankton and elementary sea life is generally accepted to have produced the oil and natural gas found today. This energy could well be available to other civilisations elsewhere and is a high density energy source. It is highly likely that a successful technology could be built up *ab initio* with this as its basic fuel.

A4.10.5. *Nuclear fission power*

This is the only energy source that is quite independent of the Sun or central star. Its source is the heaviest naturally occurring chemical elements believed to have been formed in a supernova explosion. The uranium series of radioactive elements are central to achieve high density power. There are essentially two naturally occurring isotopes of uranium $^{238}_{92}U(99.3\%)$ and $^{235}_{92}U(0.72\%)$, but there is also a third with essentially trace amounts $^{234}_{92}U(0.0058\%)$. The half-life of the three isotopes is different but in proportion to their abundances. Explicitly, the half-life of ^{238}U is 4.7×10^9 y (incidentally also essentially the age of the Solar System) while the half-life of ^{235}U is 7×10^8 y. The half-life of ^{234}U is 2.46×10^5 y. The naturally occurring proportions are consistent with the isotopes having been formed at the same time in the past.

The isotopes differ in their responses to interactions with neutrons. In general terms, fast neutrons can enter any nucleus because, the neutron having no electric charge, there is no positive electric charge repulsion to keep the neutron out and it can bludgeon its way in. This is expressed by saying that the nucleus has a large neutron cross section for fast neutrons but a small cross section for slow neutrons. This is the case with ^{238}U but ^{235}U is different. It was found (in 1938) that ^{235}U has a cross section for slow neutrons which is approximately 1,000 times that for fast neutrons. By slow neutrons is meant here neutrons with average thermal energy of 0.025 ev or room temperature. This gives the average neutron a speed of 2.2 km/s ($\sim$8,000 km/hr or 5,000 mph). The entry of a thermal neutron into a ^{235}U nucleus causes the nucleus to split into two roughly equal halves. The results of this fission are not unique, a range of elements with atomic numbers in the 50 s (very often the rare earths) being the result. These are ejected with an energy of about 200 Mev (or 3.2×10^{-11} J) which is released as heat by collision between the ejected particles and the environmental molecules. This may not sound very much energy but with many fissions proceeding at any given time (upwards of 10^{20}) it is realised that the energy potential is enormous. In practice some neutrinos are released which carry away about 5% of the energy released but the result is still a massive energy source. Nuceli with a large thermal neutron cross section are said to be fissile. Other fissile materials are known including those produced artificially by neutron bombardment of stable nuclei. Among these materials are ^{233}U with a half-life of about 160,000 y, which is made from thorium $^{90}_{232}Th$ and plutonium $^{239}_{94}Pu$ with a half-life of 24,100 y made from $^{235}_{92}U$. Very small quantities of another plutonium isotope $^{244}_{94}Pu$ with a half-life of 80 My are found in nature. This is the most massive naturally occurring element. Heavier ones have been made artificially but they are all strongly radioactive and so very unstable lasting very short times when made.

These various materials have formed the basis of a nuclear power industry able to provide electrical power for the global community for a period running into several thousand years using the uranium cycle. More than this. It has been found possible for the plutonium side to breed fuel to give an essentially endless supply of energy. There are drawbacks of course. The ash of burning is still highly radioactive with very long half-life (going into many thousands of years) and this presents problems of security in waste disposal. Again, the materials have also been the basis of a military industry and the theft of radioactive materials raises fears of terrorist activities. These associations have given the industry a questionable name on Earth but the importance of this approach to energy provision cannot be over emphasised. Certainly it would be expected that any advanced technological society would also have developed these means of energy production especially so if its fossil fuel resources are limited.

The development of breeder reactors is especially significant. Here the reactor provides energy and produces more fissile fuel than it uses. Uranium and thorium can each be used for this purpose.

A4.10.6. *Nuclear fusion power*

The idea here is to apply the nuclear reactions that we spoke of earlier as providing energy for the stars to provide power for everyday use. There already is another application of this principle apart from the stars and that is the hydrogen bomb. These two cases represent opposite extremes. A star through its very high mass provides an enormous equilibrium central density and temperature which lasts at a steady level for several tens of billion years. All the energy processes take a long time: it takes of the order of 10^7 y for the energy to pass from the source at the centre to the surface and beyond. This is a totally controlled system in equilibrium. The nuclear bomb represents the other extreme. Here the fuel is contained in a small space (the size of a bomb) and not in a massive structure. The required density and pressure are achieved by explosive firing and the whole process lasts for only the merest fraction of a millionth of a second. The huge amount of energy released during this time represents but a fraction of that actually available. The system is in total disequilibrium once fired and so is very dangerous — but it is a weapon of war. These two examples of fusion need to be modified to meet the requirements for the social use of a stable, reliable and safe energy source. The need is for a source that will fit into a space of about the same size as a conventional fossil fuelled station. This requirement is more easily said than done. There are two very substantial problems that must be overcome — establishing the physics for such a system and then establishing the consequent engineering for running the station and supplying the power.

The problem of physics is to maintain a sufficiently high density/ temperature within an appropriate volume of gas without using all the released energy (or more) to achieve this initial configuration. To this point the required temperature has been achieved under laboratory conditions as has the required pressure but not the two together.

There are a number of choices of possible fuel but it is generally accepted that a process based on deuterium, ^{2}H $\equiv$ D, and tritium, ^{3}H $\equiv$ T, is likely to be the most effective eventually. Remembering n represents a neutron this is

$$D + T \rightarrow {}^4He + n + 14.1\,Mev.$$

The helium carries an energy of 3.5 Mev. Tritium is a very rare isotope of hydrogen with a half-life of 12.34 days and will need to be manufactured. Possible sources are the isotopes of lithium ^{6}Li and ^{7}Li according to the schemes

$$n + {}^6Li \rightarrow T + {}^4He,$$
$$n + {}^7Li \rightarrow T + {}^4He + n.$$

The second scheme produces a neutron which can make another reaction so this could be a self-sustaining source of tritium. Other reactions are possible but seem unable to provide the amount of energy of the D/T reaction.

The deuterium–tritium mixture would be in the form of an ionised plasma with the electrons stripped from the atoms. An ionised plasma in this form is found to be extremely unstable and a range of configurations have been tried to contain the wiggles that are observed. The most promising topology so far is the torus with an associated strong magnetic field to control the surface shape. This is the basis of the original Tokomak machine pioneered originally by the Russian scientists... Tamm and Sokolov in the 1950s. It would seem likely to form the basis for any future successes. The problems of plasma instability are central to the application of fusion and the success or otherwise in controlling it will very likely be central to any application of this approach in the future. Such a reactor machine could present a serious environmental hazard if not properly controlled.

The problems of engineering would probably be even more difficult than the problems of physics. To begin with the technology is of very high pressures and very high temperatures far in excess of anything we have got now.

The development of new techniques together with the associated safety procedures would present formidable problems. There must be materials problems in containing such inhospitable conditions and the development of new materials would be a central issue.

A new technology would be the containment of substantial quantities of tritium gas. Hydrogen is notoriously difficult to contain in a precise way but is it likely to offer an environmental hazard? Although of low grade the tritium radioactivity would be constant and surround every power station continually. Again neutron bombardment of the metal casing of the reactor would also be the source of radioactivity and this could be of longer half-life.

It should be recognised that the deuterium supply on Earth is limited for long-term purposes. This must put a limit on the long-term hopes of fission as an ultimate source of energy. The breeder fissile reactor would seem a better investment for the long-term future. If an advanced exo-civilisation were to be found it is most likely that its domestic energy sources would be based on breeder reactors.

A4.11. Summary

Energy is the essential entity in a changing Universe. This is just as true for living organisms as for the Universe itself. The living organisms is essential that energy sources must be under control and stable. In this chapter, we have reviewed the many energy sources relevant for the happy living and exploring of life. This will be as relevant to extraterrestrials, should they exist, as to Homo sapiens.

1. The Appendix began with an account of the general conditions within a star like the Sun. The source of energy is at the centre of the star and the effects were studied of the passage of photons of energy from the centre to the surface. The black body radiation curve characterises the process.

2. The actual energy sources in solar-type stars were next recognised and studied. These are the proton–proton cycle and the carbon–nitrogen cycle. The carbon–nitrogen cycle is believed to be the major energy source of our Sun. The magnitudes of various quantities are deduced such as the proportion of hydrogen that can be converted to helium and the time it will take for the hydrogen to be used up giving the life time of a star.

3. Next the discussion moved to the behaviour of the star when the usable energy is used up. This gives ultimately a white dwarf star after a period of instability when a range of gases and dust is ejected into space. They often have the most beautiful and complex patterns which are only slightly understood.

4. Next the energy sources of heavier stars were explored. A succession of stages of atomic burning is present in more massive stars, the hierarchy being that of increasing stellar mass. Hydrogen gives helium atoms which can be heated to form carbon atoms which can be heated to form neon atoms which can be heated to form oxygen atoms, and these form silicon atoms which then can form silicon atoms which finally will yield iron atoms. At this point the process stops because beyond iron requires heat energy to be put into the process.

5. The death of stars heavier than the main sequence as considered. The associated processes were seen to be fully able to provide the heavy element abundance of matter. This leads to a detailed account of nuclear synthesis.

6. Next the brown dwarf was treated as an example of deuterium burning.

7. The emission of photons with energies in the γ-ray range is considered next.

8. The arguments next turn to animate matter when the ATP battery is described. This is so effective an energy source that it is difficult to imaging that it would not have developed elsewhere if life had formed.

9. Space travel raises its own special energy problems. These will be equally shared between us on Earth and any extraterrestrials that might be else-where. It is interesting that these developments have been made on Earth only during the last 80 y although science fiction writers had considered the problems rather earlier.

10. The chapter ends with a consideration of domestic energy sources and those that would be necessary for the development of any advanced technological society anywhere. There comes a realisation that the problems we face on Earth would also be faced by any extraterrestrials that there might be, or have been.

Appendix 5

The Language of Science: Developing Mathematics

The development of science and from it advanced technology has been possible because of the parallel development of mathematics. This will also be the case for extraterrestrial civilisations (should they exist or have existed) and it is therefore of interest to see how this has come about here and why it should have been necessary. Of some interest is the time scale for achieving advanced status although centres around other stars may be more efficient at advancing than we have been. It will be seen in what follows that there was considerable duplication in the earlier stages of the mathematics story largely due to the effects of different civilisations. The various aspects of the "global village" of the modern world, which is increasingly providing a single way of life, are having their effects on mathematics which has now become essentially a single universal activity. We can do no more here than give very brief indications of when the new discoveries were made and a little account of the mathematical methods themselves hoping this might tempt the reader to delve further and explore this fascinating subject in more detail.

Why do we need a special language for science? It perhaps is not obvious why normal language is not adequate for scientific purposes. Certainly it is an obligation that practicing scientists must accept to explain what is happening in science in terms that everyone can understand. Why is not this sufficient? There are two reasons.

First we must have a language that will not change. Verbal language changes quite often and in unexpected ways. Take English as an example but other languages will give the same effect. It is obvious that modern English is very different from, say, Shakespearean English of 500 ya. To be simple then meant to be devout in religion and honest and straightforward — now it means to be a simpleton and rather backward, easily cheated. Even during the last 50 y the meanings of words have changed very dramatically.

One example is the word gay — then a gay time meant a happy, carefree evening or week end at a party of with friends but now it has very different connotations. The way to avoid this ambiguity of meaning is to construct a language of symbols where every member is defined in a quite precise way and where the definitions do not change. This is best kept separate from the changing face of everyday language. One can read today the symbols in Isaac Newton's Principia with exactly the same meanings as he meant them to have when he wrote the manuscript 400 ya. One can read Einstein's paper on general relativity now with precisely the meaning he set down some 90 ya.

A *second* insufficiency in verbal language is that there are so many of them and they are all changing all the time. If science is to be of international interest it is best to choose a symbolic language that can have international validity. Since it is meant to represent the real world it must have flexibility and be numerate. It must be possible to add to it as new situations arise but in such a way that what has been defined and constructed before is still entirely valid. It is possible to read what *Isaac Newton* or *Laplace* wrote knowing that one can read precisely what they wrote and not a modified version of it. These are the basic properties of mathematics. Its aim is to describe what is observed in the Universe and especially to explore the underling structure. The mathematics in this way takes the form of the material it is describing but there is more. Mathematics is a logical language which can itself have hidden structure that can be unfolded. This structure is associated with what we observe and may lead to a new understanding enhanced by experiments to reveal more about the world. There is, then, a linked structure between mathematics and the real world. Both are affected by what we observe: both are able to reveal new structures and patterns not thought of before.[1] And, like physics, mathematics is changed by new discoveries. This places mathematics as a practical experimental subject like physics itself. It is the language of discovery developed by our way of viewing the world.

The beginnings of mathematics. The very earliest use of numbers for adding and subtracting has been lost in the mist of time but it was clear that the earliest evidence we have got presumed a facility of both these processes. The very earliest mathematical writings we have discovered so far date back to about 6,000 BC at the time of the Egyptian Old Kingdom. Mathematics then had the very practical task of aiding the administration of the Kingdom and the ordering of religious activities. Taxes had to be raised on land and goods and this was the earliest reference we have got to mathematics.

---

[1] For a technical discussion of this special relationship see: J.P. Killingbeck and G.H.A. Cole, 1976, *Mathematical Techniques and Physical Applications*, Academic Press, New York.

This was elementary geometry and the beginnings of number theory. An important task was calculating how much material was required to build a pyramid and how to plan its shape and structure. There are two papyri, the Chid and the Horus, which give some information about these things although most information has been irretrievably lost. From these, it is clear that the Egyptians knew the special case of a triangle with sides 3, 4 and 5 units enclose a right angle. They were not able to generalise this knowledge into a general theorem — this was to come later in the work of Pythagoras.[2] The materials on which the mathematics was written were extremely fragile so it is very surprising that any evidence from this distant past has managed to survive to the present through a turbulent history.

The earliest firm evidence of ancient mathematics is contained in four written texts. These are concerned with Babylonian mathematic about 2,000 BC (Plimpton 322), Egyptian mathematics in the Old Kingdom about 1,900 BC (the Moscow Mathematical Papyrus) and about 1,650 BC (the Rhind Mathematical Papyrus) and Indian mathematics about 800 BC (the Shulba Sutras). These earliest activities had an immediately practical aim. They were for measuring land areas for taxation and for buying and selling land, calculations for selling and buying in markets and elsewhere and similar very mundane day to day activities. The approaches were rule of thumb with no attempt to find general arguments to support the special cases. This was a mental attitude which would not give rise to a theoretical description of the physical world nevertheless it gave full numerical support to daily activities.

Mesopotamia. The *Sumerians* gave the first evidence of a written mathematics. By 3,000 BC they had developed a sophisticated system of metrology and from about 2,500 BC developed sets of multiplication tables for daily use. The majority of tablets known so far belong to the period 1,800–1,600 BC and include a wide range of topics including: extended multiplication tablets, trigonometry tables, and methods for solving linear and quadratic equations. They calculated $\sqrt{2}$ correctly to five decimal places. The *Babylonians* used base 60 for their arithmetic from which we have got the measurement of time and angles.

Indian mathematicians added enormously to knowledge over many centuries. The *Sulba Sutras* (800–850 BC) were a set of geometrical texts that introduced irrational and prime numbers, cross multiplication including the rule of three, solutions of linear and quadratic equations, and a numerical proof of what we know as the Pythagoras theorem.

---

[2]What became known as the Pythagoras' theorem seems to have been known in various special forms from the very earliest times. Pythagoras was the first person to prove it as a general statement.

A highly significant step was taken in *Jaina mathematics* (400 BC–200 AD) when, although with religious connections, mathematics was studied for itself as a subject not immediately connected with the outside world. A very wide range of results were found including: transfinite numbers (later developed by Georg Cantor in the 19th century), set theory, logarithms, indices, sequences and progressions and permutations and combinations among many others. This power house was supplemented in the *Bakhshali manuscript* (200 BC–200 AD) by the introduction of the entity zero and of negative numbers. Computations were important and irrational numbers could be included together with the square roots of numbers even as large as 1 million to at least 11 decimal places. This could perhaps be compared in power to a Turing machine.

In about 400 AD *Surya Siddhanta* displayed the trigonometric functions sin, cos and inverse sine. A little more than a century later *Aryabata* provided the first sine tables and developed the versine function. He also introduced infinitesimals and differential equations. He made very accurate astronomical measurements and, a significant point, he based his analyses on the assumption of the Sun at the centre of the Solar System, the heliocentric theory which entered Europe a millennium later. The 7th century saw the work of the gifted mathematician *Brahmagupta*. He introduced the number system that was later championed by the Arab mathematicians and often called the Hindu-Arab system. This is the system in common use today using Arabic numerals. He was particularly interested in geometric theorems and developed theorems still of value today which supplement Euclid's collection. Major steps forward were made in the 12th century by *Bhaskara*. He conceived the differential calculus with the concepts of the derivative, the differential coefficient and differentiation, investigating the derivative of the sine function. He obtained a special form of the mean value theorem (explicitly Rolle's Theorem). Various forms of analysis appeared in the 14th century when the *Madhava* School introduced series including the Taylor series and the trigonometric series with convergence tests. Finally in the 16th century, *Jyeshradeva* brought much of this work together and produced the *Yuktibhasa* which is the first known text on the differential calculus and included the concepts of the inverse processes namely the integral calculus. All this activity represents an enormous contribution to the field of mathematics. Indian mathematical developments came to a halt in the 16th century due to political upheavals and a breakdown of an environment suitable for mathematical thought.

Greek mathematics (600 BC–200 AD) added a new element to the subject. Before this the arguments had been inductive, rule of thumb accounts

of special cases. The Greeks introduced deductive reasoning where general
results could be deduced from a set of axioms.[3] Aristotle (384 BC–c355 BC)
introduced laws of logic but the outstanding example is perhaps the *Elements*
of Euclid. This set down an ordered set of geometrical theorems based on
five axioms. One (no. 5) states the truism that parallel lines meet only at
infinity and was destined to have very important consequences much later in
European mathematics. The Elements also included some results in number
theory. There is, for example, a proof of the irrational nature of $\sqrt{2}$ and a
proof that there are an infinite number of prime numbers, again by deductive logic. This pre-dated Cantor's work by two millennia. Primes could be
discovered by the Sieve of Eratosthenes (around 230 BC).

Greek mathematics is generally believed to have begun by Thales
(c624 BC–c546 BC) and Pythagoras (c582–c507 BC). There is the legend
that the inspiration came from the mathematics of Egypt, Mesopotamia but
directly from India through a visit of Pythagoras to the India holy scholars
but this may not be true. Among their numerous significant mathematicians was Archimedes (c287 BC–212 BC), regarded by many as the greatest
mathematician that has ever lived.

Chinese mathematics during 500 BC–1300 AD was also flourishing in the
East whilst the developments were taking place in the West. There is difficulty in finding out about the earliest work, however, because in 212 BC the
Emperor Qin Shi Huang ordered the burning of all books outside Qin state.
There was obviously earlier work which formed the basis for later activities.
The Chinese mathematicians replicated the work of their Western counter-
parts but went further in some directions. They found the binomial theorem,
the Pascal triangle, matrix methods for solving systems of linear equations,
the basis of the calculus and permutations. One member, Shen Quo applied
mathematics to find the areas of terrain necessary for a battle and considered
the logistics of a military campaign.

The *Islamic* contribution during the period c88–1500 AD again was very
significant. It was focused to a large extent on algebra but not entirely so.
Indeed, the word algebra itself is a corruption of an Arabic description of
balancing in equations — moving terms to the other side of an equation
or cancelling like terms on each side. The Islamic Empire was very wide in
the 8th century and, although not all its citizens were Arab, Arabic was
the accepted universal language of scholarship. This was the same as the
Mediaeval Western world where Latin was the common method of scholarly

[3]Although this was thought an absolute procedure for nearly 2000 y Gödel showed in 1932 through
his Incompleteness Theorem that this is not the case. The purity of logic was also dispelled.

communication. The prolific author *Muhammad ibn Mūsā al-Kwārizmī* in the 9th century helped to make Indian mathematics known to the West. Indeed, the word algorithm derives from *Algorithmi* the Latinised form of his name and the word algebra from the word al-jabr in the title of one of his books. He has been called the father of algebra and is interesting because he studied an equation for its own sake, without obvious practical need although this knowledge was available later for use.

About 1000 AD *al Fakhi* introduced the very important procedure of a proof by induction. This is to show that a proposition is demonstrated to be true in one case and from this can be extended to show the proposition is true in the next related case and so on to demonstrate its truth generally. In this way he proved the binomial theorem, Pascal's triangle[4] and the summation formula of integral cubes. In the 10th century, *Alu Wafa* translated the mathematical works of *Diophanus* into Arabic from the Greek. This made available a treasure trove of results. He is well known especially for developing the tangent function. At about the same time *Ibn al-Haytham* developed integration (the inverse of differentiation) to the point where he could find the volume of a paraboloid. He generalised his result to cover integral polynomials up to the 4th degree.

A very important contribution was made by *Omar Khayyám* (the tent maker) towards the end of the 11th century in connection with Euclid's *Elements* in his book *Treatise on the Difficulties of the Postulates of Euclid*. He drew attention to various difficulties but importantly questioned in detail the general validity of the parallel lines hypothesis in the axioms. His discussion could well have laid the foundations for both analytical geometry and for non-Euclidean geometry which was to play so important a rôle in the 20th century. He also published in 1070 a very influential *Treatise on Demonstration of Problems of Algebra* which laid down the principles of the subject. It is interesting that he was the first person to call an unknown in an equation by a name — he called it *shay* meaning *thing*. This was taken up by others and it has come down to the present day, through Spanish, as x, the unknown. He is also known for having constructed a geometrical solution of cubic equations.

Like all Persian mathematicians of the time he was also an astronomer. He was involved with forming and using an astronomical observatory with colleagues and measured the length of the solar year as 365.24219858156 days. This is correct to 6 decimal places which translate to an error of 1 s

---

[4]This is a triangular arrangement of members of the binomial expansion in which every number is the sum of the two numbers above it.

every 1.52777^r y. He was an adherent of the heliocentric model of the Solar System, an idea he may well have gained from *Aristarchus of Samos* who proposed it 1,000 y before. *Khayyam* was very much involved in calendar reform and was a member of the Committee which provided the *Jalali calendar* (after the Sultan) which was used in Persia and then Iran up to the 20th century. It proved more accurate than the Gregorian calendar. The *Jalali* calendar is the basis of the modern calendar used today in Iran and Afghanistan. He is also known as a poet (see page v).

Islamic mathematics had achieved a great deal during its time of power especially in the more abstract matters of algebra and trigonometry. It was a virile arm of a very extensive Empire which at its height bit into India on the one side and into Europe via Spain on the other. This mathematical activity came to an end in the 16th century with the rise of the Otterman Empire and the sword of mathematics passed back to Europe for the first time since it was the held by the Ancient Greeks.

European mathematics had essentially laid dormant between the time of the ancient Greeks and the 12th century. Europe was in the restrictive mediæval period, but by the 12th century there was interest in what others had done and European scholars began to obtain Arab texts through Spain and Sicily. Of especial interest was algebra and Euclid's *Elements*. The first consequences of this new knowledge came from *Fibonacci of Pisa* (c1170–c1250) who wrote several books including *Liber Abaci* (published 1202 and revised in 1254) which is a book on what might be called commercial arithmetic. He particularly introduced the Hindu-Arab numerals into Europe but it took a further two centuries for this to become universally accepted.[5] Well known is the Fibonacci sequence in which each number after the first is the sum of the two numbers that precede it. Explicitly, the sequence is 1, 1, 2, 3, 5, 8, 13, 21, 34, 55, ... The interesting thing about the series is that the ratio of two consecutive numbers approaches the ratio 1:1.618 in the limit of high members. This is called the Golden Ratio. Mathematics of a fashion had begun in Italy.

European mathematics became alive during the 15th and 16th centuries. This was a period of great intellectual and emotional turmoil and a period of exploration away from Europe itself. It is time of the Protestant break from Rome, the explorations at sea to find new and more effective trade routes, the discovery of the Americas and the various wars that were associated with

---

[5] Before this Roman numerals were generally used. There was no way that any substantial calculations could be achieved this way — try multiplying 342 by 264 using only Roman numerals. Division was almost impossible. One can see why there were no Roman mathematicians. Their engineering seems to have either used the Greeks (applying geometry) or be based on trial and error.

it all. These activities had mathematical implications. The printing press was invented and the techniques developed to provide the universal distribution of knowledge. One of the earliest books was *Theorieae nova planetarium* by *Peurbach* published in 1472. Very soon after came *Treviso Arithmetic* and *Euclid's Elements* published by *Ratdolt* in 1482. Maps became of central importance which led to developments of trigonometry.[6] The publication that gave the word was *Trigonometria* by *Bartholomaeus Pitisens* in 1595. A little before that, in 1533 in fact, *Regionontanus* published tables of sines and cosines using the Hindu-Arabic numerals.

European thinkers became interested in the motion of particles at this time which involved problems of instantaneous movement. The reason for this is two-fold. One was an interest in the motion of the planets that had been measured most accurately by *Tycho Brahe*[7] and can be treated as particles because any structure they might have cannot be discerned. The other was the development of the cannon as a serious weapon of war. Barricades were erected to prevent it having a direct line of fire. It was important, therefore, to be able to calculate trajectories that could go over the obstructions and strike behind them. It turned out that both these problems involve gravity but this was not realised at first.

Galilei Galileo tackled the ballistics problem by realising that the projectile had two separate motions. One was vertical and the other horizontal. The vertical problem was falling under gravity. It was already realised that the Earth holds everything to it through the action of the force called gravity. *Galileo* demonstrated in his well-known experiments from the Tower of Pisa that all bodies fall to Earth from a given height in accelerated motion taking the same time which can be stated alternatively that the acceleration of gravity is independent of the particle mass. The height reached in the projectile problem then requires the body to be raised to an appropriate height. The horizontal motion is a steady motion forward (forgetting resistance from the air). Putting these two components together gives a parabolic path whose profile is determined by the initial velocity of the projectile at the cannon mouth and the angle at which the cannon is set.

Galileo also experimented with particles moving down an inclined plane for different values of the incline including the very smallest angles. This very ingenious set of experiments led him to his *Law of Particle Motion* that

---

[6] The development of maps led to the abstract areal view which had implications for fine art.
[7] This was before the refracting telescope which was invented in the very early 17th century. There are indications that a form of reflecting telescope had been constructed in England in the second half of the 16th century and had been available to observe the arrival of the Spanish Armada in 1588. It is alleged to have been hidden as a state secret then.

a body will remain at rest or move with a constant speed unless it is acted on by a force. This is the *Law of Inertia* which was to play so important a rôle in classical and relativistic mechanics very much later. This was extended later by *Isaac Newton* as the 1st of *three laws of particle motion*. The 2nd law recognised a force as affecting inertial motion and the 3rd accounting for the response of inertia to the action of the force.[8] One of the problems that Newton attacked with his laws of motion was the motion of the planets around the Sun.

Johannus Kepler was a mathematical assistant of *Tycho Brahe* and was given the task of making sense of *Tycho*'s very accurate observations of planetary motion. Mars was selected as the object to study first. During the 1590s and very early 1600 he showed (by a very ingenious method) that the path is a smooth curve in a plane and that the curve is an ellipse with the Sun at one of the foci. This applied also to Earth and Venus.[9] The success in these cases led him to extend the result to the major planets that were then known. *Galileo* showed the overall correctness of this approach by his telescopic observations of Jupiter and its system of four principal moons. There was no way of visualising the actual shape of *Kepler*'s curves until *René Descartes* (1596–1650) invented his Cartesian co-ordinates for the purpose. Incidentally, *John Napier* (1571–1630) investigated natural logarithms in the hope it might help calculations such as those of planetary orbits. Incidentally, he also invested a system known as Napier's bones for making easy the tasks of arithmetic (which in some ways pre-dated the slide rule). Here was a modern approach to planetary motion which did not depend on the assumption of circular motion (although the circle is only as ellipse with major and minor axes equal) but was based on pure observation. How could *Newton*'s laws apply?

The essential elements came when *Newton* realised that the acceleration of the Moon in its orbit (it follows an elliptic curve and not a straight line) is a scaled down version (depending on the inverse square of the distance) of that holding things to the surface of the Earth. Gravity is the key which he provided through his *law of universal gravitation*. *Kepler*'s laws follow immediately by assuming motion under the Sun's gravity constrained by the conservation of energy and of angular momentum. Here for the first time is a natural phenomenon explained fully on the basis of natural laws and

---

[8]The logical content of *Newton's laws* has been debated by many mathematicians and physicists but we will not delve into this now.

[9]Fortunately no data were available for the innermost planet Mercury (it is close to the Sun and is difficult to observe). Its path is anomalous and would probably have confused Kepler perhaps even into not accepting the full validity of his results. Scientific data can be too accurate on occasion!

forces.[10] Newton largely pursued his arguments using Cartesian geometry but realised the need to have an analytical tool for detailed study. This led him to develop his theory of fluxions: *Gottfried Leibniz* (1646–1716) also saw this need and devised his calculus in parallel which was very much the same thing. It is the *Leibniz* form that is used today.

The beginnings of analysis and the calculus brought out new problems and one that intrigued thinkers in the 17th century was that of finding the maximum or minimum of paths. The first problem was devised by *John Bernoulli* in 1696. He asked: what is the figure of a track which allows a particle to descend by sliding without friction from one level to one below in the shortest possible time? This became known as the problem of the brachistochrone from brachistos, the shortest, and chrone, the time. Both brothers *Bernoulli*, *Newton* and *Leibniz* all found the correct solution: the answer is along a cycloid curve sometimes called a brachistochrone. It was also found that the smallest area enclosing a given constant volume is a sphere.

This interest began an era covering the 18th and 19th centuries of the most beautiful analyses by many eminent mathematicians beginning with the giant Swiss mathematician *Leonhard Euler* (1707–1783). He made major contributions to many branches of mathematics and invented some himself. Through his teaching books he introduced the notation $f(x)$ for a function of x, introduced the modern notation for the trigonometric functions and e for the base of the natural logarithms. It is the only function that is equal to its own derivative and was first explored by *Napier*. This is a transcendental number and so irrational. He defined e through a limiting process:

$$e = \lim_{n \to \infty} \left(1 + ds\frac{1}{n}\right)^n = 2.7182818284\ldots \quad \text{and} \quad e^x = \sum_{n=0}^{\infty} \frac{x^n}{n!}.$$

He introduced the notation $i = \sqrt{-1}$ and Greek Σ for summation. He made advances in trigonometry, algebra, fluid mechanics (continuum theory), number theory and geometry. His work led to a recognition of the γ function. He even contributed to the theory of the lunar orbit (a problem still of interest resulting from a rather complicated orbital motion). He is distinguished by having one of his equations[11] called "the jewel of mathematics". One

---

[10] Newton realised that the law of gravitation had its problems and required an infinite universe for stability. He thought of a modification such as an exponential dependence on the distance.

[11] This can be derived as follows although there are other ways. First express $e^{i\theta}$, $\sin\theta$ and $\cos\theta$ (with θ in radians) as series expansions in θ. Notice that the expression for $e^{i\theta}$ is the sum of the series for $\cos\theta$ plus the sum for the series $i\sin\theta$ which gives $e^{i\theta} = \cos\theta + i\sin\theta$. Set $\theta = \pi$ and remember that $\cos\pi = 1$ and $\sin\pi = 0$. This gives $e^{i\pi} = 1$ which is expressed in the conventional form $e^{i\pi} - 1 = 0$ with zero on the right-hand side of the equation. Everything in this equation was developed by Euler himself.

important development in the calculus involved replacing the analysis of a function by the analysis of functionals. This gave rise to the calculus of variations which proved to be of the greatest importance to later developments in physics.

d'Alembert (1717–1783) made a very powerful reformulation of Newton's laws of motion which led to a new approach to mechanics. Newton envisaged particles with specified locations being acted on by forces providing what *Bernard Shaw* called the Bradshaw universe. *d'Alembert* made a very simple reformulation of the 2nd law applied to a system of particles of mass m_i with acceleration a_i due to the action of the force F_i as $F_i - m_i a_i = 0$ so that an expression of dynamical motion becomes a statement of statics if the inertia force is introduced in the form $I_i = -m_i a_i$. This means it is subject to the Principle of Virtual Work which, for a virtual displacement δr_i, takes the form $\sum_{i=1}^{\infty} [F_i - m_i a_i].\delta r_i = 0$. From this Principle, the equations of motion can be deduced. Of especial significance in this approach is the emergence of two scalar quantities (with magnitude but no direction) for systems for which the force can be derived from a potential. If the kinetic energy is denoted by T and the potential energy is denoted by V then the first quantity is the Lagrangian, $L=T-V$ while the second quantity is the Hamiltonian $H = T + V$. These two scalar quantities[12] have played a central rôle in the development of modern theories of matter. They have led to the generalised definitions of particle momentum and force with the equations of motion having a generalised form called the Euler- Lagrange equations. Their most direct derivation is through the calculus of variations which was very largely the achievement of Euler and Lagrange. This approach was to prove of great utility in the 20th century including applications to quantum mechanics. Indeed the variation principle involving an assumed Lagrangian has shown many basic applications in describing physical systems.

This was a time of important developments in algebra including that in the complex plane, the theory of differential equations and of the application of the various mathematical developments to problems of physics which spilt over into the 19th century with growing momentum. One important development was the reconsideration of Newtonian mechanics by *Joseph-Louis Lagrange* (1736–1813) where the direct vector approach was replaced by and analytic analysis using the scalar kinetic and potential energies as basic entities. The use of potential functions came naturally. This more modern approach was explained by *Lagrange* in two important books: *Méchanique*

---

[12]They were named after *Joseph Lagrange* (1736–1813) and *William Hamilton* (1805–1865) who made central contributions to the subject.

Analytique giving the general theory and *Méchanique Céleste* which applied the work specifically to astronomical problems which meant orbits of the planets and the Moon. These studies involved a second differential of a function which vanishes first discussed by *Laplace* — this is the famous *Laplace equation*. The variation principles show that this is the condition for the gradient of the quantity to be a smooth as possible. This invites some interpretation of nature along the lines of minimum energy/effort which in fact has played a part in future work. The solution of the Laplace equation provides the so-called harmonic functions which have played a very important part in applications to physics and astronomy. It was shown by *Gauss* (1777–1855) that the solutions of the Laplace equation have the property that the average value over a spherical surface is equal to the value at the centre of the sphere. This is his so-called harmonic theorem. The solutions must be constrained by boundary conditions and there are two. One is names after Dirichlet and specifies the function at the boundary of the region while the other requires the solution to conform to its gradient at the boundary and it named after Newman. The harmonic functions can be added together to also provide a solution (it is a linear equation) and the solution can be achieved by separating the variable and solving for each separately. This proved a very powerful technique in analysis. The solution of many physical problems was found to involve the solution of modified Laplace equations such as the equation of Helmholtz and the equation of *Poisson* (1781–1840). Such problems covered fluid mechanics, electricity and planetary orbits among others.

One problem that interested was the stability of the Solar System against gravitational perturbations. Lagrange showed using a perturbation approach that the planetary orbits are stable to a first approximation. The problem was taken up by Siméon Poisson that the stability extended to the second approximation of perturbation. No one has moved to the analysis of a third approximation because the perturbation technique is often very cumbersome. One interest of many workers was the discovery of perturbation expansions in particular cases which converge quickly so that perturbations after the first or perhaps the second provide the final answer to high approximation.

The behaviour of fluids under forces began to be studied and *Claude Navier* (1785–1836) first published the consequences of applying the conservation requirements of energy and momentum to the flow of an ideal fluid without viscosity or thermal conduction. He obtained a first-order nonlinear equation which was not easy to solve in the general case. It seems that *Siméon Poisson* independently arrived at the same formulation but seems not to have published it. A real fluid shows viscosity so the *Navier* expression is incomplete. The effects of viscosity had been explored in the laboratory but the difficulty was to find an expression which was in a closed form. This was solved by *Sir George Stokes* accounting for shear viscosity with a simple but ingenious form. It was later extended to include the bulk viscosity but none of these extensions made it possible to solve analytically. It had to wait until numerical methods were available using a computer. There was one further development. A boundary condition for the fluid is that the fluid is at rest in contact with a rigid surface whatever its composition. Reconciling this requirement led to the development of the boundary layer and boundary layer theory by Prandtl. This led to a procedure which combines both the ideal equation coupled with that containing viscosity and so to a mathematical procedure for dealing with problems of flight.

There were a number of particularly significant mathematical developments in the 19th century with any one perhaps no more fundamental than the others. We do not take them in a particular order. The first which we will consider involves geometry and especially the parallel postulate of Euclid's axioms which can be stated as saying that parallel lines meet at infinity.[13] An alternative statement is that: there is one, and only one, line through a point not only a chosen line that is parallel to the given line. The statement that there are many lines that are parallel was made by the Russian *Nicolai Lobachevski* (1792–1856) and a similar result was concluded independently by the Hungarian *János Bolyai* (1802–1860). This defines a hyperbolic non-Euclidean geometry, the "straight lines" being a set of hyperbolae of two sheets about a central axis. Put crudely, far from parallel lines meeting at infinity in this hyperbolic geometry they do not meet at all. There is an immediate way of recognising this geometry — draw a triangle on the surface and you will find the sum of the three angles is less than 180°. This contrasts with Euclidean geometry where the sum of the three angles of a triangle are exactly 180°. We can ask if there is a geometry which has the three angles are greater than 180°? There is — it is the geometry of the sphere. Here the parallel lines (the longitude lines) meet at the poles and so provide a case where parallel lines meet after a finite distance. There is, indeed, a finite volume to the geometry. Here the sum of the three angles of a triangle drawn on the surface is greater than 180°. It is realised that the spherical case approaches the Euclidean, flat, case as the radius of the sphere increases indefinitely. The hyperbolic geometry also has the flat geometry as

---

[13]The parallel lines postulate has caused considerable disquiet in the past. Several workers had tried to derive it from the other with no success. It was viewed as a rogue member and so it turned out but a very important rogue for is held the key to different geometries.

a limiting case as the hyperbolæ become ever closer to straight lines. These three must be the only possibilities.

A systemised and generalised approach to geometry was proposed by the German mathematician *Bernhard Riemann* (1826–1866) using the absolute differential calculus sometimes called tensor calculus. The essential feature here is an independence of particular reference systems. It is a generalisation of the vector calculus and encompasses the case where the medium is nonisotropic and nonhomogeneous. This is catered for by allowing the component quantities each to have many components. This complicates the analysis to some degree, but does allow the analysis to mirror the physical world with better accuracy.

Riemann introduced a number of new concepts. He spoke of a metric as the "space" for the geometry. He accepted the Pythagorian relation between the coordinates but modified it in two ways — first to any number of dimensions and next by constraining each dimension by a quantity (which is called the metric tensor) which defines the type of geometry.[14] An important consequence was the defining of the curvature of a surface by a quantity involving only the g_{ij} and called the Riemann curvature tensor. This was later to play an important part in the structure of the field equations of the theory of general relativity.

A second major development was that of set theory by *Georg Cantor* (1845–1918). This has changed the nature of mathematics in a basic way and is now an essential mathematical tool. It especially allowed the concept of infinity to be formalised and showed that there are many infinities of different size. He asked whether the infinities formed an ordered set which can be regarded as his continuum hypothesis but this has never been proved or disproved. The problem of infinity has worried mathematicians since the earliest times. These problems arose from practical observations. Greek thinkers wondered "How many sand grains are there on all the beaches of the world?" Galileo realised there were problems associated with treating a particle trajectory as an indefinitely large number of straight lines. If such an approximation is applied to a circular path of given radius then an infinite number of elementary steps would be required to complete the approximation of the circle. But what is the situation for a circle with larger radius? — it would require more elementary steps to complete the approximation. Are

---

[14]To gain a slight flavour of Riemann's approach — the Pythagoras elementary distance ds between two points in three-dimensional space is $ds^2 = dx^2 + dy^2 + dz^2$. Riemann wrote the distances as dx^i and defined the *interval* ds between two points as $ds^2 = \sum_{i=1}^{n} \sum_{j=1}^{n} g_{ij} dx^i dx^j$ where the g_{ij} are the metric tensors defining the metric of the geometry. For Euclidean space, $g_{ij} = 1$ for $i = j$ but $= 0$ for $i \neq j$.

there infinities of different size? He took the matter no further but *Cantor* proved there are. A problem of infinities leads also to a problem of singularities for a singularity follows when an infinity is inverted. The circle problem suggests that a singularity can have an infinity of components and that there can be singularities of different sizes. This follows if one looks at the centre of the circle. There must be an infinite number of "spokes" coming from the centre to match the infinity required to form the circular surface. It would follow that circles of different radius would have different numbers of spokes which leads to different sized infinities. These problems are still with us — for instance the theory of the big bang as the origin of the Universe involves an origin in an indefinitely small volume.

The 19th century saw other fundamental advances in mathematics. One example is the introduction of a non-commutative algebra by *William Hamilton* (1805–1865). In inventing quaternions he aimed to generalise the concept of the vector by adding a scale quantity to the vector form. Unfortunately the qaurternion cannot be developed beyond four dimensions.[15] The subject of mechanics was advanced considerably when he introduced the variation principle for the deduction of the equations of motion of a particle which involves the Hamiltonian. This was to play an important part later in the development of quantum mechanics.

Other important results can be quoted. *George Boole* (1815–1864) conceived the algebra named after him which contains only the numbers 0 and 1. This was developed and used extensively in the 20th century because it forms a starting point for mathematical logic.[16] It was realised later to be valuable for the development of computer analysis. Group theory has been an important development and the foundations were laid by two gifted mathematicians each dieing very young — Niels Abel (1802–1829) and Évariste Galois (1811–1832). This forms the basis for the study of symmetries which have been a pre-occupation in physics as well as mathematics over recent years. Perhaps the most significant result in the logic of mathematics is the Incompleteness Theorems discovered by Kurt Gödel (1906–1978). They have been expressed in a number of equivalent ways but a simple statement is that any finite number of axioms in *Peano* arithmetic will lead to at least one statement that is true but cannot be proved to be true from the axioms. This result means that mathematics cannot be regarded as a closed subject and will itself contain things that cannot be proved and therefore

---

[15]Hamilton was the first person to propose an equivalence between mass and energy although he did not develop the idea and it was left to Einstein (and Poincaré independently to some extent) to formalise the equivalence into the famous formula $E = mc^2$.
[16]This algebra contains the celebrated result $1 + 1 = 1$.

may be right or wrong. The ideas have been extended to formal logic which again contains statements that cannot be proved. *Gödel's* result showed that the self-consistency of mathematics proposed by *David Hilbert* (1862–1943) cannot be achieved. It follows that mathematics is an empirical study like physics and should be treated as such. The recognition that there is no absoluteness in the world was a 20th century achievement. Turing discovered the Halting Problem when he attempted to reduce the *Gödel* theorems to more concrete terms. It seems that not all computing problems will have a solution where the computer indicates that the problem is complete by stopping. Some problems will never stop and so cannot reach a final conclusion. The interesting feature is that there is no way of knowing before you start whether a particular problem will halt, which will finally finish with a result, or not. This is the Halting Problem.

Confusion came in wave mechanics when it was realised that atomic particles show *both* wave and particles properties described as a duality through the Heisenberg Uncertainty Principle. This problem is not one of mathematical semantics — the approach of Heisenberg based on a matrix formulation was shown by Schrödinger to be equivalent to the differential equation approach based on the classical theory of Hamilton and Jacobi (1804–1851). The unified approach of bra and ket vectors by P.A.M. Dirac (1902–1984) systematised the two aspects but accepted them as simultaneous realities. The development of relativity in both special and general theories emphasises the relative nature of space and of time. There is a widespread recognition of the imaginary nature of the absolute and the simultaneous — it seems they do not exist as general conditions.

One development can be taken as characteristic of the latest mathematical thinking is numerical computation. This was first used simply as a calculation tool[17] but has quickly become a method of employing mathematical logic. Quickly used as a computational aid it has now developed into a powerful method of logical control for operations of a very wide kind. These are now an indispensable aid to daily living and are central to the many wider applications. They are vital to the current space programme. The manufactured control "silicon chips" are now very small and of immense power. They are capable of allowing a space vehicle to control

---

[17]The first computer in the modern terminology was constructed by Tommy Flowers (1905–1998) and colleagues of the British Post Office Research Laboratory in London. It was made for the Code and Cipher Establishment at Bletchley Park near London to break, successfully, the believed to be unbreakable German Enigma and other codes during the Second World War. The development of codes has since developed into a most elaborate science using quantum concepts and the task of breaking them has become enormously greater.

itself and mount quite considerable protective capability against unforeseen circumstances. They do not endow the machinery with thought but can contain pre-arranged defensive actions in a compact ordered form.[18] This capability must be central to extended space missions both to us and to any extraterrestrials that might be attempting the same things. Computing and the Internet have made our world a global village. The future will prove very interesting.

Mathematics: Good or Bad? In the brief account that we have given of mathematics so far there has been no explicit reference to the society sustaining the mathematicians or of the effects of the work of mathematicians on the structure and development of society. There have been references to the response of mathematicians to the requirements of business and commerce and the way that these requests for assistance have given mathematics a direction for advance.

Mathematics has always been a minority sport and probably always will be although the number of practitioners has increased enormously since World War II. Today manipulating mathematics and making applications of it in a wide range of fields has become a profession in itself and the number of people earning a living from mathematics is quite large. This has followed the rise of modern science (and particularly physics) and engineering. It is true, however, that the number of active mathematicians researching and being involved in the innovative activities of invention is still very small. Typically, a half dozen or so more of less isolated group of people spread around the world will form the spearhead for a new advance. Much of the technical content of their work will remain known only to themselves and not in any detail even to fellow mathematicians working in another field. This much seems not to have changed much over history. What has changed is the importance of the applications. We have seen how Archimedes designed weapons of war to thwart the Roman attacks on Syracuse and how Galileo studied the motion of projectiles to enhance the power of the newly developed canon. These gave power to soldiers and politicians. This association has continued. Mathematicians and scientists invented and improved all the articles of technology to the world we know today including those of war. The culmination came during World War II when the efforts of mathematicians were directed to the breaking of codes and so to become involved in a very pure activity. Now, using quantum principles, codes are becoming unbreakable.

---

[18]There has been considerable discussion asking whether it will be possible to produce thinking machines. Linked with this is the question whether the human brain is a very complex machine. If so, it will be subject to the restrictions of Gödel's theorem. A truly thinking machine has not yet been devised.

Mathematicians joined with physicists and engineers in the 1940s and 1950s to develop the ultimate weapons of war — the nuclear weapons. These are applications of the same principles that power the stars themselves. Politicians now have the ultimate power of being able to destroy the Earth and all advanced life on it. Presumably any technologically capable civilisation beyond the Solar System could also have developed the same weapons. Would tactical weapons of this type be taken on a mission of deep space exploration? The quest for knowledge has taken us to the point where we can take ourselves and others into oblivion. This gives an uncertain age for the existence of our, or any, technologically capable society.

Further Reading

1. There are a number of books treating the history and development of mathematics but one very good relatively recent one is

   Davis P. J. and Hersh, R. 1983, *The Mathematical Experience*, Pelican Books, reprinted 1988.

2. A very good general survey written by a range of experts is

   L'Universe des Nombres, 1999 Août, La Recherche Hors, Serie No. 2, La Société d'Editions Scientifiques, Paris.

Compendium

This compendium contains a wide range of entries the greater majority of which are referred to in the main text. Others are also included as a help to the reader who might wish to pursue matters further. It is hoped that this compendium will be extended by the reader and so form the basis of a comprehensive study.

Abyssal plain is the ocean floor away from the continental margin. The plain is generally flat but with a slight slope. The continental margin is markedly sloped to join the continent to the deep abyssal plain.

Accrete is the process of adding mass to a larger mass. In geology, it refers to the addition of terrains to a land mass. In astronomy, it refers to the joining of one small mass to a larger one (planet or star) in the process of accretion.

Acheulean is the name given to the stone tool industry that prospered in Africa during the Stone Age. The characteristic product was stone axes which gained a particular elegance in Africa after about 600,000 ya. Stone axes and tools generally are found especially with the remains of *Homo erectus* and were introduced into Europe and Asia. The name was based on the suburb of Amiens in N. France called Saint Acheul.

Adaptation is the process of change which is the result of natural selection.

Adult (hood) is the mature stage of an organism where it can reproduce.

Advanced life is multicellular life able to move independently and live an independent life perhaps collectively in tribes.

Amber is fossilised resin. Sometimes small animals have been trapped and so are preserved in their entirety.

Amoeboid is a unicellular life form without a specific shape. They move by changing shape (cytoplasmic projections called pseudopods — false feet) and feed by engulfing their prey. An example is human white blood cells.

Amniotic egg is distinguished by the presence of a fluid-filled amniotic sac that protects and cushions the developing embryo. Its essential feature is that it can be laid on land.

Anagensis is a direct evolutionary change along a lineage without branching.

Ancestor is an organism, a population or a species from which some other organism, population or species is descended by direct reproduction.

Andesite is igneous volcanic rock. It is less mafic than basalt but more mafic than dacite. It might be said to be the volcanic equivalent of diorite.

Animate matter is matter associated with living organisms.

Anthropology is the scientific study of the human species.

Anticline is a set of rock layers concave upwards.

Ape is a tailless monkey including gorilla, chimpanzee, orangutan and gibbon.

Archean era is the period of the Earth's history between 3,800 Mya and 2,5000 Mya.

Archipelago is a group of islands in an expanse of water.

Asexual reproduction involves only one parent and usually produces genetically identical offspring.

Asteroid is a member of a very numerous group of small silicate/ferrous bodies the majority of which orbit the Sun in the ecliptic plane in the region between about 1.5 AU and 4.5 AU, between the orbits of Mars and Jupiter. A small minority orbit outside these limits.

Astronomical unit (AU) is a unit of measurement based on the distance of the Earth from the Sun. It is the radius of a hypothetical circular planetary orbit for which the time to complete a single circuit is equal to a mean Earth year. It is equal to $1\,\text{AU} = 1.49598 \times 10^{15}$ m. This is often approximated by 1.5×10^{15} m.

Atmosphere is the gaseous envelope surrounding the solid centre of a planet or planetary body. It particularly describes such an atmospheric layer which is thin in relation to the radius of the planet. For the Earth the thickness of the atmosphere is rather less than 1% of the radius.

Atom is the smallest component of a chemical element. It is a complex entity being composed of protons and neutrons (in the central nucleus) and electrons (enclosing the nucleus).

Atomic fission is the breaking up of a heavy nucleus into two less massive components as the result of a collision with another atomic particle (e.g. a neutron). This also involves the release of the binding energy so fission is the basis of the release of energy on an industrial scale.

Atomic fusion is the joining together of two light atoms to form a more complex one (e.g. combing four protons to form a helium nucleus). This process releases energy up to the formation of iron, but requires the absorption of energy for greater masses.

ATP (adenosine triphosphate) is a general high energy molecule which acts as a source of energy within a cell. The energy is released when the phosphate groups are released. The "battery" is recharged by the collection of oxygen and phosphorus atoms to reform the phosphate group(s) and so reproduce the initial ATP structure.

Australopithecus afarensis (nicknamed *Lucy*) was a bipedal link with humans which lived 3.9 to 3.0 mya.

Austalopithecus africanus was a bipedal hominid which lived between 3.0 and 2.0 mya.

Avalonia was a separate crustal plate in the early Palaeozoic which contained much of Northern Europe, Newfoundland, Nova Scotia and coastal parts of New England.

Avogadro number ($N_A = 6.02214 \times 10^{23}\,\mathrm{mol^{-1}}$) is the number of elementary particles (atoms or molecules) in one mole. One mole is defined as the number of atoms in 12 gms of carbon-12.

Bacteriophage is a virus which attacks its bacterial host and in most cases destroys it. Some phages can lie dormant for a long period of time having inserted their DNA into the host. Some phages have for this reason become of great importance in genetic engineering.

Baltica was a crustal plate in early Palaeozoic which contained United Kingdom, Scandinavia, Central Europe and European Russia.

Banded iron formation is rock consisting of layers of chert alternatively dark (very iron rich) and light (less iron rich). These resulted from the reaction of the iron with oxygen to form iron oxide between 3.8 and 1.7 thousand million (billion) years ago.

Barred spiral galaxy is a spiral galaxy in which the trailing arms are linked to a central bar rather than being joined directly to the central nucleus.

Basalt is a highly mafic igneous volcanic rock. It is fine grained and dark in colour.

Basin is a large depressed area of land in which sediments are deposited.

Bioluminescence is light emitted by the compound *luciferin* due to the action of the enzyme *luciferase*.

Binding energy is the energy binding atomic nuclei together.

Bipedal describes an animal that habitually walks on two legs.

Black body is one which absorbs all frequencies of the electromagnetic spectrum that fall on its surface Equally all frequencies are emitted when the body is heated.

Boltzmann constant ($k = 1.38065 \times 10^{-23}$ J K^{-1}) is the ratio of the gas constant ($R = 8.31447$ JK^{-1} mol^{-1}) to the Avogadro number ($N_A = 6.02214 \times 10^{23}$ mol^{-1}). It forms a link between the microscopic and macroscopic worlds. It can be regarded as a gas constant per particle.

Brown dwarf is a body composed primarily of hydrogen but with a mass between the specific range of 2.8×10^{28} kg and 2×10^{29} kg. It has an energy source for the fusion burning of deuterium providing a low surface temperature.

Burgess shale is an extraordinary area in British Columbia, Canada, where an underwater landslide some 520 mya preserved an ecosystem which has now fossilised.

Calcite (calcspar) is a crystalline form of calcium carbonate ($CaCO_3$). It is the basic component in chalk, limestone and marble.

Calcium carbonate ($CaCO_3$) is the compound used by many marine invertebrates to construct exo-skeletons.

Calcareous is the term used to describe a structure rich in $CaCO_3$. It is almost always associated with a dead organism such as the shell of a bivalve.

Caldera is the name given to a large circular depression at the top of a volcano usually due to the collapse of the material.

Cambrian explosion occurred at the end of the Cambrian era about 530 mya. It saw an enormous explosion of life with all the characteristics present that were to develop later.

Capsid is the protein cover of a free virus particle enclosing the genetic material.

Carbon–nitrogen cycle is a variant cycle for the conversion of hydrogen into helium in a stellar interior. Carbon and nitrogen act as catalysts that facilitate the process and so speed the process up.

Carbohydrates are a class of compounds which includes chitin, starch, steroids and sugars.

Carbonate is a mineral composed principally of calcium and carbonate ions. It will most likely also contain traces of other minerals including magnesium and iron.

Casts are fossils formed by water containing minerals having seeped into a cavity as a mould caused by an organism. The water evaporates leaving the shape in the deposited minerals. These are often of feet marks.

Catalyst is a chemical component of a chemical process that facilitates it and so speeds it up.

Cell is the fundamental unit of all living things. It is contained within a plasma membrane which separates it from the environment. The interior contains the genetic DNA material and the fluid cytoplasm.

Cell cycle is the sequence of processes that must be performed by a cell to reproduce itself. Most of the process is taken up by the replication of DNA.

Cell membrane is the outer cover which isolates the cell contents from its environment. It is sometimes called a plasma membrane.

Cell wall is the rigid structure outside the cell membrane. This is of cellulose for plants, and chitin for fungi.

Cellulose is a polyhydrate polymer of glucose. It is used widely in the cell walls of plants, eukaryotic algae, diatoms and green algae. It is probably the most abundant compound produced by terrestrial living organisms.

Cenozoic era covers the history of the Earth during the period from 65 mya to the present day. It includes the Tertiary and Quaternary epochs.

Cephalon is the name given to the head shield of a trilobite which carries the eyes, mouth and antennæ.

Character of an organism is the heritable trait which characterises the organism.

Chert is a hard and dense sedimentary rock. It is composed of quartz crystals and sometimes amorphous silica which could be opal. The silica usually has a biological origin.

Chirility is right or left handedness: an object is chiril if it cannot be super-posed on its mirror image.

Chitin is a carbohydrate polymer offering great strength. It is used for support and protection particularly in the exo-skeletons of arthopods.

Chlorophyll is the basis of photosynthesis. It is a green coloured pigment that absorbs and stored the energy from electromagnetic radiation in the optical region. It produces glucose.

Chromosome is a linear portion of eukaryotic DNA. It is often bound by specialist proteins (called *histones*)

Clade is a group of organisms which contains all the ancestors up to the present time. It can alternatively is a monophyletic group.

Clast is an individual grain of rock. A clastic rock is one composed mainly of pieces of pre-existing rocks.

Clone is an identical copy of an organism.

Collagen is composed of long proteins wound in a triple helix giving great strength. It is a major component of mammalian hair.

Colonial is a term used when many unicellular organisms live together in a linked group keeping their individuals identities.

Comet is a small ice/dust body orbiting the Sun and having its origin in the Kuiper belt or Oort cloud. Comets with orbital period less than $200\,y$ are called short period otherwise they are classified as long period comets.

Compactions are fossils that have undergone flattening to become flat so losing their original three-dimensional form. The fossil is called a *compression*.

Complex cells are those with internal structure with specific functions.

Complex number, denoted by z, is the combination $z = a + ib$ where a and b are real numbers and $i = \sqrt{-1}$ is the imaginary unit.

Compound eye is composed of a close packed array of simple eyes each with its own lens and nerve receptors.

Compton scattering is characterised by a loss of energy by an electron when it interacts with mater. Under extreme conditions, the inverse is possible to give *Inverse Compton Scattering*.

Conglomerate is coarse grained sedimentary rock composed of clasts. These must have a dimension greater than $2\,mm$.

Congo craton was a separate crustal late that contained most of north central Africa and which rifted from the super-continent Rodinia in the late Pre-Cambrian.

Continents are land masses on a planet. For the Earth the area of the continents has grown continually with time from very small beginnings to involve now about 30% of the surface above the oceans. The distribution over the surface has varied between a range of small continents to a single land mass (Pangea). The present distribution involves essentially two concentrated masses (Eurasia/Africa) and the Americas together with Australasia and Antarctica.

Convergence describes the circumstance of similarities, often close, of features shared by two or more unrelated organisms which have arisen independently in the different cases. This arises from a shared gene.

Coprolites are fossilised faeces.

Core (of the Earth) is the central region of the Earth (very closely half the radius) composed of iron and other elements like sulphur. The inner core is about one half the core again and is composed of solid ferrous elements. It contains the source of the intrinsic magnetic field.

Craton is a part of a continent which has survived vicissitudes for a very long time. Over time other land forms around it ultimately becomes the centre of a continental land mass.

Crust is the outermost layer of the Earth. Its thickness varies from about 5 km under the oceans to as much as 70 km under continental mountains.

Cosmic abundance. This is the universal abundance of the different chemical elements. Some 75% is hydrogen with about 23% helium. The heavier elements make up the rest.

Cosmical constant is a constant of nature that first became significant in connection with the theory of general relativity. Its effect is to oppose the attraction due to gravity between bodies. It is the leading contender as the cause of the recently recognised acceleration of the expansion of the Universe.

Cytoplasm is all the contents of a cell but excluding the genetic material.

Cytoskeleton is the internal coordinated system of molecules which give a specific outward shape and internal organisation to eukaryotic cells. One example is red blood cells — they would be spherical rather than disc shaped without the cytoskeleton.

Dacite is a fine grained, light coloured, igneous volcanic rock.

Dark energy is an energy which appears to cause the acceleration of the expansion of the Hubble expansion of the Universe. Its nature is unknown.

Dark matter is the invisible component of matter which is found empirically to be associated with visible matter. Although it does not respond to electromagnetic energy it does respond to gravitation. It is not composed of the usual (baryonic) matter and its nature is as yet unknown. It appears to pervade the Universe.

Darwinius masillae, or IDA, is an extremely well preserved fossil from the Messel pit of a small furry lemur-like creature that live 47 mya. Its special importance is that it appears to form a (missing) link in the early development of mammals which became the very early ancestors of humans. The description of the fossil was made public on 2009 May and more information is awaited.

Diapsid is a vertebrate with a skull having two pairs of openings in the skull behind the eyes. Snakes, crocodiles, dinosaurs and pterosaurs are examples of such animals.

Differentiated cells are cells in which the different component processes are separated in specific regions involving organelles.

Dinosaurs which ruled the surface of the Earth from 227 mya to 65 mya were a most successful genus of reptiles. They ranged in size from the very small (cat sized) to the extremely large. Indeed the largest were among the largest creatures that have ever lived. Over 300 species have been described.

Diversity is a term used to describe number of taxa.

DNA (deoxyribonucleic acid) is a nucleic acid containing all the genetic instructions for the development and functioning of living organisms. This applies to all living organisms on Earth together with some viruses. It is the primary component of chromosomes.

Dolomite is a sedimentary rock composed of more than half of the mineral calcium-magnesium carbonate with chemical formula $CaMg(CO_3)^2$.

Domestic energy is energy from sources for use in the home.

Doppler effect is the modification of a beam of radiation received by an observer due to the relative motion of the source. The received frequency is increased as the source moves away but is reduced as the source move towards the observer. In the case of a light beam, the colour become redder as the source moves away but more blue as it approaches the observer.

Dwarf planet is a small, intrinsically spherical, body orbiting the Sun. Ten have been designated at the present time, all but one with orbits beyond that of Neptune.

Egg can mean a large gamete without flagella, sometimes called an ovum, or it can refer to a complex multicellular entity in which an animal embryo develops.

Electron is a fundamental component of atoms. It has a mass of 9.110×10^{-31} kg and radius that it too small to measure (a point particle). It carries a negative electric charge of 1.6022×10^{-19} C.

Elementary life is simple life not necessarily with mobility but with elementary internal structure and at most of very limited consciousness. The most elementary form is micro-organisms.

Elliptic galaxy a large collection of stars forming a separate entity of elliptical shape. There is very little dust and the stars are generally older.

Embryo is a zygote which is undergoing cellular division.

Endosymbiosis describes the occurrence of one organism taking permanent residence within another making the two a single functioning organism. As an example, mitochondria are believed to have been involved in this process long in the past.

Energy: domestic is energy for the home.

Energy: nuclear is the energy released when a large nucleus breaks up or when two light nuclei fuse.

Energy: sources cover a very wide range from simple coal, oil, gas, wind and waves to nuclear. The nuclear can only be developed by an advanced technological society and relies on the presence of the more elementary forms to allow the society to be developed along the way.

Energy spectrum is the proportional of different frequencies in the energy emitted by a body.

Energy: stellar is supplied by the fusion of the light elements up to iron.

Enzyme is a complex protein which speeds particular biochemical reactions. The action is usually quite strongly temperature dependent.

Epicentre (of an earthquake) is the point on the surface directly above the source or *focus* of the earthquake.

Escape velocity is that speed required by a mass to overcome the gravity of a body. For the most dense matter it can be indefinitely close to the speed of light.

Eukaryote is an organism whose DNA is contained in a nucleus separated from the other components within the cell. Its cells are supported by cytoskeletons. This category is very wide and includes, plants, animals and fungi. The current word is that eukaryotes are constructed from prokaryotes that came together in a symbiotic relationship.

Euroamerica was a super-continent that existed from the late Silurian to the Devonian. It formed by the collision of Baltica, Laurentia and Avelonia.

Evolution is, according to Charles Darwin, descent with modification.

Evo-devo (evolutionary developmental biology) is the recently established study of attempts to determine the ancestral relations between organisms. It was the natural outcome of the discovery of genes which regulate the embryonic development of organisms.

Exaptation is the change in the function of a trait through evolution. An example is bird feathers — originally developed to retain heat in an animal they were used later to help a different animal to fly.

Expansion of the Universe is the observed general motion of matter (and especially galaxies) away from the observer with a speed directly proportional to the distance away. This is the Hubble law. Although it was expected that the expansion would be steady or slowing with time it has been found recently that it is, in fact, accelerating.

Extinctions are occurrences when at least a major proportion of a species declines in number. Alternatively, members of a clade or taxon die. In some cases many, or very many, species have become extinct at the same time.

Exo-skeleton is an external covering that provides protection and support for the whole body. The skeleton is most often hard.

Eyespot is a light sensitive organelle. It is found in many protozoa and in some metazoa. It formed the first elementary eye in the ladder of evolution.

Felsic is an igneous rock containing light coloured feldspar and silica.

Fermi question: if life exists elsewhere in the Universe why have we not become aware of it?

Flagellin is a protein which is the essential component of prokaryotic flagella.

Flagellum is a tail like structure used to give locomotion to a cell. It is joined to the cell at a specific point but pointing backwards. There may be one or many. Some act by rotation producing whirl pools (the cell rotating in the opposite direction) while others act through a thrashing motion (like a swimmer's legs).

Focus of an earthquake is the point where the energy is released, immediately below the epicentre.

Fossil is any incontrovertible evidence of past life on Earth.

Fossil record is the accumulated evidence of all fossils relating to the earlier history of the Earth.

Gamete are reproductive cells which fuse to form a zygote. Gametes can be differentiated into egg and sperm in animals.

Gene is a section of an organism's DNA which defines one single trait. It is, therefore, the basic unit of heredity of the organism.

Genetic diversity of a species is the number of genetic characteristics in its genetic makeup. *Genetic variability* is the tendency of these characteristics to vary.

Geological divisions are used to divide the history of the Earth into distinct period based on the rocks' characteristic of the period.

Geological map shows the types of rock is an area together with their ages.

Gödel Incompleteness Theorem was proved and enunciated by Gödel in the form: for any finite group of axioms there is always at least one associated fact that, although known to be true, cannot be proved to be. This means that a set of axioms is incomplete in relating the consequences that follow from them.

Gondwana (Gonwanaland) was a super-continent that existed from the Cambrian to the Jurassic epochs. It contained South America, Africa, Madagascar, India, Antarctica and Australia.

Goldilocks zone is the unofficial name given to the zone surrounding a star in which the temperature is such that water can exist as a liquid. This is taken as crucial for the existence of living matter. The name comes from the nursery rhyme of the three bears: the zone is neither too hot, nor too cold but is just right.

Haden is the era in the history of the Earth between 4,550 mya and 3,800 mya. It covers the period of the formation of the Earth and the certain appearance of life.

Head is the part of a body containing brain, mouth and most sensory organs and is at the "front".

Hertzsprung–Russell diagram is the plot of the surface temperature versus the luminosity of the stars. It is found that stars are not spread at random across the plot, but fit into specific regions which can be identified with the precise evolution of a star.

Histones are proteins attached to the DNA of eukaryotes. They facilitate the collecting together of DNA to form chromosomes.

Homeobox is a DNA sequence of genes which includes the Hox genes that determine the regulation of patterns in the reproduction of animals, plants and fungi.

Hominid is a member of a group of creatures which evolved to form Modern Man.

Homo antecessor probably lived between 900,000 and 600,000 ya and indications suggest that it was right-handed. This distinguished it from a gorilla.

Homo erectus lived between about 1.3 and 0.3 mya being the first known hominoid outside Africa.

Homo ergaster lived between 1.9 and 1.4 Mya and is distinguished in having the same proportion of longer legs and shorter arms as modern humans.

Homo sapiens is the group of Hominids which, with Homo neanderthal, includes Modern Man.

Hubble expansion is the name given to the observed expansion of the Universe.

Hydrophilic (water loving) compounds are easily dissolved in water.

Hydrophobic (water hating) compounds do not dissolve easily if at all in water. Oils are an example.

Hydrothermal vent is a small region on the sea floor where warm to superheated mineral water is ejected. This is usual associated with a spreading centre and may support a varied range of fauna.

Iapetus Ocean was a relatively small ocean that existed from the Late Precambrian to the Devonian and separated the continents Avalonia, Baltica and Laurentia.

Ichnology is the study of trace fossils.

Icy bodies are bodies composed primarily of ice which may be water ice or others.

Imaginary unit, written $\sqrt{-1}$ and denoted by the symbol i, is a useful mathematical devise. It can have several equivalent interpretations, but a simple one is for it to denote an anti-clockwise rotation through $90°$. Starting with a horizontal x-axis, placing $\sqrt{-1}$ in front of a number places it on the perpendicular y-axis. Thus, the location a on the x-axis and b on the y-axis can be written more concisely as $z = a + ib$. The combination $z(a, b)$ is called a complex number plotted in the complex (x, iy) plane (the sheet of paper). The real part is the number a while the imaginary part is the number b.

Ion is an atom or small molecules that carries an electric charge, usually positive but not necessarily so.

Irregular galaxy is a galaxy without obvious structure.

Island arc is a curved chain of islands off a continent which rise from the sea floor. The concave side usually faces the continent

Isotope of a chemical element is a variant atom with the same chemical properties. Isotopes differ in the number of neutrons in the nucleus, the number of protons which define the chemical nature remaining the same.

Jupiter is the largest planet of the Solar System. Its bulk is taken as the standard for quoting the masses and radii of the observed exo-planets. It is named after the Roman King of the Gods and also God of rain.

Kenyanthropus platyops was a bipedal hominoid which lived between 3.5 and 3.2 mya.

Kepler's laws of planetary motion describe the motion of a single planet of low mass in an elliptic orbit under gravity about a body (such as a star) of high mass. The motion takes place under the constraints of the conservation of energy and angular momentum of the system. Several orbiting planets disturb the strictly elliptic shape of the orbit of each planet due to the gravitational interactions between them.

Kuiper belt is a wide region of icy bodies which lie between some 30 and 55 AU from the Sun with orbits broadly in the ecliptic plane. They are believed to be the source of short period comets.

Laurasia was a super-continent that existed from the Jurassic to the Early Tertiary Epochs after splitting away from Pangea. It was composed of Avalonia, Baltica and Laurentia. In the Tertiary Epoch the Atlantic Ocean opened by the splitting of Laurasia into Eurasia and North America.

Laurentia was a continental plate of the Late Pre-cambrian to the Silurian. It contained most of North America, northwest Ireland, Scotland Greenland and portions of Norway and European Russia.

Lava is any molten material (other than water) that is extruded from the crust. The rock that results is said to come from a molten extrusive.

Light year (ly) is the distance a ray of light would travel in one year in a vacuum in free space. In numerical terms this distance is 1 light year $= 9.4605284 \times 10^{12}$ m. It is also equal to 0.3066018 pc.

Lipid is a class of compounds including fats, oils and waxes.

Lysosome is the organelle which carry the digestive enzymes in an eukaryote.

Mafic is a measure of the iron and magnesium minerals in a dark coloured igneous rock.

Magma is molten rock formed within the Earth.

Main sequence contains stars which burn hydrogen to produce helium as their energy source.

Major planet of the Solar System is one of Jupiter, Saturn, Uranus and Neptune. They are gaseous bodies with orbital distances which lie in the range 5–30 AU from the Sun.

Mantle is that part of the Earth which lies between the central core and the crust.

Mass is the property of a body which resists the action of a force. It is a measure of the quantity of matter contained by the body.

Mean free path is the distance one particle travels between collisions with other particles.

Membrane is the semi-fluid material that binds a cell. It also partitions the interior of the cell. It is composed of two lipid layers with proteins dissolved in them.

Mesozoic era covers the period 248 mya and 65 mya. It includes the sub-divisions Triassic, Jurassic and Ctretaceous.

Metallicity is the proportion of chemical elements other than hydrogen and helium that are present in a star.

Metamorphism is the process of forming one rock from others through the action of pressure, temperature or shearing.

Metazoans are multicellular animals having differentiated cells, a nervous system and a digestive cavity. It includes all animals except protozoans.

Meteor is a small particle of matter that is caused to glow through collisions with atoms as it enters the Earth's atmosphere.

Meteorite is a fallen meteor often called a stone because of its composition.

Mitochondrion is a complex organelle which is the site of the major energy production within the cell. It is found in most eukaryotes and requires oxygen to work. Its origin appears to have been a symbiotic relationship between a primitive bacteria and a primitive eukaryote.

Micro-lensing is the use of the gravitational effects of matter on electro-magnetic waves passing them to reveal details of the hidden matter behind the gravitating matter.

Mineralisation is the process of replacing the original contents of an organism by minerals.

Minor planet is a category recently introduced by the International Astronomical Union to describe a small spherical body which (a) orbits the Sun beyond the orbit of Jupiter, (b) has small radius and (c) has nonsilicate, ferrous, hydrogen composition. This removes Pluto, previously the outer "planet", to the category of a minor planet. Other bodies of broadly Pluto size have also been discovered to exist in the same outer region.

Model universes are general descriptions of universes that are theoretically possible within the current observational constraints of the Universe.

Modern Humans (MH) are us today.

Monophyletic is a term applied to a group of organisms which includes the most recent common ancestor of all its members and the descendents of the common ancestor.

Multicell creatures are creatures composed of more than one cell. This includes all creatures that can be seen with the naked eye and many that are microscopic.

Gravitation is a universal force of attraction between bodies that is due entirely to their mass. The force decreases with distance from the body according to the inverse square power of the distance.

Neanderthals were a group of Hominids who lived from about 200,000 ya to about 30,000 ya. They spread throughout Europe and Western Asia from their origins in Africa. They developed the ability to survive in the most hostile conditions.

Neptune is the outermost planet of the Solar System. It was discovered in 1845 following mathematical calculations by Adams and Le Verrier.

Neurons are cells within the nervous system that transmit and process information. They use electrochemical signalling to maintain communication in a complex neuron network.

Neutron is an elementary constituent of atomic nuclei with a mass of 1.674967×10^{-27} kg and a dimension of order 10^{-15} m. It does not carry an electric charge.

Notochord is a central cartilaginous rod that extends the full length of the body.

Nuclear fission is the spontaneous breaking up of the heavier chemical nuclei into two roughly equal halves. The ejection of neutrons and/or electrons/positrons accompanies the disintegration together with high energy photons. The process releases binding energy into a useable form.

Nuclear fusion is the joining together of two light chemical nuclei to form a new element, for instance hydrogen nuclei to form a helium nucleus. Energy is released in the process.

Nuclear synthesis is the process of producing one atomic nucleus from another. The heavier nuclei in the Universe were produced this way.

Nucleic acid is a class of compounds which includes RNA and DNA. These compounds are among the largest molecules known.

Nucleotide is a unit in the construction of nucleic acids. It is composed of a phosphate group, a sugar and an organic base. ATP is an example.

Ocean crust is the portion of the Earth's crust under the oceans which is formed at the mid-ocean ridges. It is between 5 and 10 km thick with a mean density of about 3,000 kg/m^3.

Ockham's razor is the statement that, of a set of alternative explanations of a natural event, nature will have chosen the simplest form.

Orbital eccentricity is the degree of ellipticity of the elliptic orbit of a body around the Sun or a planet.

Oort cloud is an outer essentially spherical casing of ice particles which encompasses the Solar System beyond the Kuiper belt. Its inner radius is thought to be about 20,000 AU and its outer radius in excess of 50,000 AU. There is as yet only indirect evidence for its existence.

Organelle is a region of a cell enclosed by a membrane which has got a specific function within the cell, for example to produce energy.

Orrorin tugenensis was a transitional hominoid of 6–7 mya.

Palaeo-Tethys Ocean existed from the Ordovcian to the Jurassic Epochs. It originally lay between eastern Gondwana, Siberia, Kazakhstan and Baltica.

Panthalassic Ocean was a vast ocean from the late Pre-cambrian to the Jurassic. It encircled the Earth and connected to smaller oceans which developed.

Palaeoanthropology is the scientific study of the fossil record of humans.

Palaeontology is the scientific study of ancient life forms. It relies heavily on fossil evidence.

Palaeozoic era covers the period 543–248 mya. It is sub-divided by the epochs Cambrian, Ordovician, Silurian, Devonian, Carboniferous and Permian.

Pangea (Gk. all land) was a super-continent that existed between the end of Permian to the Jurassic Epochs. It was assembled from the large land masses Euro-America, Gondwana and Siberia and also some smaller land masses.

Pannotia was a super-continent existing in the late Pre-cambrian. In the Cambrian Epoch it gave rise to the continents Baltica, Gondwana, Laurentia and Siberia.

Panspermia is the concept that life entered the Earth's atmosphere from outside rather than forming directly on Earth.

Parallax is the apparent change in the position of an object which arises from the change of position of the observer.

Paraphyletic is a term which describes a group of organisms including its most recent ancestor. It does not include all the descendents of the most common ancestor.

Parsec (pc) is a unit of astronomical measurement of distance. It is the distance from the Earth to an astronomical object for which the parallax angle is one arc sec ($1''$). The radius of the Earth's orbit is used as the unit of measurement. $1\,\mathrm{pc} = 3.08568025 \times 10^{16}\,\mathrm{m} \approx 3.26156\,\mathrm{ly}$. For larger distances there is the kiloparcec (1,000 pc or 3.26×10^3 ly), the megaparsec (10^6 pc or 3.26×10^6 ly) and the gigaparsec (10^9 pc or 3.26×10^9 ly).

Pax genes control the development and control of specific functions of an organism. They are divided into nine categories.

Peat is a deposit in a wet place of partly decayed organic material containing more than 50% carbon.

Phanerozoic is the geological eon covering the time period from approximately 543 mya until the present time. It is the time of the fossils.

Phosphate is an ion composed of one phosphorus atom and four oxygen atoms.

Photosynthesis is the process by which absorbed light energy is absorbed by chlorophyll and converted into sugar as a source of energy.

Pigment is any coloured compound used to absorb sunlight or to block it altogether. It is used in animals for sexual displays.

Planet is a body orbiting a star or a stellar remnant: it is of sufficient mass to have assumed an intrinsically spherical shape but not of a large enough mass to support ionisation in its interior. Its composition will contain the range of chemical elements but not ice as a major component.

Planetary body is a spherical planet orbiting a central star or a spherical satellite orbiting a planet.

Planetary nebula is the end point of certain stars, such as the Sun, when the fusionable material providing the energy source is used up. Quantities of gas and dust are ejected giving the most beautiful patterns.

Plate tectonics is the study of the rigid plate system that forms the crust of the Earth.

Pluton is any mass of igneous rock that is formed below the Earth's surface.

Plutoid is a dwarf planet with an orbital semi-major axis greater than that of Neptune. Nine are known at the present time including Pluto with its satellite Charon. They are scattered bodies from the Kuiper belt.

Polypholetic refers to a group of organisms for which the common ancestor does not share the character shared by the present members of the group.

Polysaccharide is a macromolecule polymer composed of a large number of monosaccharide molecules joined by glycosic bonds. They are amorphous and insoluble in water.

Positron is a point particle elementary constituent of matter and is an exact copy of an electron with an electric charge of 1.6022×10^{-19} C except that it is now positive.

Possible universes are model universes which are all compatible with the laws of physics.

Pre-cambrian eon is the division of the Earth's history between 4,550 mya and about 543 mya. It covers the first 88% of the history of the planet. Life was primitive and essentially microbial.

Preoterozoic era covers the history of the Earth between 2,500 mya and 543 mya.

Prokaryotes are members of a group of organisms (mainly unicellular) that do not possess a cell nucleus. They are divided into two domains, bacteria and archæa.

Protein is a class of compounds made from amino acids. They may be reactive as in enzymes but they may be structural as hair and cartilage.

Proton is an elementary constituent of the nuclei of atoms. It has a mass of 1.67262×10^{-27} kg and a dimension of the order 10^{-15} m. It carries a positive electric charge of magnitude 1.607×10^{-19} C which is the same as that of the positron.

Protist is a member of a diverse set of fungus like micro-organisms associated generally with water. An example is algae. Other examples cause malaria and sleeping sickness.

Proton–proton cycle is the process of the conversion of hydrogen to helium found in main sequence stars.

Protoplasm is all the contents of a cell.

Quantum of energy is the smallest quantity of energy that can be transferred which is nonzero. It is of special importance in atomic studies.

Quantum theory is the development of the consequences of the quantum of energy in interactions between physical systems. It concerns especially atomic systems.

Radiation (biological) is an event of rapid emergence of new species (cladogenesis). This also allows organisms to move into a new niche or habitat for further evolution which is called adaptive radiation.

Radio carbon dating is the method of determining the age of a specimen by determining the quantity of the carbon isotope ^{14}C that it contains. It is accurate within the range 500–700,000 y.

Radio galaxy has got an unusually high level of radio emissions.

Renewable energy is that obtained from continuing natural sources such as the Sun, the wind, tidal energy and so on.

Reversible process is a process that can be reversed completely by an infinitesimal reversal of the forces acting on it. It follows that reversible processes require an infinite time to complete.

Rift is a long narrow crack in the surface of the Earth which penetrates the entire crust as it is pulled apart. There is a rift valley in East Africa.

RNA (ribonucleic acid) It is the nucleic acid which carries the message from DNA into the cell for interpretation and use.

Rodinia (Russian for homeland) was a super-continent of the late Pre-cambrian. It is the earliest super-continent for which there is a good record. It pre-dated the super-continent Pannotia.

Rubisco is a protein which fixes carbon in photosynthetic processes by binding CO_2 to a five-carbon molecule.

Satellite (natural) is a body orbiting a planet. There are over 60 in the Solar System, all but three orbiting the major planets.

Saturn is the second largest planet of the Solar System. It was the furthest planet from the Sun known to the ancient astronomers who had not got telescopic aids.

Seed is an entity encapsulating the embryo of a plant.

Selection (biological) is the favouring of one feature in some members of a population over other features of the population as a whole. This one feature then gives a significant evolutionary advantage.

Semi-major axis is half the long axis of an ellipse. As the shape of the ellipse widens the semi-major axis becomes the radius of the resulting limiting circle.

Sexual dimorphism describes the state of a species where the female is consistently smaller than the male. This is not true in all creatures.

Siberia in palaeo-studies was a single crustal plate that existed from the very late Pre-cambrian to the Carboniferous. It now carries the modern continent.

Silica (silicon dioxide, SiO_2) is used as a structural component by many animals.

Single cell organisms are organisms composed of a single cell. The cell may be simple or complex.

Solar activity is the collective name for the many processes and emissions associated with the surface of the Sun. Flares and the ejection of plasma are examples. The intensity goes through a cycle from maximum to minimum in about 11 y.

Solar constitution is 98% hydrogen and helium and the remainder in the remaining chemical elements.

Solar luminosity is the radiant energy emitted by the Sun and has the magnitude 3.839×10^{26} W.

Solar radiation is that emitted continuously by the Sun.

Solar System: The collection of the Sun (an average star), eight planets and an unknown number of minor planets, satellites, meteoroids, comets and dust which contains our Earth.

Spiral galaxy: The prototype is a lens shaped collection of some two hundred thousand million evolving stars (2×10^{11} stars) together with gas clouds, forming stars and dead stars. It is a few hundred thousand light years across and a few tens of light years at its thickest. It generally contains an unknown quantity of dark matter. It rotates about its axis once in a few hundred million years.

Spore is a single cell that is scattered as a single reproductive unit.

Starch is a complex polymer of glucose. It is used by plants to store sugar for later use.

Stellar evolution is the life history a star from birth to death.

Stellar motion is the motion of a star. It will have an ordered component (for instance its motion around the centre of its galaxy) and a disordered component caused by the gravitational interaction with other stars.

Stratum is a layer of sedimentary rock — several layers form a strata.

Structure of DNA is a helical double helix composed of four acids and sugars.

Sugar is one of a group of carbohydrates able to provide energy. Glucose is an example.

Sun is our parent star whose gravitation dominates that of the orbiting planets and other bodies and materials.

Super planet is a body which shows pressure ionisation in its interior but which is not massive enough for thermonuclear processes to begin.

Symbiosis is the permanent union of two organisms which, although perhaps once independent, now dependent on each other entirely for their existence.

Syncline is a set of rock layers concave downwards.

Syngamy is fertilisation — the process of the union of two gametes.

Synapis a vertebrate with a skull having one pair of openings in the side of the head behind the eyes. Mammals are like this.

Taxon is any named group of organisms. It need not be a clade.

Technologically capable life is living creatures able to develop a technological life style which moulds their environment in significant ways. It will inevitably involve the control of metals.

Tectonic forces cause the movement and deformation of the Earth's crust on a large scale.

Tethys Ocean was a small ocean that existed from the Triassic to the Jurassic. A region developed westwards as Pangea split into Gondwana and Laurasia called the Tethys Sea. Its remains now form the Mediterranean Sea.

Terrain is a tract of land as viewed in physical geography.

Terrestrial planet is one of the group Mercury, Venus, Earth and Mars. These lie within 1.5 AU of the Sun.

Tetrapod is a vertebrate that has got four limbs with digits but no fins.

The Galaxy is that particular galaxy which contains our Solar System. This lies some 25,000 ly from the centre and takes part in the general galactic rotation completing one revolution in about 200 my. It is now known to be a barred spiral.

Thermodynamic equilibrium is the condition in a physical system where thermal equilibrium, mechanical equilibrium and all other relevant equilibria are present simultaneously. It is associated with the condition of maximum entropy.

Thermonuclear processes are the interactions between atomic nuclei. These involve fission in the lightest nuclei but fission in the heavier ones.

Trace fossils are the evidence of fossils left due to the fossil movement such as footprints, burrows and so on.

Transit describes the passage of one body in front of another as seen by an observer.

Uranus is the penultimate planet from the Sun and the least massive of the Major Planets. It was discovered using a telescope by Sir William Hershel in 1781.

Vicariance is a speciation resulting from the separation and then isolation of a portion of the original population.

Volcanism is the process by which magma and gases are ejected or extruded from below the Earth's crust to the surface or the atmosphere. It is part of the mechanism by which the Earth cools down by losing heat from the interior to the outside.

White dwarf is the end point of a star of solar mass ($\approx 2 \times 10^{30}$ kg). It will usually have a significant component of carbon.

Zygote is the result of gamete fusion. This is the single cell that develops to become the new individual.

Name Index

Subject Index